W9-DGM-442

DEFINITIONS, CONVERSIONS, *and* CALCULATIONS *for* OCCUPATIONAL SAFETY *and* HEALTH PROFESSIONALS

SECOND EDITION

DEFINITIONS, CONVERSIONS, *and* CALCULATIONS *for* OCCUPATIONAL SAFETY *and* HEALTH PROFESSIONALS

SECOND EDITION

Edward W. Finucane, PE, QEP, CSP, CIH

Library of Congress Cataloging-in-Publication Data

Finucane, Edward W.
Definitions, conversions, and calculations for occupational safety and health professionals / Edward W. Finucane. -- 2nd ed.
p. cm.
Includes index.
ISBN 1-56670-248-8 (alk. paper)
1. Industrial hygiene--Handbooks, manuals, etc. 2. Industrial safety--Handbooks, manuals, etc. 3. Industrial hygiene--Problems, exercises, etc. 4. Industrial safety--Problems, exercises, etc.
I. Title.
RC963.3.F56 1998
620.8′6—dc21

98-15356
CIP

International Standard Book Number 1-56670-248-8
Library of Congress Card Number 98-15356
Printed in the United States of America 1 2 3 4 5 6 7 8 9 0
Printed on acid-free paper

With Pride & Gratitude,

this work is dedicated to my family:

to my Wife, **Gladys**,

who deserves to be identified

as the book's Co-Author;

and to my two Sons, **Phillip** and **Ryan**,

who, every day, make me proud to

be their Father.

Preface

This book is intended to serve several purposes:

1. To function as a ready Desk Reference for the Occupational Safety and Health Professional, the Industrial Hygienist, and/or the Environmental Engineer. Such an individual, in the normal development of his or her career, will likely have specialized in some relatively specific sub-area of one of these overall disciplines. For such an individual, there will likely be occasions when a professional or job related problem or situation will arise, one that falls within the general domain of Occupational Safety and Health, Industrial Hygiene, or the Environment, but is outside of this individual's area of principal focus and competence, and is, therefore, not immediately familiar to him or her. For such cases, this Reference Source will, hopefully, provide a simple path toward the answer.
2. To function as a useful Reference Source, Study Guide, or refresher to any individual who is preparing to take either the Core or the Comprehensive Examination for Certification as an Industrial Hygienist, a Safety Professional, an Environmental Engineer, or an Environmental Professional.
3. Finally, to assist Students who have embarked on a course of study in one of these disciplines. As a fairly concise compilation of most the various important mathematical relationships and definitions that these Students will be called upon to utilize as they progress in their profession, it is hoped that this group, too, may find this work to be of some value.

This book, as a Reference Information Source and Example Problem Workbook, contains virtually every Mathematical Relationship, Formula, Definition, and Conversion Factor that any Professional in any of these overall disciplines will ever need or encounter. Every effort has been made to be certain that the information and relationships in it reflect the very best of the current thinking and technological understanding, as these concepts are currently used in the field.

Each of the Problem Solutions in this book contains carefully prepared step-by-step procedures that were followed in developing the requested answer. In addition, these Solutions contain explanations of the reasons and factors that had to be considered and used in completing each step. The underlying goal in generating these very detailed Solutions was that they would constitute a very complete road map that leads from the Problem Statement, itself, all the way to its eventual Solution. It is hoped that the various Problems, most having been developed out of the real life professional experiences of the Author and some of his colleagues, will prove to be representative of the actual situations that a professional in any of these fields might routinely encounter in the normal conduct of his or her professional life; because of this, it is hoped that they, too, will be of special value to both the professional and the prospective professional, alike.

Two final comments for the individual who has chosen to follow, or check out, each specific mathematical step shown in any of the Problem Solutions.

1. Each result — as it is developed and presented in its final "boxed" format — will have been adjusted so as to contain the correct number of significant digits.
2. In many of the Problem Solutions, there will, of necessity, be separate steps involving calculations that develop "intermediate" results — as an example, please refer to Problem #3.12, from Page 3-27. In this problem there are four major, but separate, sub-results that must be calculated [see the Solution on Pages 3-54 through 3-57] in order to develop the final result that is asked for in the Problem Statement. In order, for the Solution of this Problem, these four steps are described below [labeled i, ii, iii, & iv]:
    i. Determination of the set of four mass-based concentration TLV Standards from the set of four volumetric based TLV Standards that were provided in the Problem Statement;

ii. Determination of the overall $TLV_{effective}$ for the entire four component mixture, considered as a whole;

iii. Determination of the theoretical individual vapor phase concentrations [mass-based] for each of the four components in this mixture — i.e., what these concentrations would have to be in order for the previously calculated overall mixture $TLV_{effective}$ to apply to the vapor phase; and

iv. Finally, a determination of the four volume-based equivalent concentrations corresponding to each of these calculated mass-based concentrations.

For every multi-step Solution, each of the "intermediate" results will also have been reported in an appropriate number of significant digits; however, each subsequent calculation that makes use of any of these "intermediate" results will employ the "unrounded" number value that has been retained in the math coprocessor of the computer or the calculator that is being used to perform the calculations.

Because of this, any individual who methodically checks every step of any Problem Solution in this Text will almost certainly develop "intermediate" results that differ numerically from those in this section. This will be true if the answers that he or she obtains, in this overall stepwise process, were developed from the rounded, rather than the unrounded, "intermediate" values. To understand this situation better, please consider the following specific, in-depth example:

I would like to discuss two specific calculation steps that are presented in the Solution to Problem #3.12 [Page 3-27], as shown on Pages 3-54 through 3-57. The two steps are listed on Pages 3-55 & 3-56 — they involve the determination of the mass-based concentration of Methylene Chloride. This "intermediate" result was calculated to be 189.557+ mg/m^3, and reported — in its rounded form — as 190 mg/m^3. In each subsequent calculation step in the Solution to this Problem, the value of this concentration appears to have been used in its rounded form; however, such was never the case. Its unrounded equivalent always remained in the math coprocessor (where it had been carried out to a precision of many decimal places) and in every similar case each such value was always this value — namely, 189.557+ mg/m^3 — that was used in each subsequent calculation, rather than the indicated rounded 190 mg/m^3 value.

To continue with this specific example, consider one of the expressions used to calculate the $TLV_{effective}$ for the entire mixture. The expression to which I refer is shown below:

$$TLV_{effective} = \left(\frac{1}{\frac{0.25}{6,175} + \frac{0.55}{8,361} + \frac{0.15}{2,084} + \frac{0.05}{190}} \right)$$

The final term in the denominator of the overall expression for the $TLV_{effective}$, which is taken from the set of calculations referred to directly above, is listed as:

$$\frac{0.05}{190}$$

If the individual who is carefully checking out each step of each solution carries out this mathematical operation, using these values, on any calculator, he or she will obtain as a result, 2.632×10^{-4}. Clearly, this differs slightly from the listed value of 2.64×10^{-4}; however, the difference is certainly not great. This latter value derives from using the ratio listed below, since it is the unrounded 189.557+ mg/m^3 value that had been maintained in the calculator or computer that is being used. Thus the ratio actually employed in making this math calculation was:

$$\frac{0.05}{189.557481931+} = 2.6391145649 \times 10^{-4}$$

It is this second ratio, in contrast to the first one, that produces the slightly different 2.64×10^{-4} *value listed above. Analogous slight deviations will likely occur in every Problem Solution where there are any "intermediate" numerical results, and the Reader should be aware of this possibility.*

Should any reader wish to pass on comments or suggestions as to any aspect of the contents of this volume, I can be reached at any of the following locations and/or listings:

High Tech Enterprises
P. O. Box 7835
Stockton, CA 95267-0835

Telephone:	(209) 473-1113
Fax:	(209) 473-1114
E-Mail:	ted@hi-tech-ent.com
Home Page:	http://www.hi-tech-ent.com

Finally, I would like to compliment and thank any reader who has taken the trouble to wade through all the foregoing commentary. I hope it will be helpful as you progress in your studies or your career. Good luck as you put this volume into practical use.

Edward W. Finucane, PE, QEP, CSP, CIH

AUTHOR

Edward W. Finucane was born in San Francisco, and raised in Stockton, California. He has earned degrees in Engineering from Stanford University, and in Business from Golden Gate University. Professionally, Mr. Finucane has been involved in both the Environmental and the Occupational Safety and Health fields for more than 30 years. During the last eighteen, he has operated his own professional consulting company, High Tech Enterprises, out of offices in Stockton, California. Mr. Finucane is a Registered Professional Engineer [**PE**], a Qualified Environmental Professional [**QEP**], a Certified Safety Professional [**CSP**], and a Certified Industrial Hygienist – Comprehensive Practice [**CIH**].

He has had extensive experience in the areas of: ambient gas analysis, gas analyzer calibration, indoor air quality, ventilation, noise and sound, heat and cold stress, health physics, and in the general area of hazardous wastes.

For the past several years, he has served on the faculty of the twice yearly course, *Comprehensive Review of Industrial Hygiene*, offered jointly by the Center for Occupational and Environmental Health [University of California at Berkeley, California] and the Northern California Section of the American Industrial Hygiene Association.

Edward W. "Ted" Finucane's E-Mail Address is: **ted@hi-tech-ent.com**, and High Tech Enterprises' Home Page can be accessed at: **http://www.hi-tech-ent.com**.

ACKNOWLEDGMENTS

In authoring this text, I would like to thank a number of people without whose insights and guidance this work would still not be complete. Included in this group are professional associates, colleagues, friends, and even family members — in each case, individuals whose perspectives and opinions were very important to me.

First, I would like to thank Kenneth P. McCombs, Acquisitions Editor, and his associate, Susan Alfieri, Production Manager, both of whom work with my publisher, LEWIS PUBLISHERS/CRC PRESS, in New York City, NY. Their patience and understanding with me and my very deliberate efforts to complete this work in a timely manner were remarkable. In addition, Mimi Williams, Quality Assurance Editor — also with LEWIS PUBLISHERS/CRC PRESS, however, in their Boca Raton, FL, office — did an amazing job of proofreading my manuscript. I was not able to disagree with anything she found to be incorrect!

For sharing his significant experience in gas analyzer calibrations and standards, I would like to thank Wayne A. Jalenak, Ph.D., of Maynard, MA. His comments and insights on the material in the chapter covering Standards and Calibrations were invaluable.

For their contributions on the chapter covering Ionizing and Non-Ionizing Radiation, I would like to thank my brother, James S. Finucane, Ph.D., of Bethesda, MD, and David Baron, PE, of Minneapolis, MN. Jim's help with the overall structure of this section and Dave's unique contributions in the area of non-ionizing radiation were, in each case, absolutely vital to the development of the information in this chapter.

For sharing his expertise in the area of Statistics and Probability, I would like to acknowledge the contributions of William R. Hill of Albuquerque, NM. Bill's comments and suggestions were vital in clarifying the various difficult relationships that are discussed in this chapter.

David L. Williams of Santa Clara, CA, and Joel E. Johnson of Wilmington, DE, each provided their very valuable perspectives on the content of Appendix A, the section that covers the Atmosphere.

For the knowledge and inspiration he provided me, I would like to acknowledge my teacher, Professor Andrew J. Galambos, whose work in the physical and volitional sciences has provided me with the principal foundation upon which my own professional life has been based.

Last, but most certainly not least, I would like to acknowledge and thank my wife, Gladys. In spite of the fact that her formal education included neither the environment nor the area of occupational safety and health, she proofread the entire text, and in doing so was able to identify numerous areas where my descriptions required clarification, where I had omitted important data, etc., etc. Needless to say, to the extent that the material in this book is understandable to its readers, much of the credit must go to her.

ACKNOWLEDGMENTS

Table of Contents

Chapter 1: The Basic Parameters and Laws of Physics & Chemistry

DEFINITIONS, CONVERSIONS, AND CALCULATIONS

Chapter 6: Sound and Noise

Chapter 8: Statistics and Probability

Chapter 1

The Basic Parameters and Laws of Physics & Chemistry

The basic parameters or measurements of the physical sciences [including all of the most common and widely used units that apply to each] will be identified and described in this chapter. In addition, the fundamental laws that find significant usage in the overall study and practice of industrial hygiene/occupational safety and health will also be covered in detail.

RELEVANT DEFINITIONS

Basic Units

There are seven basic or fundamental units of measure in current use today throughout the world. The most widely recognized set of these units, known as the **International System of Units** (*SI*), was initially adopted in 1960, and is reviewed and amended, as deemed necessary, at one of the General Conferences on Weights and Measures, an international meeting that convenes periodically. In addition, there are two common "metric" systems, referred to in the text that follows as the **MKS System** [**m**eters, **k**ilograms, & **s**econds], and the **CGS System** [**c**entimeters, **g**rams, & **s**econds], as well as the "non-metric" **English System** [obsolete almost everywhere on Earth, except in the United States]. Each of these Systems of Units will be covered.

Length

LENGTH is the extent or distance from one end of an object to the other, or a distance in space from one clearly identified point to any other such point.

In the International System of Units, the basic unit of LENGTH is the **meter**, which has been defined as the length of path traveled by light in a vacuum during a time interval of 1/299,792,458 of a second.

In the MKS System, the basic unit of length is the **meter**. In the CGS System, the basic unit of length is the **centimeter**. In the English System, the basic units of length can be either the **foot**, the **inch**, or the **yard**.

Mass

In physics, MASS is the measure of a body's resistance to acceleration. The MASS of any object is different than, but proportional to, its weight — weight is a force of attraction that exists between the object being considered and any other proximate massive object [i.e., the earth].

In the International System of Units, the basic unit of MASS is the **kilogram**, which has been defined as being equal to the mass of the international prototype of the kilogram.

In the MKS System, the basic unit of mass is the **kilogram**. In the CGS System, the basic unit of mass is the **gram**. In the English System, the basic unit of mass is the **slug**. In addition, although its definition as a MASS is very confusing [largely because this same unit also serves as the basic unit of FORCE for this system], a commonly used unit of MASS is the **pound mass**, which has been defined to be that mass having an exact weight of 1.0 pound force when measured on the earth at sea level [i.e., a one pound mass would exert a downward force of 1.0 pound force on a scale situated at sea level].

Time

TIME is the interval that occurs or exists between any two clearly identified events. In contrast to the situation with respect to length and mass, the basic unit of time is the same for all Systems of Units.

Until recently, the basic unit of TIME was defined to be the length of a *mean solar day*. Now, however, the basic unit of TIME is the **second**, which had been previously defined to be 1/86,400 of one *mean solar day*, but is now more precisely defined and quantified, according to the International System of Units, the MKS System, the CGS System, and the English System as the duration of 9,192,631,770 periods of the radiation corresponding to the transition between the two hyperfine levels of the ground state of the ${}^{133}_{55}Cs$ atom.

Temperature

Likewise, TEMPERATURE was not considered to be a basic unit of measure until more recent times. It has now become regarded as one of the fundamental seven units of measure. Simply, TEMPERATURE is a measure of the relative "hotness" or "coldness" of any object or system.

In the International System of Units, the basic unit of TEMPERATURE is the **degree Kelvin**, which — as the unit of thermodynamic TEMPERATURE — is a fraction [specifically, 1/273.16] of the thermodynamic temperature of the triple point of water.

Quantifications of this parameter commonly occur using either an absolute or a relative system of measurement. In both the MKS and the CGS Systems, the basic unit of TEMPERATURE is the **degree Kelvin** or the **degree Celsius**. The magnitudes of these two units are identical — i.e., a temperature difference between two states or conditions would have the identical numerical magnitude, whether expressed in **degrees Kelvin** or **degrees Celsius**. A temperature of 0 Kelvin, or 0 K, has been defined to be Absolute Zero; thus the Kelvin Scale is the absolute scale for these two Systems. A temperature of 0° Celsius, or 0°C, is the temperature at which water freezes; thus the Celsius Scale is the relative scale for these two Systems of Units. Please note that 273.16 K = 0°C.

In the English System, the basic unit of TEMPERATURE is the **degree Fahrenheit** or the **degree Rankine**; as was the case with the relative and the absolute units of measure in the MKS & CGS Systems, the magnitudes of these two English System units are also identical; i.e., a temperature difference between two states or conditions would have the identical numerical magnitude, whether expressed in **degrees Fahrenheit** or **degrees Rankine**. A temperature of 0° Rankine, or 0°R, has been defined to be Absolute Zero; thus the Rankine Scale is the absolute scale for the English System. A temperature of 32° Fahrenheit, 32°F, is the temperature at which water freezes; thus the Fahrenheit Scale is the relative temperature scale for the English System. Please note that 491.67°R = 32°F.

Electrical Current

In all the Systems of Units, the basic unit of ELECTRICAL CURRENT is the **ampere**, which has been defined to be that constant flow of electricity which, if maintained in two straight parallel conductors of infinite length, each having negligible circular cross section, and placed 1.0 meter apart in a vacuum, would produce — between these conductors and normal to the direction in which these conductors are positioned — a repulsive force equal to 2×10^{-7} newtons per meter of conductor length.

The Amount of any Substance

In all the Systems of Units, the **mole** is the basic measure of the AMOUNT OF ANY SUBSTANCE. The **mole** has been defined to be the precise number of elementary entities, as there are atoms in exactly 0.012 kilograms [12.0 grams] of ${}^{12}_{6}C$. When the **mole** is used, the specific elementary entities must be specified; however, they may be atoms, molecules, ions, electrons, protons, neutrons, other particles, or any specified groupings of such particles. In general, one **mole** of any substance will contain Avogadro's Number, $\mathbf{N_A}$, of atoms, molecules, or particles of some sort. Avogadro's Number is 6.022×10^{23}.

Luminous Intensity

In all the Systems of Units, the **candela** is the unit of LUMINOUS INTENSITY. The **candela** has been defined to be the luminous intensity, in a given direction, of a source that emits monochromatic radiation of a frequency equal to 5.40×10^{14} hertz, and that has a radiant intensity in that same given direction that is a fraction [specifically, 1/683] of 1.0 watt per steradian.

Supplemental Units

Plane Angle

A PLANE ANGLE is a figure composed, in the simplest sense, of two different rays having a common endpoint. By definition these rays will always lie in a single plane.

As a supplemental ***SI*** Unit, the basic unit of measure for a PLANE ANGLE is the **radian**. 1.0 **radian** is the PLANE ANGLE formed when the tip, or end, of a rotating vector [the generator] of unit length, moving in a plane, has traveled a circular path of LENGTH equal to the LENGTH of the unit vector.

PLANE ANGLES are dimensionless quantities, since they are defined in terms of:

$$\frac{\text{LENGTH}}{\text{LENGTH}}.$$

Clearly by this definition, there will be a total of 2π **radians** in one complete circle. In the MKS, the CGS, and the English Systems of units, PLANE ANGLES are measured in **radians** and also frequently in **degrees**. Note that a PLANE ANGLE of 360 **degrees** [written as 360°] = 2π **radians**, or 1.0 **radian** = 57.296°.

Solid Angle

A SOLID ANGLE is that part of the space bounded by a moving straight line [the generator] issuing from a single point [the vertex] and moving to all points on an arbitrary closed curve. It characterizes the angle of "seeing" from which this curve is "seen".

As another supplemental ***SI*** Unit, the basic unit of measure for SOLID ANGLES is the **steradian**. 1.0 **steradian** is the SOLID ANGLE [i.e., the area] cut out of a unit sphere by the tip of a rotating unit radius vector [the generator], having the center of the sphere as its vertex and producing a PLANE ANGLE [in <u>any</u> plane through this vertex] of 1.0 **radian** — i.e., a "cone" with its vertex angle = 1.0 **radian**.

SOLID ANGLES are dimensionless quantities. They are defined as:

$$\frac{\text{AREA}}{\text{AREA}} = \left[\frac{(\text{LENGTH})^2}{(\text{LENGTH})^2}\right].$$

In the MKS, the CGS, and the English Systems of units, SOLID ANGLES are measured in **steradians**.

Derived Units

Area

AREA is the measure of the size or extent of a surface. Its dimensions are:

$$\text{AREA} = [\text{LENGTH}]^2$$

In the MKS System, AREA is measured in: **meters²**.

In the CGS System, AREA is measured in: **centimeters²**.

In the English System, AREA is measured in: **feet²**, or frequently **inches²**.

Volume

VOLUME is the measure of the size or extent of any three-dimensional object or region in space. Its dimensions are:

$$\text{VOLUME} = [\text{LENGTH}]^3$$

In the MKS System, VOLUME is measured in: **meters³** or **liters** [with 1 meter³ = 1,000 liters].

In the CGS System, VOLUME is measured in: **centimeters³**, which is numerically identical to this parameter when expressed in **milliliters**.

In the English System, VOLUME is measured in: **feet³**, **inches³**, or frequently **gallons**.

Velocity or Speed

VELOCITY or SPEED is the distance traveled by an object during each unit interval of time. Its dimensions are:

$$\text{VELOCITY} = \text{SPEED} = \left[\frac{\text{DISTANCE TRAVELED}}{\text{TIME}}\right] = \left[\frac{\text{LENGTH}}{\text{TIME}}\right]$$

In the MKS System, VELOCITY or SPEED is measured in:

$$\frac{\textbf{meters}}{\textbf{second}}.$$

In the CGS System, VELOCITY or SPEED is measured in:

$$\frac{\textbf{centimeters}}{\textbf{second}}.$$

In the English System, VELOCITY or SPEED is measured in:

$$\frac{\textbf{feet}}{\textbf{second}}, \text{ or occasionally } \frac{\textbf{feet}}{\textbf{minute}}.$$

Acceleration

ACCELERATION is the time rate of change of SPEED or VELOCITY. Its dimensions are:

$$\text{ACCELERATION} = \left[\frac{\text{CHANGE in SPEED or VELOCITY}}{\text{TIME}}\right] = \left[\frac{\text{LENGTH}}{\text{TIME}^2}\right]$$

In the MKS System, ACCELERATION is measured in:

$$\frac{\textbf{meters}}{\textbf{second}^2}.$$

In the CGS System, ACCELERATION is measured in:

$$\frac{\textbf{centimeters}}{\textbf{second}^2}.$$

In the English System, ACCELERATION is measured in:

$$\frac{\textbf{feet}}{\textbf{second}^2}.$$

Force

FORCE is the capacity to do work or to cause change. It is a vector quantity that tends to produce an acceleration in the body on which it is acting, and to produce this acceleration in the direction of the application of that FORCE. Its dimensions are:

$$\text{FORCE} = [(\text{MASS})(\text{ACCELERATION})] = \left[\frac{(\text{MASS})(\text{LENGTH})}{\text{TIME}^2}\right]$$

In the International System of Units, FORCE is measured in **newtons**, which have been defined as:

$$1.0\ \textbf{newton} = 1.0\ \frac{(\text{kilogram})(\text{meter})}{\text{second}^2}.$$

In the MKS System, FORCE is also measured in:

$$\textbf{newtons} = \frac{(\text{kilograms})(\text{meters})}{\text{second}^2}.$$

In the CGS System, FORCE is measured in:

$$\textbf{dynes} = \frac{(\text{grams})(\text{centimeters})}{\text{second}^2}.$$

In the English System, FORCE is measured in:

$$\textbf{pounds of force} = \frac{(\text{slug})(\text{feet})}{\text{second}^2}.$$

Pressure

PRESSURE is the relative magnitude of a FORCE per unit AREA through which that FORCE is acting. Its dimensions are:

$$\text{PRESSURE} = \left[\frac{\text{FORCE}}{\text{AREA}}\right] = \left[\frac{\text{FORCE}}{(\text{LENGTH})^2}\right] = \left[\frac{(\text{MASS})}{(\text{TIME})^2(\text{LENGTH})}\right]$$

In the International System of Units, PRESSURE is measured in **pascals**, which have been defined as:

$$1.0\ \textbf{pascal} = 1.0\ \frac{\text{newton}}{\text{meter}^2}.$$

In the MKS System, PRESSURE is also measured in:

$$\textbf{pascals} = \frac{\text{newtons}}{\text{meter}^2} = \frac{\text{kilograms}}{(\text{meters})(\text{second})^2}.$$

In the CGS System, PRESSURE is measured in:

$$\frac{\textbf{dynes}}{\textbf{centimeter}^2} = \frac{\text{grams}}{(\text{centimeters})(\text{second})^2}.$$

In the English System, PRESSURE is often measured in:

$$\frac{\textbf{pounds of force}}{\textbf{foot}^2} = \frac{\text{slugs}}{(\text{foot})(\text{second})^2},$$

or, alternatively, and even more frequently, in:

$$\frac{\textbf{pounds of force}}{\textbf{inch}^2} = \frac{(\text{slug})(\text{feet})}{(\text{inch})^2(\text{second})^2} = \frac{1}{12}\left[\frac{\text{slug}}{(\text{inch})(\text{second})^2}\right],$$

often abbreviated as **psi**.

PRESSURE is also frequently measured or characterized in a variety of other ways:

1. It is expressed in terms of the height of a column of some reference fluid that would be required to exert some identifiable FORCE on the unit AREA on which this column of fluid is resting — i.e., [**millimeters of Mercury**], commonly written as **mm Hg**, or [**feet of Water**], commonly written as **ft H_2O**;
2. It is expressed in terms of its ratio to the average pressure of the Earth's atmosphere when measured at Mean Sea Level — i.e., [**atmospheres**], commonly written as **atms**;
3. It is expressed in terms of its ratio to the accepted ***SI*** unit of pressure, the **pascal** — i.e., [**torrs**], commonly written as **torrs**, of which 1.0 torr = 133.32 pascals; [**bars**], commonly written as **bars**, of which 1.0 bar = 10^5 pascals; and [**millibars**], commonly written as **mb**, of which 1.0 mb = 100 pascals.

Energy, **Work**, or **Heat**

ENERGY or WORK is the work that any physical system is capable of doing, as it changes from its existing state to some well-defined different state. In general, ENERGY can be any of the following three types: potential, kinetic, and/or rest. WORK is a measure of the result of applying a FORCE through some identified DISTANCE. In the case of either ENERGY or WORK, the dimensions are identical and are:

$$\text{ENERGY, WORK, or HEAT} = [(\text{FORCE})(\text{LENGTH})]$$

$$\text{ENERGY, etc.} = [(\text{MASS})(\text{ACCELERATION})(\text{LENGTH})] = \left[\frac{(\text{MASS})(\text{LENGTH})^2}{(\text{TIME})^2}\right]$$

In the International System of Units, ENERGY or WORK is measured in **joules**, which have been defined as:

$$1.0\ \textbf{joule} = 1.0\ (\text{newton})(\text{meter}).$$

In the MKS System, ENERGY or WORK is measured also in **joules**, where:

$$\textbf{joules} = (\text{newtons})(\text{meters}) = \frac{(\text{kilograms})(\text{meters})^2}{\text{second}^2}.$$

In the CGS System, ENERGY or WORK is usually measured in **ergs**; however, in the context of chemical reactions, etc., it is more frequently measured in **calories**. Dimensionally, an **erg** is equal to:

$$\textbf{ergs} = (\text{dynes})(\text{centimeters}) = \frac{(\text{grams})(\text{centimeters})^2}{\text{second}^2}.$$

A **calorie** [abbreviated "**cal**"] has been defined to be:

$$1.0\ \textbf{calorie} = 1.0\ \textbf{cal} = \left[\begin{array}{l}\text{the amount of heat energy required to raise the tem-}\\ \text{perature of one gram of water by one degree Kelvin}\end{array}\right].$$

In the English System, ENERGY or WORK is measured in either **foot pounds** or **British Thermal Units**, where:

$$\textbf{foot pounds} = (\text{pounds of force})(\text{feet}) = \frac{(\text{slugs})(\text{feet})^2}{\text{second}^2}.$$

The **British Thermal Unit** [abbreviated "**BTU**"] has been defined to be:

$$1.0\ \textbf{BTU} = \begin{bmatrix}\text{the amount of heat energy required to raise the temperature} \\ \text{of a mass of one pound of water by one degree Fahrenheit}\end{bmatrix}$$

In the micro-scale world [i.e., in the world of the atom, the electron, and other sub-atomic particles, etc.], ENERGY or WORK is measured in **electron volts** [abbreviated "**ev**"], where this unit has been defined to be:

$$1.0\ \textbf{electron volt} = 1.0\ \textbf{ev} = \begin{bmatrix}\text{the energy necessary to accelerate a single elec-} \\ \text{tron through a potential difference of one volt}\end{bmatrix}$$

$$1.0\ \text{ev} \approx 1.6022 \times 10^{-19}\ \text{joules}$$

Power

POWER is the TIME rate of performing WORK. The dimensions of POWER are:

$$\text{POWER} = \left[\frac{\text{WORK}}{\text{TIME}}\right] = \left[\frac{(\text{FORCE})(\text{LENGTH})}{\text{TIME}}\right]$$

$$\text{POWER} = \left[\frac{(\text{MASS})(\text{ACCELERATION})(\text{LENGTH})}{\text{TIME}}\right] = \left[\frac{(\text{MASS})(\text{LENGTH})^2}{(\text{TIME})^3}\right]$$

In the International System of Units, POWER is measured in **watts**, which has been defined as:

$$1.0\ \textbf{watt} = 1.0\ \frac{\text{joule}}{\text{second}}.$$

In the MKS System, POWER is also measured in **watts**, where:

$$\textbf{watts} = \frac{\text{joules}}{\text{second}} = \frac{(\text{newtons})(\text{meters})}{\text{second}} = \frac{(\text{kilograms})(\text{meters})^2}{\text{second}^3}.$$

In the CGS System, POWER is measured in $\frac{\textbf{ergs}}{\textbf{second}}$, where:

$$\frac{\textbf{ergs}}{\textbf{second}} = \frac{(\text{dynes})(\text{centimeters})}{\text{second}} = \frac{(\text{grams})(\text{centimeters})^2}{\text{second}^3}.$$

In the English System, POWER is measured in $\frac{\textbf{foot pounds}}{\textbf{second}}$, **horsepower**, and/or $\frac{\textbf{BTUs}}{\textbf{second}}$. Of these three general English System units, the only one that is not inherently defined by its description is the **horsepower** [abbreviated "**hp**"], which has been defined to be:

$$1.0\ \textbf{hp} = 550\ \frac{\text{foot pounds}}{\text{second}}$$

Electric Charge

ELECTRIC CHARGE is a measure of the quantity of electricity or charge which, in a time period of one second, passes through a section of conductor in which a constant current of one **ampere** happens to be flowing. The dimensions of ELECTRIC CHARGE are:

$$\text{ELECTRIC CHARGE} = [(\text{CURRENT})(\text{TIME})]$$

In both the MKS and the English System, ELECTRIC CHARGE is measured in **coulombs**.

In the CGS System, ELECTRIC CHARGE is measured in **microcoulombs**.

The **coulomb** is a relatively large unit of ELECTRIC CHARGE. A more "natural" unit would probably be the negative charge carried by an electron [or, alternatively, the equivalently large positive charge carried by a proton]. The ELECTRIC CHARGE carried by either of these basic atomic particles is equivalent to 1.6022×10^{-19} **coulombs**.

Electrical Potential or Potential Difference

ELECTRICAL POTENTIAL is a measure of the work required to move a quantity of charge in an electrostatic field. Specifically, it is a measure of the potential energy per unit electrical charge that is characterized by any point in an electric field. The dimensions of ELECTRICAL POTENTIAL are:

$$\text{ELECTRICAL POTENTIAL} = \left[\frac{\text{WORK}}{\text{CHARGE}}\right] = \left[\frac{(\text{MASS})(\text{LENGTH})^2}{(\text{CURRENT})(\text{TIME})^3}\right]$$

In the MKS System, ELECTRICAL POTENTIAL is measured in:

$$\frac{\textbf{joules}}{\textbf{coulomb}}, \text{ or } \textbf{volts} \text{ — where: } 1.0 \textbf{ volt} = 1.0 \frac{\textbf{joule}}{\textbf{coulomb}}.$$

In the CGS System, ELECTRICAL POTENTIAL is measured in **millivolts** or **microvolts**.

In the English System, ELECTRICAL POTENTIAL is also measured in **volts**, even though in this system, the unit of work is not the **joule**.

Capacitance

CAPACITANCE is the ratio of [the charge on either of two conductors] to [the electrical potential difference between these two conductors]. For all practical applications, it is used only to refer to the charging capacity characteristics of a "capacitor", which is a common component in many electrical circuits. The dimensions of CAPACITANCE are:

$$\text{CAPACITANCE} = \left[\frac{\text{CHARGE}}{\text{ELECTRICAL POTENTIAL}}\right] = \left[\frac{(\text{CURRENT})^2(\text{TIME})^4}{(\text{MASS})(\text{LENGTH})^2}\right]$$

In the MKS System, CAPACITANCE is measured in **farads**, where:

$$1.0 \textbf{ farad} = 1.0 \frac{\textbf{coulomb}}{\textbf{volt}}.$$

In the CGS System, CAPACITANCE is measured in **microfarads** or **picofarads**.

In the English System, CAPACITANCE is also measured in **farads**.

Density

DENSITY is a measure of the MASS of anything measured per the unit VOLUME that is occupied by that MASS. The dimensions of DENSITY are:

$$\text{DENSITY} = \left[\frac{\text{MASS}}{\text{VOLUME}}\right] = \left[\frac{\text{MASS}}{(\text{LENGTH})^3}\right]$$

In the MKS System, DENSITY is measured in:

$$\frac{\textbf{kilograms}}{\textbf{meter}^3}\text{, or occasionally in } \frac{\textbf{kilograms}}{\textbf{liter}}.$$

In the CGS System, DENSITY is measured in:

$$\frac{\textbf{grams}}{\textbf{centimeter}^3} = \frac{\textbf{grams}}{\textbf{milliliter}}.$$

In the English System, DENSITY is measured in:

$$\frac{\textbf{pounds of mass}}{\textbf{foot}^3}, \frac{\textbf{pounds of mass}}{\textbf{inch}^3}\text{, or occasionally in } \frac{\textbf{pounds of mass}}{\textbf{gallon}}.$$

Concentration

CONCENTRATION is a measure of: (1) the amount or quantity of any substance per the units of VOLUME that is occupied by that amount or quantity of material, (2) the mass of any substance per the units of VOLUME that is occupied by that mass, (3) the ratio of the amount, quantity, or volume of any substance to the total amount, quantity, or volume of all the substances present in the overall VOLUME being considered, and/or (4) the ratio of the MASS of any substance to the total MASS of all the substances present in the overall VOLUME being considered. Let us consider each of these four CONCENTRATION scenarios in order:

1. The first of the four listed categories of concentration is most commonly used to express the concentration of some material, substance, or chemical as a component in a solution. Its dimensions are:

$$\text{CONCENTRATION}_1 = \left[\frac{\text{MOLES}}{\text{VOLUME}}\right]$$

In the MKS System, CONCENTRATION_1 is usually measured in:

$$\frac{\textbf{moles}}{\textbf{liter}}\text{; however, it can occasionally be measured in } \frac{\textbf{equivalents}}{\textbf{liter}}.$$

In the CGS System, CONCENTRATION_1 is measured in:

$$\frac{\textbf{millimoles}}{\textbf{milliliter}} = \frac{\textbf{millimoles}}{\textbf{centimeter}^3};$$

however, it can occasionally also be measured in:

$$\frac{\textbf{milliequivalents}}{\textbf{milliliter}} = \frac{\textbf{milliequivalents}}{\textbf{centimeter}^3}.$$

The English System does not have a commonly used unit of CONCENTRATION_1.

2. The second of the four listed categories of concentration is most commonly used to express the concentration of a particulate, an aerosol, a mist, a dust, or a fume in the air. Its dimensions are:

$$\text{CONCENTRATION}_2 = \left[\frac{\text{MASS}}{\text{VOLUME}}\right]$$

In the MKS System, CONCENTRATION_2 is measured in:

$$\frac{\textbf{milligrams}}{\textbf{meter}^3}.$$

In the CGS System, CONCENTRATION_2 is measured in:

$$\frac{\textbf{micrograms}}{\textbf{centimeter}^3} = \frac{\textbf{micrograms}}{\textbf{milliliter}}.$$

The English System does not have a commonly used unit of CONCENTRATION_2.

3. The third category of CONCENTRATION is most commonly used to express the concentration of some vapor or gas in the air. This category of CONCENTRATION is considered to be "Volume-Based Concentration". Its dimensions are:

$$\text{CONCENTRATION}_3 = \left[\frac{\text{MOLES}}{\text{MOLE}}\right], \text{ or more commonly } \left[\frac{\text{MOLES}}{\text{MILLION MOLES}}\right]$$

In all of the systems of units, namely: the MKS, the CGS, and the English Systems, CONCENTRATION_3 is measured in:

$$\frac{\textbf{parts}}{\textbf{million parts}}, \text{ or, simply, } \textbf{ppm(vol)}.$$

4. The fourth and final category of CONCENTRATION is most commonly used to express concentrations of some mixture of materials, or solid sample of some sort. This category of CONCENTRATION is considered to be "Mass-Based Concentration". Its dimensions are:

$$\text{CONCENTRATION}_4 = \left[\frac{\text{MASS}}{\text{TOTAL MASS}}\right]$$

In the MKS and the CGS Systems, CONCENTRATION_4 is measured in:

$$\frac{\textbf{milligrams}}{\textbf{kilogram}}, \text{ or, more simply, } \textbf{ppm(mass)}.$$

The English System does not have a commonly used unit of CONCENTRATION_4.

Luminous Flux

LUMINOUS FLUX is the product of the LUMINOUS INTENSITY and the SOLID ANGLE over which the illumination being considered is being emitted. LUMINOUS FLUX is measured in **lumens**, and its dimensions are:

$$\text{LUMINOUS FLUX} = [(\text{LUMINOUS INTENSITY})(\text{SOLID ANGLE})]$$

By definition, a point light source radiating light energy in all directions and having a LUMINOUS INTENSITY of 1.0 **candela** will produce a total LUMINOUS FLUX of 4π **lumens**.

The basic unit of LUMINOUS FLUX is the same in all systems of units, and is the **lumen**.

Frequency

FREQUENCY is a measure of the number of cycles per unit TIME for any periodically repeating physical phenomenon — i.e., a sound, or an electromagnetic wave, each of which exhibits a form of simple harmonic motion. The dimensions of FREQUENCY are:

$$\text{FREQUENCY} = \frac{\text{CYCLES}}{\text{TIME}} = \frac{1}{\text{TIME}}$$

The basic unit of measure for FREQUENCY is the same in all systems of units, and is the **hertz**, which has been defined to be:

$$1.0\ \textbf{hertz} = \frac{\text{one complete cycle}}{\text{second}}$$

Radioactive Activity

The basic event that characterizes any radioactive element or nuclide is the transformation or decay of its nucleus into the nucleus of some other species. The number of such transformations per unit time is known as the RADIOACTIVE ACTIVITY of the element or nuclide being considered. The basic unit of RADIOACTIVE ACTIVITY is the **becquerel**, which is defined to be one transformation per second. Its dimensions are:

$$\text{RADIOACTIVE ACTIVITY} = \frac{\text{DECAY EVENTS}}{\text{TIME}} = \frac{1}{\text{TIME}}$$

Its dimensions are, at least from the perspective of pure dimensionality, identical to those of FREQUENCY, namely, $\frac{1}{\text{TIME}}$; however, there is a clear difference as the seemingly dimensionless "numerator" for each of these two parameters is totally different.

The basic ***SI*** unit of RADIOACTIVE ACTIVITY is the **becquerel**.

This is also the basic unit of RADIOACTIVE ACTIVITY for the MKS, the CGS, and the English Systems. In each of these systems, additional widely used units of RADIOACTIVE ACTIVITY are: the **curie**, the **millicurie**, the **microcurie**, and the **picocurie**, 1.0 **curie** = 2.22×10^{12} **becquerels**. The **becquerel** is a very small unit, while the **curie** is extremely large. Because of this, one most commonly encounters the **millicurie** [1/1,000 of **curie**], the **microcurie** [1/1,000 of a **millicurie**] and the **picocurie** [1/1,000 of a **microcurie**].

Absorbed Radiation Dose

The basic quantity that is used to characterize the amount of energy [in the form of some type of radiation] that has been imparted to matter is the ABSORBED RADIATION DOSE. The ABSORBED RADIATION DOSE is the ratio of [the radioactive energy imparted to the matter in that region] to [the mass of the matter in that region]. The dimensions of ABSORBED RADIATION DOSE are:

$$\text{ABSORBED RADIATION DOSE} = \frac{\text{ENERGY}}{\text{MASS}} = \left[\frac{(\text{LENGTH})^2}{(\text{TIME})^2}\right]$$

The basic ***SI*** unit of ABSORBED RADIATION DOSE is the **gray**.

This is also the basic unit of ABSORBED RADIATION DOSE for each of the other systems of units, namely, the MKS, the CGS, and the English Systems. In these latter systems additional very widely used units of ABSORBED RADIATION DOSE are the **rad**, the **millirad**, and the **microrad**. By definition:

$$1.0\ \textbf{gray} = 1.0\frac{\text{joule}}{\text{kilogram}} = 100\ \textbf{rads}.$$

$$\text{Therefore, } 1.0\ \textbf{rad} = 100\frac{\text{ergs}}{\text{gram}}.$$

The energy of most radioactive particles is most frequently expressed in millions of electron volts [MeV]; therefore, it is useful to quantify these ABSORBED RADIATION DOSES in terms of this unit of energy, thus:

$$1.0\ \textbf{rad} = 6.24 \times 10^{7}\ \frac{\text{MeV}}{\text{gram}},\ \&\ 1.0\ \textbf{millirad} = 62{,}400\ \frac{\text{MeV}}{\text{gram}}.$$

In general, harmful levels of ABSORBED RADIATION DOSE are expressed in **rads**, whereas the **millirad** [often abbreviated, **mrad**] is most frequently used to specify permissible dose rates. Occupational radioactivity exposures should never exceed a few **mrads** per hour.

Radiation Dose Equivalent or Radiation Dose Equivalent Index

The basic quantity used in radiation protection, as it applies to humans, is the RADIATION DOSE EQUIVALENT. Injuries produced by an individual's exposure to radioactivity depend not only on the total of the radioactive energy imparted to that person, but also on the nature of the radiation that produced the exposure. The dimensions of the RADIATION DOSE EQUIVALENT are:

$$\text{RADIATION DOSE EQUIVALENT} = \frac{\text{ENERGY}}{\text{MASS}} = \left[\frac{(\text{LENGTH})^2}{(\text{TIME})^2}\right]$$

Note the dimensional identity between the ABSORBED RADIATION DOSE and the RADIATION DOSE EQUIVALENT. We must use this additional parameter because of differences in the harmful effects one type of radiation can produce when compared to other types. As an example, a 1.0 **mrad** ABSORBED RADIATION DOSE from neutrons would produce a far greater injury to an exposed individual than would the same ABSORBED RADIATION DOSE from beta particles. Because of this variability in harmful effects, it is clear that if we use only the simple ABSORBED RADIATION DOSE parameter, we will fail to provide a completely accurate and unambiguous picture when evaluating the injuries that result from exposure to various different types of radiation. Although, as stated above, the dimensionalities of ABSORBED RADIATION DOSE and RADIATION DOSE EQUIVALENT are identical, for the sake of clarity, there is a different unit for RADIATION DOSE EQUIVALENT — this unit is the **sievert**. In the MKS, the CGS, and the English Systems, an additional unit of RADIATION DOSE EQUIVALENT is also in wide use. This unit is the **rem**.

The "effectiveness" in producing injury for any one category of radiation relative to any other can vary widely depending on a number of factors and circumstances. Included among these are: (1) the nature of the biological material exposed, (2) the length of time during which the exposure occurred, (3) the type of effect being considered, and (4) the ability of the exposed body to repair the injuries produced by the radiation. Therefore, when considering aspects of radiation protection, an occupational safety and health professional must evaluate the manner in which various different forms of ionizing radiation impart energy to the tissue onto which it has fallen.

The physical measure that is used for evaluating the "effectiveness" of equal ABSORBED RADIATION DOSES from different types of radiation — insofar as the ability of the radiation to produce injury — is the linear energy transfer, or *LET*. The greater the *LET* for any type of radiation, the greater will be the injury produced for any given ABSORBED RADIATION DOSE. The quantitative factor that expresses the relative "effectiveness" of a given type of radiation, based on its *LET*, is the *Quality Factor*. Numeric values of these

Quality Factors, as functions of the radiation type and energy, have been developed from animal experimental data, and are given in **Table 1.1**, below.

Although the issue is even more complex than has been indicated to this point, this discussion will go no further. When the ABSORBED RADIATION DOSE is evaluated for the purposes of radiation protection, this dose number must be multiplied by the appropriate *Quality Factor* in order to provide an accurate RADIATION DOSE EQUIVALENT. Specifically, the following relationships provide the method for obtaining accurate values of the RADIATION DOSE EQUIVALENT for any measured ABSORBED RADIATION DOSE, each as a function of the specific units of these two similar parameters:

$$\mathbf{rems} = \left[(\mathbf{rads})(\mathbf{QF})\right], \ \&$$

$$\mathbf{sieverts} = \left[(\mathbf{grays})(\mathbf{QF})\right].$$

As stated above, a tabulation listing some of these *Quality Factors* — from data published in 1987 by the National Council on Radiation Protection — is shown below in **Table 1.1**.

Table 1.1

— Approximations to a Variety of Different *Quality Factors* —

Particle Type	Parameters Related to Energy	*Quality Factor*
X-Rays & Gamma Rays	any	1
Beta Particles	any	1
Heavy Ionizing Particles	$LET \leq 35$ MeV/cm	2
"	35 MeV/cm < LET < 70 MeV/cm	2 to 4
"	70 MeV/cm < LET < 230 MeV/cm	4 to 10
"	230 MeV/cm < LET < 530 MeV/cm	10 to 20
"	530 MeV/cm < LET < 1,750 MeV/cm	20 to 40
"	$LET \geq 1{,}750$ MeV/cm	40
Neutrons	thermal	5
"	0.0001 MeV	4
"	0.01 MeV	5
"	0.1 MeV	15
"	0.5 MeV	22
"	1.0 MeV	22
"	10 MeV	13
Protons	≤ 10 MeV	20
Alpha Particles	≤ 10 MeV	20
Other Multiply Charged Particles	≤ 10 MeV	20

Atmospheric Standards

Standard Temperature and Pressure

Standard Temperature and Pressure (usually abbreviated, **STP**) is the designation given to an ambient condition in which the Barometric Pressure [**P**] and the Ambient Temperature [**T** or **t**] are, respectively:

Barometric Pressure [P]		Ambient Temperature [T or t]
1 atmosphere, or		0°C, or
760 mm Hg, or		32°F, or
14.70 psia, or	AND	273.16 K, or
0.00 psig, or		491.67°R
1,013.25 millibars, or		
760 Torr		

Normal Temperature and Pressure

Normal Temperature and Pressure (usually abbreviated, **NTP**) is the designation given to an ambient condition in which the Barometric Pressure [**P**] and the Ambient Temperature [**T** or **t**] are, respectively:

Barometric Pressure [P]		Ambient Temperature [T or t]
1 atmosphere, or		25°C, or
760 mm Hg, or		77°F, or
14.70 psia, or	AND	298.16 K, or
0.00 psig, or		536.67°R
1,013.25 millibars, or		
760 Torr		

Ventilation-Based Standard Air

The conditions that are characteristic of ventilation-based **Standard Air** differ slightly from those of either **S**tandard **T**emperature and **P**ressure or **N**ormal **T**emperature and **P**ressure listed above. **Standard Air** is the designation given to ambient air for which the following three specific conditions must be met, namely:

Barometric Pressure [P]	Ambient Temp. [T or t]	Humidity Level
1.00 atmosphere, or	21.1°C, or	Relative Humidity = 0.0%
760 mm Hg, or	70°F, or	Absolute Humidity = 0.0%
29.92 inches of Hg, or	294.26 K, or	Water Conc. = 0.0 ppm(vol)
14.70 psia, or	529.67°R	
0.00 psig, or		i.e., **Standard Air** is DRY
1,013.25 millibars, or		
760 Torr		

Standard Air is also regarded as possessing the following characteristics of density and specific heat:

Density of **Standard Air** = $\rho_{\text{Standard Air}}$ = 0.075 lbs/ft^3 = 1.201×10^{-3} gm/cm^3

Specific Heat of **Standard Air** = $C_{P\text{—Standard Air}}$ = 0.24 BTU/lb/°F.

Metric Prefixes (for use with *SI* Units)

"EXPANDING" PREFIXES			"DIMINISHING" PREFIXES		
Prefix	Abbreviation	Multiplier	Prefix	Abbreviation	Multiplier
deca-	da	10^{1}	deci-	d	10^{-1}
hecto-	h	10^{2}	centi-	c	10^{-2}
kilo-	k	10^{3}	milli-	m	10^{-3}
mega-	M	10^{6}	micro-	μ	10^{-6}
giga-	G	10^{9}	nano-	n	10^{-9}
tera-	T	10^{12}	pico-	p	10^{-12}
peta-	P	10^{15}	femto-	f	10^{-15}
exa-	E	10^{18}	atto-	a	10^{-18}
zetta-	Z	10^{21}	zepto-	z	10^{-21}
yotta-	Y	10^{24}	yocto-	y	10^{-24}

RELEVANT FORMULAE & RELATIONSHIPS

Temperature Conversions

As stated earlier, temperatures can be expressed in relative or absolute units, and in the English or the Metric Systems [i.e., CGS, MKS, and/or ***SI*** Systems]. To convert from one form of temperature units to any other, the following four relationships should be used: Equation **#1-1**, to convert from relative to absolute Metric System units; Equation **#1-2**, to convert from relative to absolute English System units; Equation **#1-3**, to convert from relative Metric to relative English System units; & Equation **#1-4**, to convert temperature changes in any Metric System unit [relative or absolute] to the corresponding temperature change in English System units [relative or absolute].

Equation #1-1:

$$t_{Metric} + 273.16 = T_{Metric}$$

Where:

t_{Metric} = the temperature, expressed in relative Metric units — i.e., degrees Celsius [°C];

T_{Metric} = the temperature, expressed in absolute Metric units — i.e., degrees Kelvin [K].

Equation #1-2:

$$t_{English} + 459.67° = T_{English}$$

Where:

$t_{English}$ = the temperature, expressed in relative English units — i.e., degrees Fahrenheit [°F];

$T_{English}$ = the temperature, expressed in absolute English units — i.e., degrees Rankine [°R].

Equation #1-3:

$$t_{Metric} = \frac{5}{9}\left[t_{English} - 32°\right]$$

Where: $t_{English}$ & t_{Metric} are as defined above on this page for the two previous Equations.

Equation #1-4:

$$\Delta t_{Metric} = \frac{5}{9}\Delta t_{English} \quad \text{and/or} \quad \Delta T_{Metric} = \frac{5}{9}\Delta T_{English}$$

Where:

Δt_{Metric} = the temperature change in degrees Celsius [°C];

$\Delta t_{English}$ = the temperature change in degrees Fahrenheit [°F];

ΔT_{Metric} = the temperature change in degrees Kelvin [K]; &

$\Delta T_{English}$ = the temperature change in degrees Rankine [°R].

The Standard Gas Laws

The following Formulae make up the five Standard Gas Laws, which are, in the order in which they will be presented and discussed: Boyle's Law (Equation **#1-5**); Charles' Law (Equation **#1-6**); Gay-Lussac's Law (Equation **#1-7**); the General Gas Law (Equation **#1-8**); and the Ideal or Perfect Gas Law (Equation **#1-9**).

Equation #1-5:

The following relationship, Equation **#1-5**, is Boyle's Law, which describes how the Pressure and Volume of a gas vary under conditions of constant temperature.

$$P_1V_1 = P_2V_2$$

Where:

P_1 = the Pressure of a gas @ Time #1, measured in some suitable pressure units;

V_1 = the Volume of that same gas @ Time #1, measured in some suitable volumetric units;

P_2 = the Pressure of that same gas @ Time #2, measured in the same pressure units as P_1 above; &

V_2 = the Volume of that same gas @ Time #2, measured in the same volumetric units as V_1 above.

Equation #1-6:

The following relationship, Equation **#1-6**, is Charles' Law, which describes how the Volume and the Absolute Temperature of a gas vary under conditions of constant pressure.

$$\frac{V_1}{T_1} = \frac{V_2}{T_2}$$

Where:

V_1 & V_2 are the Volumes of the gas of interest at each of its two states, with this term as was defined for Equation **#1-5**, above on this page;

T_1 = the Absolute Temperature of a gas @ Time #1, measured in either K or °R; &

T_2 = the Absolute Temperature of the same gas @ Time #2, measured in the same Absolute Temperature units as T_1.

Equation #1-7:

The following relationship, Equation **#1-7**, is Gay-Lussac's Law, which describes how the Pressure and Temperature of a gas vary under conditions of constant volume.

$$\frac{P_1}{T_1} = \frac{P_2}{T_2}$$

Where: **P_1 & P_2** are the Pressures of the gas of interest at each of its two states, with this term as was defined for Equation **#1-5** on the previous page, namely, Page 1-17; &

T_1 & T_2 are the Absolute Temperatures of the gas of interest at each of its two states, with this term as was defined for Equation **#1-6**, on the previous page, namely, Page 1-17.

Equation #1-8:

The following formula, Equation **#1-8**, is the General Gas Law, which is the more generalized relationship involving changes in any of the basic measurable characteristics of any gas. This relationship permits the determination of the value of any of the three basic characteristics of the gaseous material being evaluated — namely: its Pressure, its Temperature, and/or its Volume — any one of which might have changed as a result of changes in either one or both of the other two characteristics.

$$\frac{P_1V_1}{T_1} = \frac{P_2V_2}{T_2}$$

Where: **P_1 & P_2** are the Pressures of the gas of interest at each of its two states, with this term as was defined for Equation **#1-5** on the previous page, namely, Page 1-17;

V_1 & V_2 are the Volumes of the gas of interest at each of its two states, with this term as was defined for Equation **#s 1-5** & **1-6** on the previous page, namely, Page 1-17; &

T_1 & T_2 are the Absolute Temperatures of the gas of interest at each of its two states, with this term as was defined for Equation **#s 1-6** on the previous page, namely, Page 1-17, & **1-7**, above on this page.

Equation #1-9:

The following relationship, Equation **#1-9**, is the Ideal Gas Law [it is also frequently called the Perfect Gas Law]. This law is one of the most commonly used Equations of State. Like the immediately preceding formula, this one provides the necessary relationship for determining the value of any of the measurable characteristics of a gas — namely, again: its Pressure, its Temperature, and/or its Volume; however, it does not require that one know these characteristics at some alternative state or condition. Unlike Equation **#1-8**, on the previous page, namely, Page 1-18, it does require that the quantity of the gas involved in the determination be known (i.e., the Number of Moles involved, or the weight and identity of the gas involved, etc.).

$$\boxed{PV = nRT}$$

Where:

P = the Pressure of the gas involved, in some appropriate units;

V = the Volume of the gas involved, in some appropriate units;

n = the number of moles of the gas involved [for more detailed information on this "number of moles" factor, see the discussion below on this and on the following pages, namely Pages 1-19 & 1-20, under Equation **#s 1-10 & 1-11**];

R = the Universal Gas Constant, in units consistent with those used for the Pressure, **P**, the Volume, **V**, and the temperature, **T**:

$$= 0.0821 \frac{(\text{liter})(\text{atmospheres})}{(\text{K})(\text{moles})};$$

$$= 62.36 \frac{(\text{liter})(\text{mm Hg})}{(\text{K})(\text{mole})};$$

$$= 1.631 \frac{(\text{feet})^3(\text{millibars})}{(^\circ\text{R})(\text{mole})};$$

= [The Universal Gas Constant can be expressed in a wide variety of units so as to provide a suitable proportionality for any application of the Ideal Gas Law.]

T = the Absolute Temperature of the gas involved, in some appropriate units of absolute temperature.

The Mole

Equation #1-10:

The following Equation, **#1-10**, involves the Mass-Based relationship that defines the **mole**. Specifically, a mole [which for this relationship is equivalent to a gram mole or a gram molecular weight] of any compound or chemical is that quantity, which when weighed will have a weight, IN GRAMS, that is numerically equivalent to that compound's or chemical's molecular weight, when that molecular weight is expressed in Atomic Mass Units [amu]. Analogously, we can also define the pound mole of any compound or chemical as that quantity, which when weighed will have a weight, IN POUNDS, that is numerically equivalent to that compound's or chemical's molecular weight, again when this molecular weight is expressed in Atomic Mass Units [amu]. The pound mole is only very rarely ever used, particularly when compared to the ***SI*** mole — thus, when a reference identifies only "a mole", it can always be assumed that it is the ***SI*** mole that is involved. A pound mole has a mass that is 453.59 times greater than that of an ***SI*** mole.

$$\mathbf{n = \frac{m}{MW}}$$

Where:

n = the number of moles of the chemical or the material being evaluated;

m = the mass, measured in grams, of the chemical or the material being evaluated; &

MW = the molecular weight of the chemical or the material being evaluated, expressed in Atomic Mass Units [amu].

Equation #1-11:

The following Equation, **#1-11**, involves the numerical count-based relationship that also defines the mole. A mole [i.e., a gram mole] of any compound or chemical will always contain Avogadro's Number, usually designated as $\mathbf{N_A}$, which equals — numerically — 6.022×10^{23} particles [i.e., molecules, atoms, ions, etc.].

$$\mathbf{n = \frac{Q}{N_A} = \frac{Q}{6.022 \times 10^{23}}}$$

Where:

n = the number of moles, as defined above on this as well as on the previous page, in the discussion of Equation **#1-10**;

Q = the actual count or number of particles [i.e., molecules, atoms, ions, etc.] of the chemical or material being evaluated; &

$\mathbf{N_A}$ = Avogadro's Number = 6.022×10^{23}

Equation #s 1-12 & 1-13:

The following two Equations, **#s 1-12** & **1-13**, incorporate a third relationship that can be used to identify the quantity of any substance. Again, these relationships also serve to define the mole. In these cases the relationships involved are volumetric rather than weight based. It is known that a mole of any compound or chemical that exists as a gas or vapor will occupy a very specific volume under the prevailing conditions of pressure and temperature at which the evaluation is being made. Under STP Conditions, the Molar Volume_{STP} = 22.414 liters; under NTP Conditions, the Molar Volume_{NTP} = 24.465 liters.

Equation #1-12:

$$\mathbf{n = \frac{V_{sample}}{V_{molar-STP}} = \frac{V_{sample}}{22.414}} \quad \text{@ } \mathbf{STP} \text{ Conditions}$$

Where:

n = the number of moles, as defined on Pages 1-19 & 1-20, in Equation **#1-10**;

V_{sample} = the volume, expressed in liters, of the sample of gaseous material being evaluated — the sample for which the specific quantity or amount of material present must be determined; &

$V_{molar-STP}$ = the Molar $Volume_{STP}$, expressed in liters, under conditions **S**tandard **T**emperature and **P**ressure — under these specific conditions of temperature and pressure, this Molar Volume is **22.414 liters**.

Equation #1-13:

$$n = \frac{V_{sample}}{V_{molar-NTP}} = \frac{V_{sample}}{24.465} \quad @ \textbf{ NTP} \text{ Conditions}$$

Where:

n = the number of moles, as defined on Pages 1-19 & 1-20, in Equation **#1-10**;

V_{sample} = the volume, expressed in liters, as defined above on this, as well as on the previous page, under Equation **#1-12**; &

$V_{molar-NTP}$ = the Molar $Volume_{NTP}$, expressed in liters, under conditions **N**ormal **T**emperature and **P**ressure — under these specific conditions of temperature and pressure, this Molar Volume is **24.465 liters**.

Other Gas Laws

Equation #1-14:

The following Equation, **#1-14**, provides a vehicle for relating the Density of <u>any</u> gas to its Pressure and Absolute Temperature characteristics.

$$\frac{\rho_1 T_1}{P_1} = \frac{\rho_2 T_2}{P_2}$$

Where:

P_1 & P_2 are the Pressures of the gas of interest at each of its two states, as was defined for Equation **#1-5** on Page 1-17;

T_1 & T_2 are the Absolute Temperatures of the gas of interest at each of its two states, as was defined for Equation **#1-6** on Page 1-17;

ρ_1 = the Density of the gas @ Time #1, measured commonly, in units of $\frac{\text{grams}}{\text{cm}^3}$; &

ρ_2 = the Density of the gas @ Time #2, also measured in units of $\frac{\text{grams}}{\text{cm}^3}$, but most importantly, in the same units as was the case for ρ_1.

Equation #s 1-15 & 1-16:

The following two Equations, #s **1-15** & **1-16**, are expressions of Dalton's Law of Partial Pressures; in general, this Law simply states that the total pressure exerted by the mixture of different gases in any volume will be, and is, equal to the sum of the partial pressures of each individual component in the overall gas mixture. Dalton's Law of Partial Pressures also states that each individual component in any gas mixture exerts its own individual partial pressure, in the exact ratio as its mole fraction in that mixture.

Equation #1-15:

$$\mathbf{P_{total} = \sum_{i=1}^{n} P_i = P_1 + P_2 + \ldots + P_n}$$

Where:

$\mathbf{P_{total}}$ = the Total Pressure of some volume or system containing a total of "**n**" different identifiable gaseous components, with all pressures measured in some suitable and consistent units; &

$\mathbf{P_i}$ = the Partial Pressure of the **i**th member of the "**n**" total gaseous components of the entire mixture being considered, again with all pressures measured in some suitable and consistent units.

The second portion of the overall statement of Dalton's Law of Partial Pressures involves the previously mentioned relationship of the partial pressure of any component to its mole fraction in the entire mixture. This relationship also provides a way to determine the volume-based Concentration of any individual gaseous component in a mixture of several other gases OR in the ambient air. This relationship can be usefully expressed in a variety of different forms:

Equations #1-16:

$$\mathbf{P_i = P_{total} m_i} \quad \text{or} \quad \mathbf{P_i = \frac{P_{total} C_i}{1,000,000}} \quad \text{or} \quad \mathbf{C_i = \frac{[1,000,000] P_i}{P_{total}}}$$

Where:

$\mathbf{P_i}$ = the Partial Pressure of the **i**th member of the "**n**" total gaseous components of the entire mixture being considered, as defined on the previous page, namely, Page 1-22, for Equation **#1-15**;

P_{total} = the Total Pressure of some volume or system containing a total of "**n**" different identifiable gaseous components, as defined on the previous page, namely, Page 1-22, for Equation **#1-15**;

m_i = the Mole Fraction of the **i**th component in the entire mixture of "**n**" total gaseous components; &

C_i = the Concentration of the **i**th component in the entire mixture of "**n**" total gaseous components, expressed in ppm(vol).

Equation #s 1-17 & 1-18:

The following two relationships, Equations **#s 1-17** & **1-18**, are known as Raoult's Law. This Law involves the relationship between the partial vapor pressure of each of the components in a solution containing two or more volatile components — i.e., the partial pressure of each component in the mixed vapor phase that exists in equilibrium with the solution — AND the solution mole fraction of each component [**NOT** the vapor phase mole fraction!!!]. The basic relationship utilizes the Vapor Pressure each component would exert if it was alone in the pure liquid state at the temperature of the solution. The first of these two relationships involves a consideration of just one single component member of the solution being considered.

Equation #1-17:

$$PVP_i = m_i VP_i \quad \text{for the } i\text{th component in the solution}$$

Where:

PVP_i = the Partial Vapor Pressure — in the vapor space above the solution — produced by the **i**th component of a solution containing "**n**" different components, with this pressure measured in some suitable and consistent units of pressure; &

m_i = the Solution Mole Fraction of the **i**th component in a solution containing a total of "**n**" different components; &

VP_i = the **Vapor Pressure** of the "pure" **i**th component of the solution, always listed as a function of its temperature, with this pressure measured in some suitable and consistent units of pressure.

Equation #1-18:

This second expression involves considering the multi-component solution as a whole.

$$\mathbf{TVP_{solution} = \sum_{i=1}^{n} PVP_i = \sum_{i=1}^{n} m_i VP_i = m_1VP_1 + m_2VP_2 + \cdots + m_nVP_n}$$

Where: $\mathbf{TVP_{solution}}$ = the Total Vapor Pressure for all the volatile components in the solution, with this pressure also measured in some suitable and consistent units of pressure; &

$\mathbf{PVP_i}$, $\mathbf{m_i}$, & $\mathbf{VP_i}$ are all as defined for Equation **#1-17**, on the previous page, namely, Page 1-23.

Equation #1-19:

The following relationship, Equation **#1-19**, constitutes the definition of Absorbance. Absorbance is the most commonly used measure of the decrease in intensity of any beam of electromagnetic radiation, so long as this decrease has been produced by the absorption of some fraction of the energy of that beam by the components of the matrix through which it is passing. Absorbance is a dimensionless parameter, but its quantitative measure is frequently referred to and identified as "Absorbance Units".

$$\mathbf{A} = log\frac{\mathbf{I_o}}{\mathbf{I}}$$

Where: $\mathbf{A}$ = Absorbance produced in a beam of electromagnetic energy produced by the absorption of some fraction of that beam by components of the matrix through which that beam has passed, measured in Absorbance Units or AUs.

$\mathbf{I_0}$ = the Intensity, or Total Energy, of the beam as it enters the matrix through which it will be passing, &

$\mathbf{I}$ = the Intensity, or Total Energy, of the beam as it exits the matrix through which it has passed.

Settling Velocity of Various Types of Particulates Suspended in Air

Equation #s 1-20 & 1-21:

These two Equations, **#s 1-20** & **1-21**, are, in order: (1) for Equation **#1-20**, the rigorous statement of Stoke's Law, which relates the terminal, or settling velocity, in air of any suspended particle [i.e., dusts, fumes, mists, etc.] to the combination of that particle's effective diameter and density, and (2) for Equation **#1-21**, the most useful form of this equation, simplified from its more precise cousin by ignoring the contribution of the numerically very small density of air. In every case, the density of the settling particle must be greater than the density of the air – which will virtually always be less than 0.0013 grams/cm^3 — in order for the particle to settle at all. Particles less dense than air are extremely rare, and because of their low density, they will never settle at all. At the other extreme, particles with an effective diameter greater than 50 to 60 microns will simply "fall" to earth, rather than "settle"; thus this approximation to Stoke's Law [Equation **#1-20**] should never be used in any case where the effective diameter of the particle being considered is at or above this size range.

Equation #1-20:

$$V_s = 0.003\left[\rho_{particle} - \rho_{air}\right]d^2_{particle}$$

Where:

V_s = the Settling Velocity in air of the particle being considered, measured in cm/sec;

$\rho_{particle}$ = the Density of the particle being considered, measured in gms/cm^3;

$\rho_{air\text{-}STP}$ = the Density of air under STP conditions, identified as 0.00129 gms/cm^3;

$\rho_{air\text{-}NTP}$ = the Density of air under NTP conditions, identified as 0.00118 gms/cm^3; &

$d_{particle}$ = the Effective Diameter of the particle, measured in microns, (μ).

Equation #1-21:

$$V_s = 0.003\left[\rho_{particle}\right]\left[d^2_{particle}\right]$$

Where:

V_s = the Settling Velocity in air of the particle being considered, measured in cm/sec;

$\rho_{particle}$ = the Density of the particle being considered, as defined above on this page for Equation #1-20; &

$d_{particle}$ = the Effective Diameter of the particle, also as defined above on this page for Equation #1-20.

In general, as stated above, the density of most particulate matter will always be far greater than the density of air in which the particles being considered are suspended; therefore, we can simply ignore the contribution to this quantitative relationship, represented by Equation **#1-20**, that results from the subtraction of the extremely small density of air from the very large density of the particulate matter being considered. Equation **#1-21** is the result of this simplification.

BASIC PARAMETERS AND LAWS OF PHYSICS & CHEMISTRY PROBLEM SET

Problem #1.1:

The boiling point of Oxygen has been determined to be –182.95°C. What is the value of this temperature when expressed as an Absolute Temperature in the MKS System of units?

Applicable Definition:	Temperature	Page 1-2
Applicable Formula:	Equation **#1-1**	Page 1-16
Solution to this Problem:	Page 1-46	

Problem Workspace

Problem #1.2:

The approximate melting point of Tungsten [Wolfram] is 6,651.3°R. To what temperature, in degrees Fahrenheit, is this very high melting point temperature equivalent?

Applicable Definition:	Temperature	Page 1-2
Applicable Formula:	Equation **#1-2**	Page 1-16
Solution to this Problem:	Page 1-46	

Problem Workspace

Problem #1.3:

A Laboratory Technician has determined that the melting point of napthalene is 176.4°F, using a calorimeter equipped with a mercury capillary thermometer that reads out temperatures in degrees Fahrenheit. Literature references list the melting point of napthalene as 80.2°C. Was the thermometer in this calorimeter accurate [i.e., was its response within a tolerance of ± 0.2°C]?

Applicable Definition:	Temperature	Page 1-2
Applicable Formula:	Equation **#1-3**	Page 1-16
Solution to this Problem:	Page 1-46	

Problem Workspace

Problem #1.4:

The element mercury has a melting point of –38.83°C, and a boiling point of 356.73°C. What is the temperature difference between the melting and boiling points of mercury when expressed in degrees Fahrenheit?

Applicable Definition:	Temperature	Page 1-2
Applicable Formula:	Equation **#1-4**	Pages 1-16 & 1-17
Solution to this Problem:	Pages 1-46 & 1-47	

Problem Workspace

Problem #1.5:

A 100-liter meteorological balloon is filled with helium at NTP. What will its volume become at an altitude of 50,000 ft where the barometric pressure is 175 millibars?

Applicable Definition:	Normal Temperature & Pressure	Page 1-14
Applicable Formula:	Equation **#1-5**	Page 1-17
Solution to this Problem:	Page 1-47	

Problem Workspace

Problem #1.6:

A spherical bladder has been filled with air at a pressure of one atmosphere. What must the external pressure on this bladder become in order to reduce its diameter by 75%?

Applicable Formula:	Equation **#1-5**	Page 1-17
Solution to this Problem:	Page 1-47	

Problem Workspace

Problem #1.7:

A one-liter balloon is filled to capacity at STP, and then placed in a commercial freezer which operates at –13°F. By how much will the volume of the balloon decrease?

Applicable Definition:	Standard Temperature & Pressure	Page 1-14
Applicable Formulae:	Equation **#1-2**	Page 1-16
	Equation **#1-6**	Page 1-17
Solution to this Problem:	Pages 1-47 & 1-48	

Problem Workspace

Problem #1.8:

A soap bubble has a volume of 46 ml at a room temperature of 77°F. When this bubble passed through the beam of a heat lamp, the temperature of the air inside it increased until it [the bubble] burst. When it burst, the bubble volume had increased to 46.5 ml. By how many degrees Fahrenheit had the air temperature in this soap bubble increased at its moment of its destruction?

Applicable Formulae:	Equation **#1-2**	Page 1-16
	Equation **#1-6**	Page 1-17
Solution to this Problem:	Pages 1-48 & 1-49	

Workspace for Problem **#1.8**

Problem #1.9:

A capped soft drink bottle has an internal pressure of 950 mm Hg at its refrigerator temperature of 3°C. If this bottle is accidentally left, unprotected, in the trunk of a car during the summer, and undergoes a temperature increase to 45°C, what will be its internal pressure (in mm Hg)?

Applicable Formulae:	Equation **#1-1**	Page 1-16
	Equation **#1-7**	Page 1-18
Solution to this Problem:	Page 1-49	

Problem Workspace

Problem #1.10:

In the morning, after a car has been parked overnight, its tires will typically be at the same temperature as the atmosphere. For one car, the morning ambient temperature was 56°F. On that particular morning, the pressure in its tires was 32 psig. What will this tire pressure be (in psig) after the car has been driven through a hot desert environment causing its tires to be heated to 175°F?

Applicable Formulae:	Equation **#1-2**	Page 1-16
	Equation **#1-7**	Page 1-18
Solution to this Problem:	Pages 1-49 & 1-50	

Problem Workspace

Problem #1.11:

A 1.0-liter cylinder containing chlorine at 5 atmospheres and 77°F is used to fill a balloon at STP. What will be the volume of the chlorine resistant balloon (in liters), after all the chlorine that was in the cylinder has been transferred into the balloon?

Applicable Definition:	Standard Temperature & Pressure	Page 1-14
Applicable Formulae:	Equation **#1-2**	Page 1-16
	Equation **#1-8**	Page 1-18
Solution to this Problem:	Page 1-50	

Problem Workspace

Problem #1.12:

A partially filled cylinder, having an internal volume of 30 liters, contained N_2O, and was located in the sunlight. It showed an internal pressure of 450 psia. This cylinder was used to provide a flow of N_2O, laughing gas, for anesthesia purposes at a flow rate of 3,000 cc/min for a time period of 4 hours and 30 minutes, or until the cylinder's internal pressure had dropped to 766 mm Hg. If this flow was delivered at a pressure of 766 mm Hg, and a temperature of 20°C, what must have been the temperature of the sun-heated cylinder, in degrees Celsius, at the start of the process?

Applicable Formulae:	Equation **#1-1**	Page 1-16
	Equation **#1-8**	Page 1-18
Solution to this Problem:	Pages 1-50 & 1-51	

Continuation of Workspace for Problem **#1.12**

Problem #1.13:

Approximately 34 grams of a refrigerant occupy 8.58 liters at a temperature of 18°C and a pressure of 0.92 atmospheres. It is known that this refrigerant is one of the following three. Please identify the correct material and justify your choice.

No.	Refrigerant	Formula	Molecular Weight
1.	Freon 12	CCl_2F_2	120.91
2.	Freon 21	$CHFCl_2$	102.92
3.	Freon 22	$CHClF_2$	86.48

Applicable Formulae:	Equation **#1-1**	Page 1-16
	Equation **#1-9**	Pages 1-18 & 1-19
	Equation **#1-10**	Pages 1-19 & 1-20
Solution to this Problem:	Pages 1-51 & 1-52	

Continuation of Workspace for Problem **#1.13**

Problem #1.14:

What is the "Effective Molecular Weight" of the Air if 4.36 liters of it weigh 5.00 grams at a temperature of 31°C and a pressure of 755 mm Hg?

Applicable Formulae:	Equation **#1-1**	Page 1-16
	Equation **#1-9**	Pages 1-18 & 1-19
	Equation **#1-10**	Pages 1-19 & 1-20
Solution to this Problem:	Pages 1-52 & 1-53	

Continuation of Workspace for Problem **#1.14**

Problem #1.15:

What will be the weight of 4.4 moles of Chile Salt Peter, $NaNO_3$? The following atomic weights may be of interest:

Element	Atomic Weight
Sodium	24.31 amu
Nitrogen	14.01 amu
Oxygen	16.00 amu

Applicable Definition:	The Amount of Any Substance	Page 1-3
Applicable Formula:	Equation **#1-10**	Pages 1-19 & 1-20
Solution to this Problem:	Page 1-53	

Problem Workspace

Problem #1.16:

The actual weight of an individual atom of the most commonly occurring isotope of copper is 1.045×10^{-13} nanograms [remember: 1 gram = 1,000,000,000 nanograms]. What is the atomic weight of this isotope of copper?

Applicable Definition:	The Amount of Any Substance	Page 1-3
Applicable Formula:	Equation **#1-11**	Page 1-20
Solution to this Problem:	Pages 1-53 & 1-54	

Problem Workspace

Problem #1.17:

The STP volume of a sample of helium gas [atomic weight = 4.00 amu] was found to be 3,400 cc. How many moles of helium were in the sample, and what did it weigh?

Applicable Definitions:	The Amount of Any Substance	Page 1-3
	Standard Temperature & Pressure	Page 1-14
Applicable Formulae:	Equation **#1-10**	Pages 1-19 & 1-20
	Equation **#1-12**	Pages 1-20 & 1-21
Solution to this Problem:	Page 1-54	

Problem Workspace

Problem #1.18:

During the decade of the '80s, one of the most commonly used hospital gas sterilants was a product known as Oxyfume 12®. This material was sold in pressurized cylinders and was blended by its manufacturer, Union Carbide, to contain 12%, by weight, of ethylene oxide [Molecular Wt. = 44.05], with the balance being Freon 12 [Molecular Wt. = 120.92]. What are the proportions of this gas mixture when expressed in % by volume? In the event you are interested you should know that % by volume = % by the actual count of molecules present.

® Registered Trademark of the Union Carbide Corporation

Applicable Definition:	The Amount of Any Substance	Page 1-3
Applicable Formula:	Equation **#1-10**	Pages 1-19 & 1-20
Solution to this Problem:	Pages 1-54 & 1-55	

Workspace for Problem **#1.18**

Problem #1.19:

It is known that 1.0 therm of a gaseous fuel will occupy a volume of 22.1 liters as it issues from a standard pipeline compressor. If the output pressure and temperature of this compressor are 225 psia and 55°C, respectively, how many moles of this combustible gas make up 1.0 therm?

Applicable Definition:	The Amount of Any Substance	Page 1-3
Applicable Formulae:	Equation **#1-1**	Page 1-16
	Equation **#1-9**	Pages 1-18 & 1-19
Solution to this Problem:	Page 1-55	

Problem Workspace

Workspace Continued on the Next Page

Continuation of Workspace for Problem **#1.19**

Problem #1.20:

In order to make Phineas Fogg's hot air balloon lift its 1,650 lbs load, it is necessary to decrease the density of the air in the balloon envelope by only 5%. If Mr. Fogg wishes to start his "80-Days-Around-The-World" odyssey on a day when the ambient air temperature is 55°F, by how many degrees Fahrenheit will he have to increase the temperature of the gas in the balloon envelope to achieve a lift off?

Applicable Formulae:	Equation **#1-2**	Page 1-16
	Equation **#1-14**	Pages 1-21 & 1-22
Solution to this Problem:	Page 1-56	

Problem Workspace

Problem #1.21:

The nominal density of air under NTP conditions is 1.182 mg/cm^3. What will its density be at the booster outlet of a turbocharged internal combustion gasoline engine, where the temperature is 1,400°F and the pressure is 3,460 mm Hg?

Applicable Definition:	Normal Temperature & Pressure	Page 1-14
Applicable Formulae:	Equation **#1-2**	Page 1-16
	Equation **#1-14**	Pages 1-21 & 1-22
Solution to this Problem:	Pages 1-56 & 1-57	

Problem Workspace

Problem #1.22:

Bone dry air contains three principal components. It is made up of 78.1% nitrogen, 20.9% oxygen, and 0.9% argon. The remaining balance of 0.1% is comprised of carbon dioxide, neon, and methane. Expressed in mm Hg, what is the partial pressure of oxygen at NTP? Of argon?

Applicable Definition:	Normal Temperature & Pressure	Page 1-14
Applicable Formula:	Equation **#1-16**	Pages 1-22 & 1-23
Solution to this Problem:	Page 1-57	

Problem Workspace

Problem #1.23:

The Lower Explosive Limit (LEL) for propane in air is 2.1%. Its Upper Explosive Limit (UEL) is 9.5%. Expressing your answer in millibars, what would the STP partial pressures of propane be at each of these concentration levels?

Applicable Definition:	Standard Temperature & Pressure	Page 1-14
Applicable Formula:	Equation **#1-16**	Pages 1-22 & 1-23
Solution to this Problem:	Pages 1-57 & 1-58	

Problem Workspace

Problem #1.24:

The vapor pressure of pure ethanol at 20°C is 44 mm Hg. 90 Proof Irish Whiskey contains 45% ethanol and 55% water, by weight. What would be the partial vapor pressure of ethanol above a shot glass containing 200 ml of 90 Proof Irish Whiskey in a closed room in which the temperature is 20°C, and the barometric pressure, 765 mm Hg? If this shot glass of Irish Whiskey were left uncovered in this room until its contents had come to a full evaporative equilibrium, what would be the ambient concentration of ethanol, in ppm? Should you have any use for the molecular weights of ethanol and water, they are: $MW_{ethanol}$ = 46.07 amu & MW_{water} = 18.02 amu.

Applicable Definition:	The Amount of Any Substance	Page 1-3
Applicable Formulae:	Equation **#1-10**	Pages 1-19 & 1-20
	Equation **#1-16**	Pages 1-22 & 1-23
	Equation **#1-17**	Page 1-23
Solution to this Problem:	Pages 1-58 & 1-59	

Workspace for Problem **#1.24**

Problem #1.25:

It is widely recognized that the specific gravity of Irish Whiskey is the same as that of water. It is desired to add distilled water to the shot glass listed in **Problem #1.24** in order to reduce the ultimate ambient equilibrium concentration level of ethanol that could be achieved in the air of the closed room. If the target ambient concentration level of ethanol is to be its PEL-TWA of 1,000 ppm(vol), how much distilled water must be added to the shot glass? ***Note:*** *It is highly unlikely that any Irishman would enjoy such a dilution of his favorite beverage!*

Applicable Definition:	The Amount of Any Substance	Page 1-3
Applicable Formulae:	Equation **#1-10**	Pages 1-19 & 1-20
	Equation **#1-16**	Pages 1-22 & 1-23
	Equation **#1-17**	Page 1-23
Solution to this Problem:	Pages 1-59 & 1-60	

Workspace for Problem **#1.25**

Problem #1.26:

Opacity measurements on the exhaust stack of any fossil fuel combustion system [i.e., a boiler, a power plant, etc.] are often used to monitor the particulate content of the combustion gas matrix emanating from the stack. An environmental engineer has determined that the stack gas matrix from a coal burning incinerator he is monitoring has reduced the intensity of a beam of visible light passing through it by 72%. What Absorbance did this environmental engineer report?

Applicable Formula:	Equation **#1-19**	Page 1-24
Solution to this Problem:	Page 1-60	

Problem Workspace

Problem #1.27:

The owner of a metallic spring manufacturing company was advised by an Industrial Hygienist that the beryllium dust concentration on his stamping floor was 0.0032 mg/m^3, which was 160% of the PEL-TWA for this material [i.e., the PEL-TWA for beryllium dust is 0.002 mg/m^3]. This IH determined: (1) that the density of beryllium particulates was 1.848 gms/cm^3; and (2) that the average diameter of the particles in question in this factory was 1.5 microns. What settling velocity, in inches/min, did this Industrial Hygienist determine for the beryllium particles in this facility?

Applicable Formula:	Equation **#1-21**	Pages 1-24 & 1-25
Solution to this Problem:	Pages 1-60 & 1-61	

Problem Workspace

Problem #1.28:

The Industrial Hygienist from Problem **#1.27** has further determined that the principal source of the beryllium dust is the facility's stamping presses. He determined that coatings of a silicone agent on the alloys to be stamped would minimize the dust generation, primarily by causing small dust particles to agglomerate into significantly larger ones. If this Industrial Hygienist has determined that he must recommend a procedure that will increase the settling velocity of the beryllium dust particles to a minimum of 72 inches/hour, will any of the following products achieve the desired result? If so, which?

Silicone Designation	Guaranteed Particle Diam. Increase/mg applied
A	1.90 X
B	2.00 X
C	2.10 X

Applicable Formula:	Equation **#1-21**	Page 1-25
Data from:	Problem **#1.27**	Page 1-44
Solution to this Problem:	Page 1-61	

Problem Workspace

SOLUTIONS TO THE BASIC PARAMETERS AND LAWS OF PHYSICS & CHEMISTRY PROBLEM SET

Problem #1.1:

The solution to this problem requires the use of Equation **#1-1**, from Page 1-16:

$$t_{Metric} + 273.16 = T_{Metric} \qquad \text{[Eqn. #1-1]}$$

Where: $t_{Metric} = -182.95°C$, therefore

$$-182.95° + 273.16 = T_{Metric}, \ \&$$

$$\therefore \ T_{Metric} = 90.21 \text{ K}$$

Problem #1.2:

The solution to this problem requires the use of Equation **#1-2**, from Page 1-16:

$$t_{English} + 459.67° = T_{English} \qquad \text{[Eqn. #1-2]}$$

Where: = $T_{English} = 6{,}651.3°R$, therefore

$$6{,}651.3° = t_{English} + 459.67°, \ \&$$

$$\therefore \ t_{English} = 6{,}191.6°F$$

Problem #1.3:

The solution to this problem requires the use of Equation **#1-3**, from Page 1-16:

$$t_{Metric} = \frac{5}{9}\left[t_{English} - 32°\right] \qquad \text{[Eqn. #1-3]}$$

Where: $t_{English} = 176.4°F$, therefore

$$t_{Metric} = \frac{5}{9}\left(176.4° - 32°\right) = \frac{5}{9}\left(144.4°\right) = 80.2°, \ \&$$

$\therefore$ $t_{Metric} = 80.2°C$ & the thermometer appears to be sufficiently accurate (i.e., $\pm 0.2°C$)

Problem #1.4:

The solution to this problem requires the use of Equation **#1-4**, from Pages 1-16 & 1-17:

$$\Delta t_{Metric} = \frac{5}{9}\Delta t_{English} \qquad \text{[Eqn. #1-4]}$$

Where: $\Delta t_{Metric} = 356.73° - (-38.83°) = 395.59°C$, therefore

$$395.59° = \frac{5}{9}\Delta t_{English}, \text{ and}$$

$$\Delta t_{English} = \frac{(9)(395.59^\circ)}{5} = 712.06^\circ F, \&$$

$$\therefore \quad \Delta t_{English} = 712.06^\circ F$$

Problem #1.5:

The solution to this problem requires the use of Boyle's Law, Equation **#1-5**, from Page 1-17:

$$P_1 V_1 = P_2 V_2 \qquad \text{[Eqn. #1-5]}$$

$$(175)V_1 = (1,013.25)(100) = 1.01325 \times 10^5$$

$$V_1 = \frac{1.01324 \times 10^5}{175} = 579$$

$$\therefore \quad V_1 = 579 \text{ liters}$$

Problem #1.6:

The solution to this problem requires the use of Boyle's Law, Equation **#1-5**, from Page 1-17:

$$P_1 V_1 = P_2 V_2 \qquad \text{[Eqn. #1-5]}$$

Also we must remember that the volume of a sphere, V_{sphere}, in terms of its diameter, d, is given by:

$$V_{sphere} = \frac{\pi d^3}{8}$$

We must now select the quantitative values that we will use for the "before" and "after" diameters of the balloon, as we attempt to develop a solution to this problem. Since the balloon is having its diameter decreased by 75%, we can select 4 units as the "before" diameter and 1 unit as the "after" diameter — a diameter decrease of 75% [i.e., from 4 units to 1 unit]. Now applying Boyle's Law:

$$(1)\left(\frac{(\pi)(4)^3}{8}\right) = P_2\left(\frac{(\pi)(1)^3}{8}\right) \quad \text{\& canceling, as appropriate:}$$

$$P_2 = 4^3 = 64$$

$$\therefore \quad P_2 = 64 \text{ atmospheres}$$

Problem #1.7:

This problem requires the use of Charles' Law, Equation **#1-6**, from Page 1-17; however, we will first have to convert the listed relative Fahrenheit temperature to its absolute Rankine equivalent, using Equation **#1-2**, from Page 1-16:

$$t_{English} + 459.67^\circ = T_{English} \qquad \text{[Eqn. #1-2]}$$

$$-13° + 459.67° = T_{English}, \ \&$$

$$T_{English} = 446.67°R$$

Next, we use this calculated absolute temperature, as stated above, in conjunction with Charles' Law, Equation **#1-6**, from Page 1-17, to develop the ultimate answer [remember that a volume of 1 liter = 1,000 ml]; thus:

$$\frac{V_1}{T_1} = \frac{V_2}{T_2} \qquad \text{[Eqn. \#1-6]}$$

$$\frac{V_1}{1,000} = \frac{446.67°}{491.67°}, \ \&$$

$$V_1 = \frac{(1,000)(446.67°)}{491.67°} = 908.5$$

$$V_1 = 908.5 \text{ ml}$$

The result that was asked for was the decrease in the volume of the balloon; therefore, the answer is 91.5 ml [1,000 – 908.5 = 91.5].

∴ The decrease in the balloon's volume = 91.5 ml.

Problem #1.8:

This problem also requires the use of Charles' Law, Equation **#1-6**, from Page 1-17; however, as was the case with Problem **#1.7** above, we must first convert the listed relative Fahrenheit temperature to its absolute Rankine equivalent, using Equation **#1-2**, from Page 1-16:

$$t_{English} + 459.67° = T_{English} \qquad \text{[Eqn. \#1-2]}$$

$$77° + 459.67° = T_{English}, \ \&$$

$$T_{English} = 536.67°R$$

Next, we use this calculated absolute temperature, as stated above, in conjunction with Charles' Law, Equation **#1-6**, from Page 1-17, to develop the information necessary to develop the ultimate answer:

$$\frac{V_1}{T_1} = \frac{V_2}{T_2} \qquad \text{[Eqn. \#1-6]}$$

$$\frac{46.5}{T_1} = \frac{46}{536.67}, \ \&$$

$$T_1 = \frac{(46.5)(536.67)}{46} = \frac{24,955.16}{46} = 542.5°R$$

Now the problem has asked for the ambient temperature increase inside the bubble at the moment of its popping, when its temperature was 542.5°R as calculated above. Since the air in this bubble started out at 536.67°R [from the initial calculation for this problem], we can see that the temperature increase was 5.83°R [542.5° – 536.67° = 5.83°R]. Since the size of a Rankine degree is exactly equal to the size of a Fahrenheit degree, we see that the answer to the problem is 5.83°F.

∴ The increase in the bubble's internal temperature ~ 5.8°F

Problem #1.9:

This problem can be solved by using Gay-Lussac's Law, Equation **#1-7**, from Page 1-18, but first the Celsius temperature provided in the problem statement must be converted to its absolute temperature equivalent using Equation **#1-1**, from Page 1-16:

$$t_{Metric} + 273.16 = T_{Metric} \qquad \text{[Eqn. \#1-1]}$$

$$3 + 273.16 = T_{Metric-1}, \text{ therefore}$$

$$T_{Metric-1} = T_1 = 276.16 \text{ K}, \&$$

$$45 + 273.16 = T_{Metric-2}, \text{ therefore}$$

$$T_{Metric-2} = T_2 = 318.16 \text{ K}, \&$$

We can now apply Equation **#1-7**, from Page 1-18, to obtain the desired result:

$$\frac{P_1}{T_1} = \frac{P_2}{T_2} \qquad \text{[Eqn. \#1-7]}$$

$$\frac{950}{276.16} = \frac{P_2}{318.16}$$

$$P_2 = \frac{(950)(318.16)}{276.16} = \frac{302,252}{276.16} = 1,094.48$$

∴ The bottle's eventual internal pressure will become ~ 1,094.5 mm Hg at 45°C.

Problem #1.10:

This problem also can be solved by using Gay-Lussac's Law, Equation **#1-7**, from Page 1-18; however, as with most of the previous problems, we must first convert two relative temperatures to their absolute counterparts, using Equation **#1-2**, from Page 1-16:

$$t_{English} + 459.67^\circ = T_{English} \qquad \text{[Eqn. \#1-2]}$$

$$56^\circ + 459.67^\circ = T_{English-1}, \text{ therefore}$$

$$T_{English-1} = T_1 = 515.67^\circ\text{R}, \text{ and}$$

$$175^\circ + 459.67^\circ = T_{English-2}, \text{ therefore}$$

$$T_{English-2} = T_2 = 634.67^\circ\text{R}, \&$$

At this point, we must note that the internal tire pressure has been given in "psig" [pounds per square inch - gauge]. To convert this "relative" pressure to its "absolute" equivalent, we must add the equivalent of one atmosphere to the listed pressure — i.e., 14.70 psia — converting this listed pressure from its "psig" units into "psia" units [pounds per square inch - absolute]. Once this has been accomplished, we can apply Equation **#1-7**, from Page 1-18:

$$\frac{P_1}{T_1} = \frac{P_2}{T_2} \qquad \text{[Eqn. \#1-7]}$$

$$\frac{(p_1 + 14.70)}{634.67^\circ} = \frac{(32 + 14.70)}{515.67^\circ}$$

$$p_1 + 14.70 = \frac{(32 + 14.70)(634.67^\circ)}{515.67^\circ} = \frac{29{,}639.09}{515.67^\circ} = 57.48\text{, therefore}$$

$$p_1 + 14.70 = 57.48, \&$$

$$p_1 = 57.48 - 14.70 = 42.78 \text{ psig}$$

Note, we have determined the requested tire pressure directly in the "relative", psig, units.

> ∴ The tire's internal pressure, when its temperature has reached 175°F, will be ~ 42.8 psig.

Problem #1.11:

This problem can be easily solved by using the General Gas Law, Equation **#1-8**, from Page 1-18, again after converting the relative temperatures to their absolute equivalents, using Equation **#1-2**, from Page 1-16:

$$t_{English} + 459.67^\circ = T_{English} \qquad \text{[Eqn. #1-2]}$$

$$77^\circ + 459.67^\circ = T_{English} = 536.67^\circ R, \&$$

$$32^\circ + 459.67^\circ = T_{English-STP} = 491.67^\circ R$$

Now, we can apply Equation **#1-8**, from Page 1-18:

$$\frac{P_1 V_1}{T_1} = \frac{P_2 V_2}{T_2} \qquad \text{[Eqn. #1-8]}$$

$$\frac{(5)(1)}{536.67^\circ} = \frac{(1)V_2}{491.67^\circ}, \&$$

$$V_2 = \frac{(5)(1)(491.67^\circ)}{(1)(536.67^\circ)} = 4.58 \text{ liters}$$

> ∴ The chlorine resistant balloon will have an STP volume of ~ 4.6 liters.

Problem #1.12:

The primary information for which this problem has asked can also be solved using the General Gas Law, Equation **#1-8**, from Page 1-18; however, as with most of the previous problems, we must first convert the listed Celsius temperature to its Kelvin equivalent using Equation **#1-1**, from Page 1-16:

$$t_{Metric} + 273.16 = T_{Metric} \qquad \text{[Eqn. #1-1]}$$

$$20 + 273.16 = T_{Metric} = 293.16 \text{ K}$$

To solve this problem, we must next calculate the total volume of N_2O that was delivered to the Operating Room. This is readily accomplished by multiplying the known flow rate of N_2O [3,000 cc/minute] by the time period [4 hours & 30 minutes = 4.5 hours] during which that flow rate was maintained, and — for the sake of using suitably sized units — we will convert the calculated volume, in cubic centimeters, to its equivalent volume in liters:

$$V_{total} = \frac{(3,000\ ^{cc}\!/_{min})(4.50\ \text{hours})(60\ ^{min}\!/_{hour})}{1,000\ ^{cc}\!/_{liter}} = 810\ \text{liters}$$

We must also convert the pressure of the N_2O cylinder from psia to mm Hg, using the following relationship:

$$P_{mm\ Hg} = \left[\frac{760\ ^{mm\ Hg}\!/_{atmosphere}}{14.70\ ^{psia}\!/_{atmosphere}}\right] P_{psia}$$

$$P_{mm\ Hg} = \frac{(760)(450)}{14.70} = 23,265.31\ \text{mm Hg}$$

We can now finally apply the General Gas Law, Equation **#1-8**, from Page 1-18, in order to obtain the desired result:

$$\frac{P_1 V_1}{T_1} = \frac{P_2 V_2}{T_2} \qquad \text{[Eqn. \#1-8]}$$

$$\frac{(23,265.31)(30)}{T_1} = \frac{(766)(810)}{293.16}$$

$$T_1 = \frac{(23,265.31)(30)(293.16)}{(766)(810)} = \frac{204,613,714.3}{620,460} = 329.78\ \text{K}$$

Finally we must convert this resultant absolute temperature back to the requested degrees Celsius, again using Equation **#1-1**, from Page 1-16:

$$t_{Metric} + 273.16 = T_{Metric} \qquad \text{[Eqn. \#1-1]}$$

$$t_{Metric} + 273.16 = 329.78\text{, and}$$

$$t_{Metric} = 329.78 - 273.16 = 56.62\ ^\circ\text{C}$$

∴ The sun-heated N_2O cylinder started out at a temperature of ~ 57°C.

Problem #1.13:

This problem can be solved through the use of the Ideal or Perfect Gas Law, Equation **#1-9**, from Pages 1-18 & 1-19, in conjunction or combination with Equation **#1-10**, from Pages 1-19 & 1-20, again after converting the Celsius temperature provided in the problem statement to its Kelvin equivalent, using Equation **#1-1**, from Page 1-16:

$$t_{Metric} + 273.16 = T_{Metric} \qquad \text{[Eqn. \#1-1]}$$

$$18 + 273.16 = T_{Metric} = 291.16\ \text{K}$$

Now we must join Equation **#s 1-9 & 1-10**, from Pages 1-18, 1-19, & 1-20, into a "combined" form that will simplify the determination of the required answer:

$$PV = nRT \qquad \text{[Eqn. \#1-9]}$$

&

$$n = \frac{m}{MW} \qquad \text{[Eqn. \#1-10]}$$

Combining these two Equations, we arrive at a single relationship that will permit us to proceed directly to the answer requested:

$$PV = \left[\frac{m}{MW}\right]RT, \;\&$$

$$MW = \frac{mRT}{PV}$$

Now substituting known values, and using a value for the Universal Gas constant in units that are consistent with those provided in the problem statement, we see that:

$$R = 0.0821\ ^{(liter)(atmospheres)}\!/_{(K)(mole)},\ \text{also from Page 1-19}$$

$$MW = \frac{(34)(0.0821)(291.16)}{(0.92)(8.58)} = \frac{812.74}{7.89} = 102.96\ \text{amu}$$

∴ Since the molecular weight of the refrigerant is approximately 102.96 amu, we can conclude that the material in question is most likely Freon 21, for which the molecular weight is 102.92 amu.

Problem #1.14:

This problem can be solved in the same manner as Problem **#1.13**, namely, through the use of the Ideal Gas Law, Equation **#1-9**, from Pages 1-18 & 1-19, again in conjunction or combination with Equation **#1-10**, from Pages 1-19 & 1-20. As before, we must again convert the Celsius temperature provided in the problem statement to its Kelvin equivalent, using Equation **#1-1**, from Page 1-16:

$$t_{Metric} + 273.16 = T_{Metric} \qquad \text{[Eqn. #1-1]}$$

$$31 + 273.16 = T_{Metric} = 304.16\ K$$

Now we must again join Equation **#s 1-9** & **1-10**, from Pages 1-18, 1-19, & 1-20, into a form that will simplify the determination of the required answer:

$$PV = nRT \qquad \text{[Eqn. #1-9]}$$

&

$$n = \frac{m}{MW} \qquad \text{[Eqn. #1-10]}$$

Combining these two Equations, we again arrive at the single relationship that will permit us to proceed directly to the answer requested in the problem statement:

$$PV = \left[\frac{m}{MW}\right]RT, \;\&$$

$$MW = \frac{mRT}{PV}$$

Let us again substitute in known values in order to obtain the required answer. Note that in this case we will use a value for the Universal Gas Constant, R, again from Page 1-19, choosing units that are consistent with the data provided in the problem statement; we see that:

$$R = 62.36\ ^{(liter)(mm\ Hg)}\!/_{(K)(mole)}$$

$$MW = \frac{(5)(62.36)(304.16)}{(755)(4.36)} = \frac{94,837.09}{3,291.8} = 28.81 \text{ amu}$$

∴ The "Effective Molecular Weight" of air is ~ 28.81 amu.

Problem #1.15:

This problem will employ Equation **#1-10**, from Pages 1-19 & 1-20:

$$n = \frac{m}{MW} \qquad \text{[Eqn. \#1-10]}$$

Transposing to a form better suited to the solution of this problem, we get:

$$m = (n)(MW)$$

Prior to performing the final calculations, we must determine the molecular weight of Chile Salt Peter, $NaNO_3$:

One atom of sodium @ 24.31 amu each	=	24.31 amu
One atom of nitrogen @ 14.01 amu each	=	14.01 amu
Three atoms of oxygen @ 16.00 amu each	=	48.00 amu
Molecular Weight of $NaNO_3$	=	86.32 amu

Therefore:

$$m = (4.4)(86.32) = 379.81 \text{ grams}$$

∴ 4.4 moles of $NaNO_3$ weigh 379.81 grams.

Problem #1.16:

To solve this problem, we must apply Equation **#1-11**, from Page 1-20. We will seek always to be dealing with a one mole quantity of these atoms of copper. Remember, in order to have one mole [i.e., for n = 1] of anything, it is clear that we must have Avogadro's Number of the items being considered. In this case, we must have Avogadro's Number of the atoms of this specific isotope of copper, namely, 6.022×10^{23} of them. If we have this number, then the weight of this total group of atoms, in grams, will be numerically equal to the atomic weight of this isotope of copper.

$$n = \frac{Q}{N_A} = \frac{Q}{6.022 \times 10^{23}} \qquad \text{[Eqn. \#1-11]}$$

Clearly, for one mole [i.e., for n = 1] we have:

$$1 = \frac{Q}{6.022 \times 10^{23}}, \ \&$$

$$Q = 6.022 \times 10^{23} \text{ atoms}$$

We are told that each atom weighs 1.045×10^{-13} nanograms. This can be converted to grams by dividing by 1,000,000,000 or, alternatively, by multiplying by 10^{-9}. This means that one atom of this isotope of copper weighs 1.045×10^{-22} grams. Therefore, Avogadro's Number of these atoms will weigh as follows [remember: this calculated weight — which will be expressed in grams — will be numerically equal to the actual atomic weight of the copper isotope, so long as this latter parameter is expressed in Atomic Mass Units].

$$W = (N_A)(\text{weight of one atom})$$

$$W = \left(6.022 \times 10^{23}\right)\left(1.045 \times 10^{-22}\right) = 62.93 \text{ amu}$$

∴ The atomic weight of this isotope of Copper is ~ 62.93 amu. As indicated in the problem statement, this is the most commonly occurring isotope of this metal [69.17% of all the copper produced on this planet], and its designation is ${}^{63}_{29}Cu$.

Problem #1.17:

To solve this problem, we must first apply Equation **#1-12**, from Pages 1-20 & 1-21, and then Equation **#1-10**, from Pages 1-19 & 1-20:

$$n = \frac{V_{sample}}{V_{molar-STP}} = \frac{V_{sample}}{22,414} \qquad \text{[Eqn. \#1-12]}$$

$$n = \frac{3,400}{22,414} = 0.152 \text{ moles}$$

We can finally now apply Equation **#1-10**, from Pages 1-19 & 1-20:

$$n = \frac{m}{MW} \qquad \text{[Eqn. \#1-10]}$$

Transposing this relationship:

$$m = (n)(MW), \&$$

$$m = (0.152)(4.00) = 0.607 \text{ grams} = 607 \text{ mg}$$

∴ There are 0.152 moles of Helium in 3,400 cc of this gas, at STP; and 0.152 moles of Helium weigh ~ 0.607 grams, or ~ 607 mg.

Problem #1.18:

To solve this problem, we must apply Equation **#1-10**, from Pages 1-19 & 1-20, to determine the actual number of moles of each component in the Oxyfume 12®, and then deal with the basic definition for volume percent of a component, as this concept relates to the actual number of moles of any component and/or the mole fraction of that component.

Let us begin by assuming that we are dealing with exactly 1,000 grams of Oxyfume 12®. Obviously, this quantity of this material will contain 120 grams of ethylene oxide and 880 grams of Freon 12. We must next determine the number of moles of each of these materials that would be in this 1,000 gram quantity; this will give us the necessary information to determine the mole fractions of each. Using Equation **#1-10**, from Pages 1-19 & 1-20:

$$n = \frac{m}{MW} \qquad \text{[Eqn. \#1-10]}$$

$$n_{ethylene\ oxide} = \frac{120}{44.05} = 2.72 \text{ moles of ethylene oxide in 1,000 grams of Oxyfume 12®}$$

$$n_{Freon\ 12} = \frac{880}{120.92} = 7.28 \text{ moles of Freon 12 in 1,000 grams of Oxyfume 12®}$$

Therefore the total number of moles present in 1,000 grams of Oxyfume 12® is 10.00 [i.e., 2.72 + 7.28 = 10.00 moles total]. From this data we can see that the mole fractions of these two components are:

For ethylene oxide: $$mf_{\text{ethylene oxide}} = \frac{2.72}{10} = 0.272$$

For Freon 12: $$mf_{\text{Freon 12}} = \frac{7.28}{10} = 0.728$$

By definition, the mole fraction for each component in a mixture or solution can be converted to its volume percentage simply by multiplying the mole fraction by 100.

> ∴ Oxyfume 12® is made up as follows:
> ethylene oxide: ~ 27.2% by volume
> Freon 12: ~ 72.8% by volume

Problem #1.19:

To solve this problem, we must apply Equation **#1-9**, from Pages 1-18 & 1-19, after first correcting the listed Celsius temperature to its Kelvin equivalent using Equation **#1-1**, from Page 1-16:

$$t_{\text{Metric}} + 273.16 = T_{\text{Metric}} \qquad \text{[Eqn. \#1-1]}$$

$$55 + 273.16 = T_{\text{Metric}} = 328.16 \text{ K}$$

Since we do not have any tabulated Universal Gas Constant, R, in units of $^{(\text{liter})(\text{psia})}\!/_{(\text{K})(\text{mole})}$, we have two choices:

(1) We can calculate a new listing of the Universal Gas Constant, converting from its listed units into units of $^{(\text{liter})(\text{psia})}\!/_{(\text{K})(\text{mole})}$ so as to be consistent with the units provided in the problem statement, or

(2) We can simply convert the pressure, listed in the problem statement in units of "psia", to a pressure unit that is consistent with the units of one of the listed choices for the Universal Gas Constant — i.e., mm Hg, atmospheres, etc. — that appear in one of the listed Universal Gas Constants, from Page 1-19.

For the sake of instruction, we will choose the slightly more difficult option, namely, Option (1):

$$\left[0.0821\,^{(\text{liter})(\text{atmosphere})}\!/_{(\text{K})(\text{mole})}\right]\left[14.70\,^{\text{psia}}\!/_{\text{atmosphere}}\right] = R = 1.207\,^{(\text{liter})(\text{psia})}\!/_{(\text{K})(\text{mole})}$$

Now with this value of the Universal Gas Constant in units consistent with those in the problem statement, we can apply Equation **#1-9**, from Pages 1-18 & 1-19:

$$PV = nRT \qquad \text{[Eqn. \#1-9]}$$

$$(225)(22.1) = n(1.207)(328.16), \;\&$$

$$n = \frac{(225)(22.1)}{(1.207)(328.16)} = \frac{4,972.5}{396.09} = 12.55$$

> ∴ 12.55 moles of this gaseous fuel make up 1.0 therm.

Problem #1.20:

To solve this problem, we must apply Equation **#1-14**, from Pages 1-21 & 1-22, after first correcting the Fahrenheit temperature listed in the problem statement to its Rankine equivalent, using Equation **#1-2** from Page 1-16:

$$t_{English} + 459.67^{\circ} = T_{English} \qquad \text{[Eqn. \#1-2]}$$

$$55^{\circ} + 459.67^{\circ} = T_{English} = 514.67^{\circ}R$$

Also, we must remember the relationship, as stated in the problem, between the starting density of the cool, unheated air, and the density of this same air, once it has been heated to a sufficiently high temperature to achieve a liftoff of the 1,650 lbs load, namely:

$$\frac{\rho_{hot\ air}}{\rho_{cool\ air}} = 0.95\text{, or}$$

$$\rho_{hot\ air} = (0.95)(\rho_{cool\ air})$$

At this point, and using this [hot air density] vs. [cold air density] relationship, we can apply Equation **#1-14**, from Pages 1-21 & 1-22, to obtain the desired problem solution:

$$\frac{\rho_1 T_1}{P_1} = \frac{\rho_2 T_2}{P_2} \qquad \text{[Eqn. \#1-14]}$$

$$\frac{(0.95)(\rho_{cool\ air})T_1}{P} = \frac{(\rho_{cool\ air})(514.67^{\circ})}{P}$$

Note finally that the atmospheric pressure, designated above as "P", both before and after the balloon liftoff is the same, and that this is the pressure that characterizes both the heated air in the balloon envelope, and the unheated ambient air that surrounds it. In addition, the density of cool air, designated as $\rho_{cool\ air}$, is also obviously equal to itself; therefore, these two terms, which appear on both sides of this equation, cancel out giving us the following relationship:

$$(0.95)T_1 = 514.67^{\circ}\text{, \&}$$

$$T_1 = \frac{514.67^{\circ}}{0.95} = 541.76^{\circ}R$$

Finally, we must determine the change in temperature, as required in the problem statement, that was necessary to achieve the liftoff:

$$\Delta T = 541.76^{\circ} - 514.67^{\circ} = 27.09^{\circ}F$$

∴ The air in the balloon envelope must undergo an increase in temperature of ~ 27.1°F in order to achieve a liftoff of Phineas Fogg's hot air balloon.

Problem #1.21:

To solve this problem, we must again use Equation **#1-14**, from Pages 1-21 & 1-22, but we must first convert the listed Fahrenheit temperature to its Rankine equivalent, using Equation **#1-2**, from Page 1-16:

$$t_{English} + 459.67^{\circ} = T_{English} \qquad \text{[Eqn. \#1-2]}$$

$$1,400^{\circ} + 459.67^{\circ} = T_{English} = 1,859.67^{\circ}R$$

We can now use Equation **#1-14**, from Pages 1-21 & 1-22, to obtain the solution to the problem:

$$\frac{\rho_1 T_1}{P_1} = \frac{\rho_2 T_2}{P_2} \qquad \text{[Eqn. \#1-14]}$$

$$\frac{\rho_1(1,859.67^\circ)}{3,460} = \frac{(1.182)(536.67^\circ)}{760}, \;\&$$

$$\rho_1 = \frac{(1.182)(536.67^\circ)(3,460)}{(760)(1,859.67^\circ)}$$

$$\rho_1 = \frac{2,194,830.03}{1,413,349.20} = 1.553 \text{ mg/cm}^3$$

∴ The density of air at the booster outlet will be 1.553 mg/cm³.

Problem #1.22:

To solve this problem, we must use Dalton's Law of Partial Pressures, Equation **#1-16**, from Pages 1-22 & 1-23. Remember that the volume percentage of any component in a gas mixture, when divided by 100, will give the mole fraction of that component; thus:

$$P_i = P_{total}\, m_i \qquad \text{[Eqn. \#1-16]}$$

Modifying this relationship to account for the percent-based information as stated in the problem, we have the following relationship:

$$P_i = \frac{P_{total}\, m_i}{100}$$

For Oxygen:
$$P_{Oxygen} = \frac{(760)(20.9\%)}{100\%} = 158.84 \text{ mm Hg}$$

For Argon:
$$P_{Argon} = \frac{(760)(0.9\%)}{100\%} = 6.84 \text{ mm Hg}$$

∴ The partial pressures of these two components of air at NTP are:
Oxygen: ~ 158.8 mm Hg
Argon: ~ 6.8 mm Hg

Problem #1.23:

To solve this problem, we must again use Dalton's Law of Partial Pressures, Equation **#1-16**, from Pages 1-22 & 1-23. For this problem, too, recall that the volume percentage of any component in a gas mixture, when divided by 100, will give the mole fraction of that component:

$$P_i = P_{total}\, m_i \qquad \text{[Eqn. \#1-16]}$$

Modifying this relationship to account for the percent-based information as stated in the problem, we have the following relationship:

$$P_i = \frac{P_{total}\, m_i}{100}$$

The Propane LEL: $$P_{LEL-Propane} = \frac{(1{,}013.25)(2.1\%)}{100\%} = 21.28 \text{ millibars}$$

The Propane UEL: $$P_{UEL-Propane} = \frac{(1{,}013.25)(9.5\%)}{100\%} = 96.26 \text{ millibars}$$

∴ The partial pressures of propane at its LEL & UEL are:
Propane LEL: ~ 21.3 millibars
Propane UEL: ~ 96.3 millibars

Problem #1.24:

To solve this problem, we must use Equation **#1-10**, from Pages 1-19 & 1-20, Raoult's Law, Equation **#1-17**, from Page 1-23, and then finally Equation **#1-16**, from Pages 1-22 & 1-23. In order to determine the mole fractions of ethanol in Irish Whiskey, we will assume that we have a 100-gram sample of this beverage. Because we know the proportions of ethanol and water by weight in Irish Whiskey, it is possible to state that our sample contains exactly 45 grams of ethanol and 55 grams of water. Now using Equation **#1-10**, from Pages 1-19 & 1-20:

$$n = \frac{m}{MW} \qquad \text{[Eqn. \#1-10]}$$

For ethanol: $$n_{ethanol} = \frac{45}{46.07} = 0.977 \text{ moles of ethanol}$$

For water: $$n_{water} = \frac{55}{18.02} = 3.052 \text{ moles of water}$$

Clearly the total number of moles in this 100-gram sample of Irish Whiskey is the sum of these two numbers, namely, 4.029 moles. The mole fraction of the ethanol in this sample is given by:

$$m_{ethanol} = \frac{0.977}{0.977 + 3.052} = \frac{0.977}{4.029} = 0.242$$

Since we now know the mole fraction of this material, we can determine its partial vapor pressure through the use of Raoult's Law, Equation **#1-17**, from Page 1-23:

$$PVP_i = m_i VP_i \qquad \text{[Eqn. \#1-17]}$$

$$PVP_{ethanol} = (0.242)(44) = 10.67 \text{ mm Hg}$$

Finally, we apply Equation **#1-16**, from Pages 1-22 & 1-23, to obtain the equilibrium concentration of ethanol in the ambient air of this room:

$$PVP_i = \frac{P_{total}\, C_i}{1{,}000{,}000} \qquad \text{[Eqn. \#1-16]}$$

Transposing this equation into a more useful form for the purposes of this problem, we have:

$$C_i = \frac{(1{,}000{,}000)(PVP_i)}{P_{total}}, \ \&$$

$$C_{ethanol} = \frac{(1,000,000)(10.67)}{765} = 13,947.7 \text{ ppm of ethanol}$$

∴ The ultimate, ambient, evaporative equilibrium concentration of ethanol that will be achieved in this room is ~ 13,950 ppm.

Problem #1.25:

To solve this problem, we will have to employ Equation **#1-16**, from Pages 1-22 & 1-23, then Equation **#1-17**, from Page 1-23. We must start by determining what partial pressure of ethanol is required if this material is to produce an equilibrium ambient concentration of 1,000 ppm(vol). To do this, we will use Equation **#1-16**, from Pages 1-22 & 1-23:

$$PVP_i = \frac{P_{total} C_i}{1,000,000} \qquad \text{[Eqn. \#1-16]}$$

$$PVP_{ethanol} = \frac{(765)(1,000)}{1,000,000} = \frac{765,000}{1,000,000} = 0.765 \text{ mm Hg}$$

We can now apply Equation **#1-17**, from Page 1-23, to obtain the mole fraction of ethanol that must exist in the liquid phase in order for the partial vapor pressure to be as calculated above:

$$PVP_i = m_i VP_i \qquad \text{[Eqn. \#1-17]}$$

Transposing into a more useful format for determining the mole fraction of ethanol, we have:

$$m_{ethanol} = \frac{PVP_{ethanol}}{VP_{ethanol}}$$

$$m_{ethanol} = \frac{0.765}{44} = 0.0174$$

Now, we must determine the total number of moles of ethanol that were in the original shot glass of Irish Whiskey. We have been told in the problem statement that the specific gravity of Irish Whiskey is the same as that of water (i.e., its density = 1.00 grams/cm^3), and we know that this glass started with 200 ml of this liquid. We can, therefore, conclude that there must have been an initial total of 200 grams of solution in the shot glass. Of this 200 grams, 90 grams (45% by weight) had to have been Irish Whiskey and 110 grams, water. We must now determine the starting number of moles of each of the two components in the shot glass; to do this, we will use Equation **#1-10**, from Pages 1-19 & 1-20:

$$n_i = \frac{m_i}{MW_i} \qquad \text{[Eqn. \#1-10]}$$

For ethanol: $$n_{ethanol} = \frac{m_{ethanol}}{MW_{ethanol}} = \frac{90}{46.07} = 1.954 \text{ moles, \&}$$

For water: $$n_{water} = \frac{m_{water}}{MW_{water}} = \frac{110}{18.02} = 6.104 \text{ moles}$$

Since we earlier calculated the mole fraction of ethanol that would be required to achieve the target ambient ethanol concentration of 1,000 ppm(vol), we can apply the following relationship to determine the mass of water that must be added to achieve the desired necessary mole fraction of ethanol:

$$m_{ethanol} = \frac{n_{ethanol}}{n_{water-initial} + n_{water-added} + n_{ethanol}}$$

$$0.0174 = \frac{1.954}{6.104 + n_{water-added} + 1.954} = \frac{1.954}{n_{water-added} + 8.058}$$

$$n_{water-added} + 8.058 = \frac{1.954}{0.0174} = 112.361\text{, therefore}$$

$$n_{water-added} = 112.361 - 8.058 = 104.303 \text{ moles of water}$$

Now that we know the number of moles of water that must be added, we can readily convert this to a volume of water, which is what was asked for in the problem statement. We do this by first determining the weight of water that must be added; and then, since the density of water is 1.000 grams/cm^3, we see that the number of grams added and/or the volume added (when measured in cm^3) are numerically the same! Thus:

$$m_{water-added} = (104.303)(18.02) = 1{,}879.53$$

∴ The volume of water that must be added to this shot glass in order to achieve the desired equilibrium ambient concentration level of ethanol will be ~ 1,880 ml, or ~ 1.88 liters! Clearly, we will have to transfer the contents of the shot glass to a fairly large pitcher before attempting to add the amount of water required.

Problem #1.26:

The solution to this problem requires the use of Equation **#1-19**, from Page 1-24. We must first assign values to both the entering beam energy, $\mathbf{I_0}$, and the exiting beam energy, **I**. We know that the beam intensity was reduced by 72% as it passed through the gas matrix in the stack; therefore, if we assign an entering beam intensity of 100 units, we see that the exiting beam intensity must have been 28 units. We do not need to be concerned with the specific energy units for these two intensities so long as the units for these two parameters are consistent from with each other. Let us now apply Equation **#1-19** from Page 1-24:

$$A = log \frac{I_0}{I} \qquad \text{[Eqn. #1-19]}$$

$$A = log \frac{100}{28} = log\ 3.571, \&$$

$$A = 0.553 \text{ Absorbance Units}$$

∴ This environmental engineer reported an absorbance of ~ 0.55 AUs.

Problem #1.27:

To solve this problem, we must use Equation **#1-21**, from Pages 1-24 & 1-25 — this is the simplified form of Stoke's Law:

$$V_s = 0.003\left[\rho_{particle}\right]\left[d^2_{particle}\right] \qquad \text{[Eqn. #1-21]}$$

$$V_s = (0.003)(1.848)(1.5)^2 = 0.012 \text{ cm/sec}$$

Now it is necessary to convert this settling velocity into units of inches/minute, as requested in the problem statement:

$$V_s \ ^{in}\!/_{hour} = (0.012 \ ^{cm}\!/_{sec})(0.394 \ ^{in}\!/_{cm})(60 \ ^{sec}\!/_{min})(60 \ ^{min}\!/_{hour}) = 17.021 \text{ inches/hour}$$

∴	It is likely that the Industrial Hygienist calculated a Settling Velocity of ~ 17.02 inches/hour (or ~ 0.12 mm/sec) for the beryllium particles on the stamping floor of this Metallic Spring Manufacturing Company.

Problem #1.28:

To solve this problem, we must again use Equation **#1-21**, from Page 1-25:

$$V_s = 0.003\left[\rho_{particle}\right]\left[d^2_{particle}\right] \qquad \text{[Eqn. \textbf{\#1-21}]}$$

To make the solution to this problem simpler, let us convert this relationship so that it provides a Settling Velocity in inches/hour, rather than cm/sec:

$$(V_s \ ^{cm}\!/_{sec})(0.394 \ ^{in}\!/_{cm})(60 \ ^{sec}\!/_{min})(60 \ ^{min}\!/_{hour}) = V_s \ ^{in}\!/_{hour} = 1,418.4(V_s \ ^{cm}\!/_{sec})$$

$$\text{therefore: } V_s \ ^{in}\!/_{hour} = (0.003)(1,418.4)(\rho_{particle})(d^2_{particle}) = 4.255(\rho_{particle})(d^2_{particle})$$

What we are interested in is the diameter of the suspended beryllium particles, so let us rearrange the foregoing relationship to give this value directly:

$$d^2_{particle} = \frac{V_s}{(4.255)(\rho_{particle})}, \ \&$$

$$d_{particle} = \sqrt{\frac{V_s}{(4.255)(\rho_{particle})}}$$

Now, since we know the density of these beryllium particulates [$\rho = 1.848$ gms/cm^3], as well as the target settling velocity [$V_s = 72$ inches/hour], we can simply substitute in and determine the required particulate diameter, thus:

$$d_{particle} = \sqrt{\frac{72}{(4.255)(1.848)}} = \sqrt{\frac{72}{7.864}} = \sqrt{9.156} = 3.026 \text{ microns}$$

Since the initial average diameter of the beryllium particles was determined to have been 1.5 microns, and since this diameter must now be increased to ~ 3.03 microns, in order to achieve the desired settling velocity, we see that we must increase the particle diameter quantitatively by the following diameter multiplier, "DM":

$$DM = \frac{3.03}{1.5} = 2.017$$

∴	It appears as if the Industrial Hygienist will be forced to recommend the "C" category of Silicone Coating. The "B" category material will <u>almost</u> work; however, if the requirement truly is for a <u>minimum</u> of 72 inches/hour settling velocity, then this IH's recommendation will have to be for the "C" material.

Chapter 2
Standards and Calibrations

This chapter will discuss the reference methods, procedures, and standards against which all field measurements must be compared. The validity of any measurement will depend, obviously, on the accuracy of the method, procedure, technique, and instrumentation that is used to make it. Factors such as the precision, accuracy, and/or repeatability of any analytical effort completed outside of the laboratory can be and frequently are called into question. The individual who has made a challenged measurement in the field or in the lab must be able to document the relationship between the result he or she has reported and an appropriate, accepted, and well-established standard.

RELEVANT DEFINITIONS

Primary Standard

A standard for any measurable parameter (i.e., time, length, mass, etc.) that is maintained by any of the international or national standards agencies, most commonly by either the United States National Institute of Standards & Technology [NIST], in Washington, DC — formerly known as the United States National Bureau of Standards [NBS] — or the International Organization for Standardization [ISO], in Geneva, Switzerland.

Secondary Standard

A standard for any measurable parameter (i.e., temperature, volume, etc.) that is maintained by any commercial, military, or other organization — excluding any of those groups referenced above, i.e., groups that maintain Primary Standards. A Secondary Standard will have been thoroughly documented as to the fact of its having been directly referenced against an appropriate and applicable Primary Standard. Common Secondary Standards include such things as balance weights, atomic clocks, etc.

Standard Reference Material

A Standard Reference Material — often abbreviated as SRM — is any material, item, etc. for which one or more important characteristics [i.e., the specific make-up of a mixture such as Arizona Road Dust, the leak rate of a gas permeation device, the purity of a radioactive chemical, the precision and accuracy of a liquid-in-glass thermometer, etc.] have been certified by well-documented procedures to be traceable to some specific Primary Standard. Standard Reference Materials can be obtained from the National Institute of Standards & Technology, or any commercial supplier. In every case the SRM will have had the specific characteristic of interest to its purchaser certified as being traceable to the appropriate Primary Standard.

Calibration

Calibration is a process whereby the operation or response of any analytical method, procedure, instrument, etc. is referenced against some standard — most likely either a Secondary Standard directly, or some mechanism that incorporates a Secondary Standard. As an example, let us consider a situation wherein the actual response of a gas analyzer that has been designed to measure some specific analyte is unknown. Such an analyzer might be challenged with a number of known concentrations of the vapor of interest — with the known gas concentrations having been generated by a system that employs a Secondary Standard as its vapor source. This type of process, known as a Calibration, will document the previously unknown relationship between the analyzer response and the specific vapor concentrations that have produced each response. Such a Calibration would result in a curve or plot showing the analyzer output vs. vapor concentration.

Calibration Check

A Calibration Check is a simple process where a previously calibrated method, procedure, instrument, etc. is challenged, most commonly with a "Zero" and a single "Non-Zero" calibration standard — this latter one again most likely either a Secondary Standard directly, or some mechanism that incorporates a Secondary Standard. Such a "Non-Zero" challenge is frequently referred to as a "Span Check"; it serves primarily to confirm that the system in question is working properly. A Calibration Check can also involve multi-point ("Zero" & multiple "Non-Zero") challenges designed to confirm that a system in question is responding properly over its entire designed operating range.

Sensitivity

Sensitivity is a measure of the smallest value of any parameter that is to be monitored that can be unequivocally measured by the system being considered. It is a function of the inherent noise that is present in any analytical system. Sensitivity is almost always defined and/or specified by a manufacturer as some multiple (usually in the range of 2X to 4X) of the zero level noise of the system being considered. As an example, if some type of analytical system were to produce a steady ± 0.1 mv output when it is being exposed to a zero level of whatever material it has been designed to measure, then one might specify the Sensitivity of this system to that analyte level that would produce a 0.2 to 0.4 mv output response.

Selectivity

Selectivity is the capability of any analytical system to provide accurate answers to specific analytical problems even in the presence of factors that might potentially interfere with the overall analytical process. Selectivity is most easily understood by considering a typical example; in this case we will consider sound measurements. Suppose we are dealing with an Octave Band Analyzer that has been set up to provide equivalent sound pressure levels for the 1,000 Hz Octave Band. Suppose further that the sounds being monitored include all frequencies from 20 to 20,000 Hz. The Selectivity of this analytical tool would be its ability to provide accurate measurements of the 1,000 Hz Octave Band while simultaneously rejecting the contributions of any other segment of the entire noise spectrum to which it was exposed.

Repeatability

Repeatability is the ability of an analytical system to deliver consistently identical results to specific identical analytical challenges independent of any other factors. Specifically, an analytical system can be said to be repeatable if it provides the same result (± a small percentage of this result) when challenged with a known level of the material for which this system was designed to monitor. Although the following listing is not necessarily complete, a repeatable system would have to perform as listed above under any or all of the following conditions: (1) different operators; (2) different times of day; (3) an "old" system vs. a "new" one, etc.

Timeliness

The Timeliness of any measurement is related to the interval of time between the introduction of a sample to an analytical system, and the time required for that system to provide the desired result. Systems are classified into one of the three following groupings, each as a function of this specific time interval, or delay time, to provide an analytical answer. These groupings are:

1. Instantaneous or Real-Time — Any system that provides its analytical output at the same time as it is presented with the sample. Instantaneous or Real-Time systems are the only types that are capable of determining true Exposure Limit Ceiling Values [see Page 3-2].
2. Slow — Any system that has a delay interval between a few seconds and 30 minutes would be called a Slow system. A gas chromatograph would fall into this category.
3. Very Slow — Any system that will typically require days to be able to provide its answer. Dosimeters of all types tend to fall into this grouping.

Accuracy

The Accuracy of any measurement will simply be the value that has been specified by the manufacturer of the instrument involved. For most analytical instruments, the manufacturers will have identified the specific unit's Accuracy as a percentage of its full scale reading. As an example, a Carbon Dioxide Analyzer that has been set up to operate in the range 0 to 2,000 ppm [0 to 0.2%] will typically have an Accuracy Specification of ± 10% of its full scale reading, or ± 200 ppm [200 ppm = 10% of 2,000 ppm]. Although it is not yet common, some manufacturers now specify Accuracies for their instruments in terms of a combination of: (1) a percentage of the analyzer's full scale reading and (2) a percentage of the actual reading, whichever of these values is less — i.e., an Accuracy Specification calling for ± 15% of the analytical reading, OR ± 10% of the analyzer's full scale, whichever is less.

Precision

The Precision of any measurement will be the smallest quantity that the analytical instrument under consideration can indicate in its output reading. As an example, if the readout of an analyzer under consideration is in a digital format [i.e., 3.5 or 4.5 digits] showing two decimal places, then that analyzer's Precision would be 0.01 units. It is important to note that an analytical instrument's Precision is most assuredly not the same as its Sensitivity, although frequently these two parameters are mistaken and/or misunderstood to be identical.

RELEVANT FORMULAE & RELATIONSHIPS

Flow Rate & Flow Volume Calibrations

Flow rate calibrations are routinely performed using a combination of a volumetric standard in conjunction with a time standard. Simply, the time interval required for the output of some source of interest — i.e., a pump, etc. — to fill a precisely known volume is carefully measured and used then to determine the flow rate of the gas source.

Equation #2-1:

$$\textbf{Flow Rate} = \frac{\textbf{Volume}}{\textbf{Time Interval}}$$

Where: **Flow Rate** = the volume of gas per unit time flowing in or out of some system, usually in units such as: liters/minute, cm^3/min, etc.;

Volume = the known standardized volume that has been filled in some known time interval, in units such as cm^3, or liters; &

Time Interval = the actual measured time required for the gas source to output the standardized volume of gas, in some compatible unit such as minutes, etc.

Equation #2-2:

The principal purpose for making flow rate calibrations is to be able to calculate — with a high degree of certainty — the total volume of air that has been pumped, over a well-defined time interval, by a calibrated pump. These data are required for any determination of the average ambient concentration of any airborne material [gas, vapor, particulate, etc.] that might be trapped in any sort of impinger, filter cassette, etc. used in conjunction with the calibrated pump. Note that this relationship is simply a rearrangement of the previous equation.

$$\textbf{Total Volume} = [\textbf{Flow Rate}][\textbf{Time Interval}]$$

Where: **Flow Rate** = the volume of gas per unit time flowing into or out of some system, as above, in units such as liters/minute;

Total Volume = the calculated volume that has been pumped in some known time interval, in units such as liters; &

Time Interval = the actual measured time interval during which the pump was in operation, in some compatible unit such as minutes, etc.

Gas Analyzer Calibrations & Calibration Checks

The process of calibrating, calibration checking, zeroing, span checking, etc. any gas analyzer is both a very necessary and relatively simple process. To accomplish this task, the individual involved must first develop a standard that contains a known and well-referenced concentration of the analyte of interest, and then use this standard to challenge the analyzer whose performance is to be documented.

Equation #s 2-3, 2-4, & 2-5:

One of the most common methods for preparing a single concentration calibration standard that is to be used to test, calibrate, or span check a gas analyzer employs a chemically inert bag into which known volumes of a clean matrix gas [usually air or nitrogen] and a high purity analyte are introduced, so as to create a mixture of precisely known composition and concentration.

The sample preparation procedure always involves a minimum of two steps. First, a known volume of some matrix gas is introduced into a bag, inflating it to between 50 & 80% of its capacity. Next, a known volume of an analyte that is to serve as the standard is introduced into the bag. There are three very specific "categories" that apply to these single concentration calibration standards. Each will be described in detail in this section.

Equation #2-3:

The first of the three equations is used when it is necessary to prepare and calculate the resultant concentration that arises from the introduction of small volumes of a pure gas into the matrix filled bag. This procedure is used whenever a low concentration level calibration standard — i.e., one in the ppm(vol) or ppb(vol) concentration range — is desired. Although the total volume in the chemically inert calibration bag will always consist of the volumes of both the matrix gas and the analyte, for calculation purposes, the analyte volume will be so extremely small that it can be ignored. This volume, which is typically measured in microliters, will be four to eight orders of magnitude smaller than the volume of the matrix gas, which, in contrast, will typically be measured in liters.

An important fundamental assumption in this overall process is that all of the gas volumes involved in every step of the preparation of the standard, and in completing the calculations that will identify the actual concentration in the standard, will have to have been normalized to some standardized set of conditions such as NTP or STP.

$$\mathbf{C} = \frac{\mathbf{V_{analyte}}}{\mathbf{V_{matrix}}}$$

Where:

$\mathbf{C}$ = the analyte concentration, in parts per million by volume;

$\mathbf{V_{analyte}}$ = the volume of gaseous analyte that was introduced into the bag, measured in microliters; &

$\mathbf{V_{matrix}}$ = the precise volume of matrix gas introduced into the bag, measured in liters. As stated above, this matrix gas may be any pure gas [i.e., air, nitrogen, etc.] that, by definition, is completely free of impurities.

Equation #2-4:

This second relationship is employed when the analyte is introduced as a gas into the bag or cylinder at sufficiently large volumes so as to produce a calibration standard, the concentration of which is most conveniently measured as a percent.

The very same important fundamental assumption that applied to Equation **#2-3**, above, also applies to this situation, namely, that all of the gas volumes involved in every step of the preparation of this standard, as well as in completing the calculations that will identify its actual concentration, will have to have been normalized to some standardized set of conditions such as NTP or STP.

$$C = 100\left[\frac{V_{analyte}}{1,000(V_{matrix}) + V_{analyte}}\right]$$

Where:

- C = the analyte concentration, in percent by volume;
- $V_{analyte}$ = the volume of gaseous analyte that was introduced into the bag, measured in milliliters; &
- V_{matrix} = the precise volume of matrix gas introduced into the bag, measured in liters. As stated earlier, this matrix gas may be any pure gas [i.e., air, nitrogen, etc.] that, by definition, is completely free of impurities.

The final relationship is used whenever the calibration standard is to be prepared by the introduction of a known volume of a pure liquid phase chemical into the matrix filled bag. As was the case for standards produced by the introduction of a gaseous analyte, there are two concentration-related specific situations that will be covered — the first for low, and the second for high concentration level standards. In each of these cases, but particularly in the second or high concentration level case, care must be exercised to ensure that the prevailing conditions of temperature and pressure are sufficient to guarantee that all the liquid analyte will, in fact, vaporize so as to produce the desired concentration in the calibration standard. The relationship involved is the same for both cases.

Equation #2-5:

$$C = \left[\frac{T_{ambient}v_{analyte}\rho_{analyte}}{16.036\left[P_{ambient}V_{matrix}\left(MW_{analyte}\right)\right] + T_{ambient}v_{analyte}\rho_{analyte}}\right]10^6$$

Where:

- C = the analyte concentration, in parts per million by volume;
- $T_{ambient}$ = the absolute ambient temperature, in K;
- $v_{analyte}$ = the volume of pure liquid analyte introduced into the bag, measured in microliters, µl;

$\rho_{analyte}$ = the density of the pure liquid analyte, measured in grams/cm^3;

$P_{ambient}$ = the ambient barometric pressure, in mm Hg;

V_{matrix} = the precise volume of matrix gas introduced into the bag, measured in liters; &

$MW_{analyte}$ = the molecular weight of the analyte, measured in Atomic Mass Units [or more precisely, in grams mass per mole].

If the calibration standard to be generated by the introduction of a liquid into the bag must have its concentration in the percent range, then great care must be exercised to ensure that the prevailing conditions of temperature and pressure are sufficient to guarantee that all the liquid introduced will, in fact, evaporate so as to produce the desired analyte vapor concentration.

N.B.: **In situations that involve the use of an inflatable bag, specific attention must be paid to the volume that the analyte — when completely vaporized from its liquid phase — will occupy. The injected volume of liquid will always be very small [i.e., it is measured in microliters]; however, the analyte volume, when vaporized, will almost certainly be at least 2.5 to 3.0 orders of magnitude greater [i.e., 10 ml volume of acetone, introduced as a pure liquid, will vaporize to produce a gaseous volume of 3,325 ml = 3.33 liters at NTP — an obvious 330+ fold increase in volume]. It is not at all uncommon, in the preparation of percentage concentration range standards by an individual who has overlooked this factor, to have a situation where the bag will burst when its capacity has been exceeded by the sum of the matrix gas and the vaporized analyte.**

STANDARDS AND CALIBRATIONS PROBLEM SET

Problem #2.1:

An Industrial Hygienist wishes to identify which of his three personal sampling pumps has a flow rate both greater than 450 cc/minute, but at the same time as close as possible to 500 cc/minute. To make this determination, he uses a bubble flowmeter whose marked interior volume of 135 ml has been certified to be traceable to an NIST volumetric standard. He makes five runs with each of his three sampling pumps, using a stop watch to time the movement of the soap bubble. His results are summarized in the following tabulation, which shows the five separate time intervals he measured during which each of his three candidate pumps delivered 135 ml of air. Which of these three pumps should this Industrial Hygienist select?

Sample Pump #1	Sample Pump #2	Sample Pump #3
16.42 seconds	15.59 seconds	15.88 seconds
16.49 seconds	15.82 seconds	16.07 seconds
16.62 seconds	15.70 seconds	16.11 seconds
16.37 seconds	15.85 seconds	15.95 seconds
16.53 seconds	15.81 seconds	16.08 seconds

Applicable Definitions:	Volume	Page 1-4
	Time	Page 1-2
Applicable Formula:	Equation **#2-1**	Page 2-4
Solution to this Problem:	Page 2-15	

Problem Workspace

Workspace Continued on the Next Page

Continuation of Workspace for Problem **#2.1**

Problem #2.2:

An Industrial Hygienist wishes to complete a calibration check on her carbon monoxide analyzer. Her instrument has been designed to provide accurate carbon monoxide concentration readouts in the range 0 to 100 ppm(vol). She decides to prepare a single component calibration standard of ~ 80 ppm(vol). To do this, she has available to her: (1) a 10-liter Tedlar bag, (2) a 1,000 microliter gas tight chromatographic injection syringe, (3) a precisely calibrated gas pump, (4) a zero air system that will produce up to 6 liters/minute of extremely clean air, and (5) a properly valved and regulated lecture bottle of high purity carbon monoxide. In addition, she has determined that she will require a minimum of 8.0 liters of calibration gas in order to completely flush and stabilize her analyzer.

As a first step, she decides to introduce a total of 8.5 liters of contaminant free air from her zero air system into her bag, using her calibrated gas pump. If she next uses her 1,000 microliter syringe, what volume of carbon monoxide must she inject into the bag so as to produce the desired calibration standard of ~ 80 ppm(vol) of carbon monoxide?

Applicable Formula:	Equation **#2-3**	Page 2-5
Solution to this Problem:	Page 2-16	

Problem Workspace

Problem #2.3:

To check the accuracy of an installed gas analyzer that was designed to record the fire suppressant concentration levels of carbon dioxide in a computer room, the Safety Manager prepared a calibration standard in a Tedlar bag. For reference, this individual was charged with the responsibility for maintaining the operability of the CO_2 Fire Suppressant System that was designed to protect the main frame computer system that was installed in this room. In preparing his standard, the Safety Manager first filled a 25 liter Tedlar bag with 15.13 liters of dry nitrogen, and then added 7.66 liters of CO_2. What was the concentration of carbon dioxide, expressed as a percent, in this Tedlar bag?

Applicable Formula:	Equation **#2-4**	Page 2-6
Solution to this Problem:	Page 2-16	

Problem Workspace

Problem #2.4:

Forane® (Isoflurane) is one of a group of halogenated ethers commonly used for human inhalation anesthesia. Although there is no established exposure limit for this material, common practice is to try never to permit its ambient concentration to exceed 2.0 ppm(vol). A long pathlength infrared spectrophotometric analyzer — having a response range of 0 to 5 ppm(vol) for Forane® — is used to monitor the ambient air in an Operating Room where this agent is to be used. It is necessary to prepare a 2.0 ppm(vol) span check standard to verify the operation of this analyzer. Calibration standards for this type of analyzer must always contain a minimum of 20 liters total volume. To prepare the standard, the Technician involved has charged a 25-liter Tedlar bag with 23.0 liters of clean air. To finish the preparation of his standard, he has available to him the following equipment and data:

1. A 1.0-μl chromatographic injection syringe. This syringe has divisions every 0.02 μl; and, by using "visual interpolation", it can be filled to a precision of 0.01 μl;
2. A 100-ml bottle of Forane®;
3. The prevailing ambient conditions and location data for this situation are as follows:

Location:	Boise, ID [Altitude = 2,739 ft above Sea Level]
Ambient Temperature:	71°F
Barometric Pressure:	690 mm Hg

4. The following are the relevant data on Forane®:

Chemical Formula:	**CF_3-CHCl-O-CHF_2**
Chemical Name:	2-chloro-2-(difluoromethoxy)-1,1,1-trifluoroethane
Molecular Weight:	184.50 amu
Melting Point:	48.5°C [liquid at room temperature]
Liquid Density:	1.452 gms/cm^3

What volume of Forane®, in μl, must the Technician inject into the partially filled Tedlar bag to produce the approximate 2.0 ppm(vol) standard that is required? What will be the actual final Forane® concentration that exists in the Tedlar bag after this injection of liquid Forane®?

Forane® is a registered trademark of Anaquest Corp.

Applicable Formulae:	Equation **#1-1**	Page 1-16
	Equation **#1-3**	Page 1-16
	Equation **#2-5**	Pages 2-6 & 2-7
Solution to this Problem:	Pages 2-16 through 2-18	

Problem Workspace

Workspace Continued on the Next Page

Continuation of Workspace for Problem **#2.4**

Problem #2.5:

A factory that manufactures molded foam polystyrene egg cartons uses n-hexane to "expand" this polymer foam into the molds that form and shape the desired end product. Because of the great flammability of n-hexane, this plant's manufacturing area is equipped with ten combustible gas detectors, each of which has been designed to provide an audible alarm whenever the measured ambient n-hexane concentration reaches 50% of the LEL [**L**ower **E**xplosive **L**imit] for this chemical. This plant's Safety Engineer wants to prepare a calibration standard that contains a concentration of n-hexane equal to 50% of its LEL. To prepare this standard, he plans to use a 50-liter Tedlar bag which he will fill — initially to 80% of its capacity — with clean, hydrocarbon-free air. To finish preparing this standard, he has available to him the following equipment and data:

1. A 2-ml liquid tight chromatographic injection syringe. This syringe has divisions at 0.05 ml intervals [0.05 ml = 50 μl]; and, by using "visual interpolation", it can be filled to a precision of ± 25 μl;
2. A 250-ml bottle of n-hexane [Spectrophotometric Grade Purity];
3. The prevailing ambient conditions and location data for this situation are as follows:

Location:	Fresno, CA [Altitude = 294 ft above Sea Level]
Ambient Temperature:	29°C
Barometric Pressure:	1,003 millibars

4. The following are the relevant data on n-hexane:

Chemical Formula:	$\mathbf{CH_3\text{-}CH_2\text{-}CH_2\text{-}CH_2\text{-}CH_2\text{-}CH_3}$
Molecular Weight:	86.18 amu
Freezing Point:	–59.8°C
Boiling Point:	68.7°C
Liquid Density:	0.655 gms/cm^3
Vapor Pressure:	190.5 mm Hg @ 29°C
Explosive Range:	1.2 to 7.7% in air

What volume of n-hexane, in μl, must the Safety Engineer inject into the partially filled Tedlar bag to produce a standard of approximately 50% of the LEL for n-hexane, and what will be the actual final concentration of n-hexane in the bag — in ppm(vol) — assuming that the precision of the injection syringe volume is no more than ± 0.25 μl? Will the volume of liquid n-hexane that is injected into the Tedlar bag evaporate fully under the stated conditions of temperature and pressure, etc.?

Applicable Definition:	Upper & Lower Explosive Limits	Page 3-4
Applicable Formulae:	Equation **#1-1**	Page 1-16
	Equation **#1-9**	Pages 1-18 & 1-19
	Equation **#1-10**	Pages 1-19 & 1-20
	Equation **#1-16**	Pages 1-22 & 1-23
	Equation **#2-5**	Pages 2-6 & 2-7
Solution to this Problem:	Pages 2-18 through 2-20	

Problem Workspace

Workspace Continued on the Next Page

Continuation of Workspace for Problem **#2.5**

SOLUTIONS TO THE STANDARDS AND CALIBRATIONS PROBLEM SET

Problem #2.1:

The solution to this problem requires that we first calculate the average times to pump 135 ml for each of the three pumps — and while we are at it, we should also calculate the sample standard deviations for all these measurements. Once this is done, we can then apply Equation **#2-1**, from Page 2-3, to obtain the answer we need, thus:

$$\text{Flow Rate} = \frac{\text{Volume}}{\text{Time Interval}} \qquad \text{[Eqn. \#2-1]}$$

Times for the Three Pumps to Move 135 ml of Air

	Sample Pump #1	Sample Pump #2	Sample Pump #3
Run #1	16.42 seconds	15.59 seconds	15.88 seconds
Run #2	16.49 seconds	15.82 seconds	16.07 seconds
Run #3	16.62 seconds	15.70 seconds	16.11 seconds
Run #4	16.37 seconds	15.85 seconds	15.95 seconds
Run #5	16.53 seconds	15.81 seconds	16.08 seconds
Total Time	82.43 seconds	78.77 seconds	80.09 seconds
Average Time	16.49 seconds	15.75 seconds	16.02 seconds
Average Time	0.275 minutes	0.263 minutes	0.267 minutes
Standard Deviation	0.10 seconds	0.11 seconds	0.10 seconds

We can now apply Equation **#2-1**, from Page 2-4, to obtain the answers we seek, thus:

$$\text{Flow Rate}_{\text{Pump 1}} = \frac{135}{0.275} = 491.2 \text{ cc/min}$$

$$\text{Flow Rate}_{\text{Pump 2}} = \frac{135}{0.263} = 514.3 \text{ cc/min}$$

$$\text{Flow Rate}_{\text{Pump 3}} = \frac{135}{0.267} = 505.6 \text{ cc/min}$$

Clearly, this industrial hygienist handled a stop watch very well — the standard deviations in his time interval measurements were all virtually the same and all very close to 0.1 seconds. For this reason, this person should fully trust the calculated flow rates and select Pump #3 which best satisfies the requirements of this problem

∴ The choice in this situation should be Personal Sampling Pump #3.

Problem #2.2:

The solution to this problem requires the use of Equation **#2-3**, from Page 2-5:

$$C = \frac{V_{analyte}}{V_{matrix}} \qquad \text{[Eqn. \#2-3]}$$

$$80 = \frac{V_{analyte}}{8.5}$$

$$V_{analyte} = (80)(8.5) = 680 \ \mu l$$

∴ This Industrial Hygienist should inject a total of 680 μl of pure carbon monoxide into the Tedlar bag (the bag that already contains 8.5 liters of clean air) — this will produce the required ~ 80 ppm(vol) carbon monoxide standard.

Problem #2.3:

The solution to this problem requires the application of Equation **#2-4**, from Page 2-6:

$$C = 100\left[\frac{V_{analyte}}{1,000(V_{matrix}) + V_{analyte}}\right] \qquad \text{[Eqn. \#2-4]}$$

$$C = 100\left[\frac{7,660}{(1,000)(15.13) + 7,660}\right] = 33.61 \ \%$$

∴ The CO_2 concentration in the Tedlar bag was 33.61%.

Problem #2.4:

The solution to this problem will require the application of a number of different equations, starting with Equation **#s 1-3 & 1-1**, which both appear on Page 1-16 [these are used to convert the temperature, which was provided in the problem statement in units of °F, to the required units of K]; and Equation **#2-5**, from Pages 2-6 & 2-7:

$$t_{Metric} = \frac{5}{9}\left[t_{English} - 32°\right] \qquad \text{[Eqn. \#1-3]}$$

$$t_{Metric} = \frac{5}{9}\left(71° - 32°\right) = \frac{5}{9}(39) = 21.67°C$$

$$t_{Metric} + 273.16 = T_{metric} \qquad \text{[Eqn. \#1-1]}$$

$$21.67 + 273.16 = T_{Metric} = 294.83 \ K$$

Next we must apply Equation **#2-5**, from Pages 2-6 & 2-7:

$$C = \left[\frac{T_{ambient} v_{analyte} \rho_{analyte}}{16.036\left[P_{ambient} V_{matrix}\left(MW_{analyte}\right)\right] + T_{ambient} v_{analyte} \rho_{analyte}}\right]10^6 \qquad \text{[Eqn. \#2-5]}$$

Then, using a number of algebraic steps, we must rewrite this relationship so that it can, in its modified form, be an expression from which we can solve directly for the injected volume of Forane®, "$V_{analyte}$", in μl, as has been asked for in the problem statement, thus:

$$\frac{1}{C} = \left[\frac{16.036\left[P_{ambient}V_{matrix}\left(MW_{analyte}\right)\right] + T_{ambient}v_{analyte}\rho_{analyte}}{T_{ambient}v_{analyte}\rho_{analyte}}\right]10^{-6}$$

$$\frac{1}{C} = \left[\frac{16.036\left[P_{ambient}V_{matrix}\left(MW_{analyte}\right)\right]}{T_{ambient}v_{analyte}\rho_{analyte}} + 1\right]10^{-6}$$

$$\frac{1}{C} = \frac{1.604\times10^{-5}\left[P_{ambient}V_{matrix}\left(MW_{analyte}\right)\right]}{T_{ambient}v_{analyte}\rho_{analyte}} + 10^{-6}$$

$$\frac{1}{C} - 10^{-6} = \frac{1.604\times10^{-5}\left[P_{ambient}V_{matrix}\left(MW_{analyte}\right)\right]}{T_{ambient}v_{analyte}\rho_{analyte}} = \frac{1 - C\times10^{-6}}{C}$$

$$\frac{C}{1 - C\times10^{-6}} = 62,361\left[\frac{T_{ambient}v_{analyte}\rho_{analyte}}{P_{ambient}V_{matrix}\left(MW_{analyte}\right)}\right]$$

and, next solving for the injected volume of Forane®, $v_{analyte}$, we get:

$$v_{analyte} = 1.604\times10^{-5}\left[\frac{P_{ambient}V_{matrix}\left(MW_{analyte}\right)}{T_{ambient}\rho_{analyte}}\right]\left[\frac{C}{1 - C\times10^{-6}}\right]$$

Substituting in all the appropriate factors either provided in the problem statement or calculated, we get:

$$v_{analyte} = \frac{\left(1.604\times10^{-5}\right)(690)(23.0)(184.50)(2.0)}{(294.83)(1.452)\left(1 - 2\times10^{-6}\right)}$$

Noting that $(1 - 2\times10^{-6}) = 0.999998 \approx 1.0$, we can ignore the third term in the denominator [i.e., we can replace this term with 1.00] and the expression then becomes:

$$v_{analyte} = \frac{\left(1.604\times10^{-5}\right)(690)(23.0)(184.50)(2.0)}{(294.83)(1.452)}$$

$$v_{analyte} = \frac{93.931}{428.093} = 0.22\ \mu l$$

Finally, we must again use the initial relationship, namely, Equation **#2-5**, from Pages 2-6 & 2-7, to develop the final requested answer:

$$C = \left[\frac{T_{ambient}v_{analyte}\rho_{analyte}}{16.036\left[P_{ambient}V_{matrix}\left(MW_{analyte}\right)\right] + T_{ambient}v_{analyte}\rho_{analyte}}\right]10^{6} \quad \text{[Eqn. \#2-5]}$$

$$C = \left[\frac{(294.83)(0.22)(1.452)}{(16.036)(690)(23.0)(184.50) + (294.83)(0.22)(1.452)}\right]10^{6}$$

$$C = \frac{(294.83)(0.22)(1.452)\left(10^{6}\right)}{(16.036)(690)(23.0)(184.50) + (294.83)(0.22)(1.452)}$$

$$C = \frac{94,180,495.20}{46,953,648.54 + 94.18} = \frac{94,180,495.20}{46,953,742.72} = 2.006 \approx 2.0$$

∴	The Technician must inject 0.22 microliters of Forane® into the partially filled Tedlar bag. This will produce a calibration standard of Forane® with a concentration of 2.0 ppm(vol) ± 0.3%.

Problem #2.5:

The eventual solution to this problem will also require the use of Equation **#2-5**, from Pages 2-6 & 2-7. Prior to applying this relationship, we must convert some of the data provided in the problem statement into the specific units that are required for this Equation. Let us begin with the barometric pressure, which must be in units of mm Hg, but has been provided in millibars. Remembering that 1.0 mm Hg = 1.33 millibars, we get:

$$\frac{1,003 \text{ millibars}}{1.33 \text{ millibars}/\text{mm Hg}} = 754.1 \text{ mm Hg}$$

We must next determine the actual target concentration of n-hexane. Since we are seeking a concentration equal to 50% of the LEL for n-hexane, and since we have been told that the LEL for this chemical is 1.2%, we see that we must seek a concentration of 0.6% = 6,000 ppm(vol) of this material in air. Clearly then C = 6,000 ppm(vol).

Next, we must apply Equation **#1-1**, from Page 1-16, in order to convert the relative temperature provided in the problem statement to its absolute equivalent, as is required in Equation **#2-5**:

$$t_{Metric} = 273.16 = T_{Metric} \qquad \text{[Eqn. \#1-1]}$$

$$29 + 273.16 = T_{Metric} = 302.16 \text{ K}$$

The final preparatory determination we must make is to determine the volume of matrix air this Safety Engineer will be injecting into his clean air partially filled Tedlar bag. The problem statement indicated that he has introduced a sufficient volume of clean air to fill the bag to 80% of its capacity. Since this capacity is 50 liters, we can assume that he has introduced a total of 40 liters of clean air into the bag; thus $V_{matrix} = 40.0$ liters.

We can now apply the previously listed relationship, namely, Equation **#2-5**, from Pages 2-7 & 2-8:

$$C = \left[\frac{T_{ambient} v_{analyte} \rho_{analyte}}{16.036\left[P_{ambient} V_{matrix}\left(MW_{analyte}\right)\right] + T_{ambient} v_{analyte} \rho_{analyte}}\right] 10^6 \qquad \text{[Eqn. \#2-5]}$$

As was true for the previous problem, namely Problem #2.4, we must now convert Equation **#2-5** so as to give a relationship that will provide a direct solution to the problem, namely, an equation that gives the value of the injected volume of the analyte, "$v_{analyte}$". We do not actually have to make this multi-step algebraic manipulation; we can simply use the relationship derived in Problem #2.4, thus:

$$v_{analyte} = 1.604 \times 10^{-5} \left[\frac{P_{ambient} V_{matrix}\left(MW_{analyte}\right)}{T_{ambient} \rho_{analyte}}\right]\left[\frac{C}{1 - C \times 10^{-6}}\right]$$

Finally, substituting in all the appropriate factors either provided in the problem statement or calculated, we get:

$$v_{analyte} = \frac{(1.604 \times 10^{-5})(754.1)(40.0)(86.18)(6,000)}{(302.16)(0.655)[1 - (6,000)(10^{-6})]}$$

$$v_{analyte} = \frac{250,179.11}{(197.915 - .994)} = \frac{250,179.11}{196.921} = 1,270.5$$

Considering now the precision limitations in the ability of this individual to read the injection syringe, we see that he must inject 1,275 μl of n-hexane — he cannot achieve any greater precision with the equipment he has available to him; therefore, $v_{analyte}$ = 1,275 μl.

Finally, we must again use the initial relationship, namely, Equation **#2-5**, from Pages 2-6 & 2-7, to develop the next requested answer:

$$C = \left[\frac{T_{ambient} v_{analyte} \rho_{analyte}}{16.036[P_{ambient} V_{matrix}(MW_{analyte})] + T_{ambient} v_{analyte} \rho_{analyte}}\right] 10^6 \qquad \text{[Eqn. \#2-5]}$$

$$C = \frac{(302.16)(1,275)(0.655)(10^6)}{[(16.036)(754.1)(40.0)(86.18) - (302.16)(1,275)(0.655)]}$$

$$C = \frac{2.523 \times 10^{11}}{41,686,119.53 - 252,341.37} = \frac{2.523 \times 10^{11}}{41,433,778.16} = 6,090.2 \text{ ppm(vol)}$$

If the injected liquid n-hexane actually does fully evaporate, then the concentration of the standard would be at ~ 6,090.2 ppm(vol), or roughly 0.61%, which in turn would be approximately 50.8% of the LEL for this chemical.

We must next check to determine: (1) whether this amount of n-hexane could actually be expected to evaporate in this Tedlar bag under the prevailing conditions of temperature and pressure, and (2) whether the remaining 10 liter, unfilled capacity of this 50-liter Tedlar bag would be sufficiently large to hold the evaporated volume of n-hexane.

The first of these determinations can be made by determining the equilibrium concentration of n-hexane at 29°C. We have been given that its vapor pressure at this temperature is 190.5 mm Hg. Using Equation **#1-16**, from Pages 1-22 & 1-23, we can calculate what the saturation vapor concentration of n-hexane would be. If this concentration is greater than the anticipated 6,090.2 ppm(vol) concentration that the standard has been targeted to achieve, then we can be certain that all of the liquid n-hexane will evaporate. Remember for this calculation, we must use the prevailing barometric pressure of 754.1 mm Hg for the total pressure term:

$$C_{n\text{-hexane (saturation)}} = \frac{[1,000,000]VP_{n\text{-hexane}}}{P_{total}} \qquad \text{[Eqn. \#1-16]}$$

$$C_{n\text{-hexane (saturation)}} = \frac{(1,000,000)(190.5)}{754.1} = 252,619 \text{ ppm(vol)}$$

In essence, this concentration amounts to just over 25%, so we can be certain that all the n-hexane will evaporate. Assuming that there is enough available space in the Tedlar bag to hold this evaporated volume, we will be able to generate the required standard. The question is, what vapor volume will this quantity of liquid n-hexane occupy?

We can obtain this answer first by determining the number of moles of n-hexane that are in 1,275 μl = 1.275 cm^3 of pure liquid chemical. Since we know the density of this n-hexane, we can calculate directly the mass represented by this volume, and with this mass, we can apply Equation **#1-10**, from Pages 1-19 & 1-20, to determine the number of moles, represented by this volume, thus:

$$m_{n\text{-hexane}} = (\rho_{n\text{-hexane}})(v_{n\text{-hexane}})$$

$$m_{n\text{-hexane}} = (0.655)(1.275) = 0.835 \text{ grams of n-hexane}$$

We can now apply Equation **#1-10**:

$$n = \frac{m}{MW} \qquad \text{[Eqn. \#1-10]}$$

$$n = \frac{0.835}{86.18} = 9.69 \times 10^{-3}$$

Now applying Equation **#1-9**, from Pages 1-18 & 1-19, we can determine the volume that this number of moles will occupy under the prevailing conditions of pressure and temperature, and if this volume is less than the available 40 liters of unused Tedlar bag space, then we are all set:

$$PV = nRT \qquad \text{[Eqn. \#1-9]}$$

Solving for the volume, "V", we get:

$$V = \frac{nRT}{P}$$

$$V = \frac{(9.69 \times 10^{-3})(62.36)(302.16)}{754.1} = 0.242 \text{ liters} = 242 \text{ ml}$$

Clearly, there will be sufficient space in the Tedlar bag to accommodate the n-hexane vapor.

∴ The Safety Engineer should inject a total of 1,275 μl of n-hexane into the Tedlar bag. This procedure should successfully produce a calibration standard with a concentration of ~ 6,090.2 ppm(vol) ~ 0.61% of the LEL for n-hexane.

Chapter 3
Workplace Ambient Air

This chapter focuses on the wide variety of factors and aspects that characterize, or are characteristics of, ambient air. Particular emphasis is placed on those factors and/or conditions that must be mitigated in order to make the workplace safe. The principal components of any ambient matrix to be considered will be its temperature, pressure, and volume as well as the vapors, particulates, and/or aerosols that may be resident in it.

RELEVANT DEFINITIONS

Ambient Concentration Categories

Threshold Limit Values

The **T**hreshold **L**imit **V**alue [usually abbreviated, **TLV**] refers to an ambient airborne concentration of some substance of interest, and represents a condition under which it is believed that nearly all workers may be repeatedly exposed, day after day, without adverse effect. This concentration limit can be, and is, commonly expressed in one of three forms: as an 8-hour Time Weighted Average (TLV-TWA); as a Short Term Exposure Limit (TLV-STEL); and as a Ceiling Value (TLV-C). The overall ambient concentration concept designated by the term or phrase, "Threshold Limit Value", was introduced and promulgated by the **A**merican **C**onference of **G**overnment **I**ndustrial **H**ygienists [the **ACGIH**]. Currently established TLVs are always under review, and individual listings are modified whenever relevant new information dictates that this be done.

Permissible Exposure Limits

The **P**ermissible **E**xposure **L**imit [usually abbreviated, **PEL**] is an ambient airborne concentration of some substance of interest which the **O**ccupational **S**afety and **H**ealth **A**dministration [**OSHA**], which is a branch of the **U**.**S**. **D**epartment **o**f **L**abor [**USDOL**], had adopted — largely from data furnished earlier by the ACGIH in the development of this organization's listing of Threshold Limit Values. The initial listing of these Permissible Exposure Limits was made in the Z-Tables of Title 29, Code of Federal Regulations [usually abbreviated, CFR], Part 1910.1000, as published in the *Federal Register* on January 19, 1989. Currently published PELs are always under review; established values for specific materials are modified whenever additional information indicates that this should be done. In addition, PELs for previously unlisted materials are added as data on the effect of these materials are developed and become accepted. Like TLVs, PELs are commonly expressed in one of three forms: as an 8-hour Time Weighted Average (PEL-TWA); as a Short Term Exposure Limit (PEL-STEL); and as a Ceiling Value (PEL-C).

All PELs exist as enforceable statutes; whenever an employee or worker is exposed to a listed material at a combination of: (1) an ambient concentration, and/or (2) a duration of exposure that exceeds any of these specific standards, that individual's employer can be cited and fined by OSHA.

Recommended Exposure Limits

The **R**ecommended **E**xposure **L**imit [usually abbreviated, **REL**] is still another ambient airborne concentration of some substance of interest which, in this case, the **N**ational **I**nstitute for **O**ccupational **S**afety and **H**ealth [**NIOSH**] has researched and developed. Like its two previously listed counterparts, the TLV and the PEL, RELs are commonly expressed in one of the three standard forms: as an 8-hour Time Weighted Average (REL-TWA); as a Short Term Exposure Limit (REL-STEL); and as a Ceiling Value (REL-C). Currently established RELs are always under review, and individual listings are modified whenever relevant new NIOSH research dictates that this be done.

Maximum Concentration Values in the Workplace

The **M**aximum **A**rbeitsplatz **K**onzentration [usually abbreviated **MAK** — translated into English as the Maximum Concentration Value in the Workplace] is a TLV, PEL, & REL analog; and, as such, is also an ambient airborne concentration of some substance of interest which, in this case, the **D**eutsche **F**orschungs**g**emeinschaft [**DFG**], Commission for the Investigation of the Health Hazards of Chemical Compounds in the Work Area, as a branch of the Federal Republic of Germany's central government, has developed, adopted, and promulgated. Exactly like its three U. S. counterparts, MAKs are commonly expressed in one of the three standard forms: as an 8-hour Time Weighted Average (MAK-TWA), as a Short Term Exposure Limit (MAK-STEL), and as a Ceiling Value (MAK-C); and exactly like their U.S. counterparts, MAKs are always under review, and individual listings are modified whenever relevant new information dictates that this be done.

Time Weighted Averages

The **T**ime **W**eighted **A**verage [usually abbreviated as a "suffix", **-TWA**; thus: TLV-TWA, PEL-TWA, REL-TWA, and/or MAK-TWA] is the employee's average airborne exposure in <u>any</u> 8-hour work shift of <u>any</u> 40-hour work week to which nearly all workers may be repeatedly exposed, day after day, without suffering any adverse effects. It is a value that should never be exceeded.

Short Term Exposure Limits

The **S**hort **T**erm **E**xposure **L**imit [usually abbreviated as a "suffix", **-STEL**; thus: TLV-STEL, PEL-STEL, REL-STEL, and/or MAK-STEL] is the concentration to which workers can be continuously exposed for short periods of time without suffering from:

1. irritation;
2. chronic or irreversible tissue damage; or
3. narcosis of sufficient degree to increase the likelihood of accidental injury, impair self-rescue, or materially reduce work efficiency — provided also that the corresponding **TWA** has not been exceeded.

STELs are usually 15-minute (except for those materials for which an alternative time limit has been specified) Time Weighted Average exposures which should <u>never</u> be exceeded during <u>any</u> of the specified time intervals during the work day, even if the corresponding TWA has not been exceeded. In the event any time limit other than 15 minutes is specified for some material or compound, the previous definition still holds, except as modified by the different time limit.

Ceiling Values

The **Ceiling Value** concentration [usually abbreviated as a "suffix", **-C**; thus: TLV-C, PEL-C, REL-C, and/or MAK-C] is a concentration that should <u>never</u> be exceeded, even instantaneously, at <u>any</u> time during the work day. In the event that instantaneous monitoring is not feasible, then the Ceiling Value can be assessed as a 15-minute Time Weighted Average exposure which should not be exceeded at any time during the work day, EXCEPT when the subject vapor can cause immediate irritation with exceedingly short exposures.

Action Levels

The **Action Level** is an 8-hour Time Weighted Average concentration for which there is only a 5% risk of having more than 5% of the employee workdays involve an exposure at a level greater than the relevant TLV-TWA, PEL-TWA, REL-TEA, or MAK-TWA. This value is most frequently set at or near 50% of the relevant TLV-TWA, PEL-TWA, REL-TWA, or MAK-TWA concentration standard.

Excursion Limits

An **Excursion Limit** is a term that is frequently called into use in situations and for substances where no published STEL or Ceiling Value exists. It is a Short Term Exposure Limit or a Ceiling Value without any legal standing, as would be the case for OSHA published STELs or Ceiling Values. As such, the Excursion Limit is simply an "industry recognized" factor or guideline. For materials that have no published STEL, the STEL Excursion Limit is generally understood to be three times the published 8-hour Time Weighted Average standard, for no more than 30 minutes during any work day. For materials that have no published Ceiling Value, the Ceiling Value Excursion Limit is generally understood to be five times the established 8-hour Time Weighted Average standard, and is treated as a concentration that should never be exceeded at any time.

Designation of Immediately Dangerous to Life and/or Health

The **I**mmediately **D**angerous to **L**ife and/or **H**ealth (usually abbreviated, **IDLH**) concentration is a concentration level of some substance of interest from which a worker could escape in 30 minutes or less without suffering any escape-impairing symptoms and/or irreversible health effects.

Breathing Zone

The **Breathing Zone** of an individual is a roughly hemispherical volume immediately forward of that person's shoulders and face, centered roughly on the Adam's Apple, and having a radius of 6 to 9 inches (15 to 23 cm).

Ambient Dose-Response/Concentration-Response Parameters

Median Lethal Dose

The Median **L**ethal **D**ose (usually abbreviated, **LD50**) is the toxicant dose at which 50% of a population of the same species will die within a specified period of time, under similar experimental conditions. This dosage number is usually expressed as milligrams of toxicant per kilogram of body weight [mg/kg].

Median Effective Dose

The Median **E**ffective **D**ose (usually abbreviated, **ED50**) is the toxicant dose required to produce a specific non-lethal effect in 50% of a population of the same species, under similar experimental conditions. This dosage number is usually also expressed as milligrams of toxicant per kilogram of body weight [mg/kg].

Median Lethal Concentration

The Median **L**ethal **C**oncentration (usually abbreviated, **LC50**) is the concentration of toxicant in air which will cause 50% of a population to die within a specified period of time.

This factor is a concentration, not a dose, and is usually expressed either as milligrams of toxicant per cubic meter of air [mg/m^3], or as parts per million [ppm(vol)].

"Other" Dose-Response, Concentration-Response Parameters

Other common analogous Dose-Response or Concentration-Response Terms have been indicated below — in each case, the definition of each of these terms is directly analogous to the definition of the "Same Name" term on the previous page, the only difference being in the percentage figure involved:

Lethal Dose for **10%** of a population (abbreviated, **LD10**)

Lethal Dose for **90%** of a population (abbreviated, **LD90**)

Effective Dose for **10%** of a population (abbreviated, **ED10**)

Effective Dose for **90%** of a population (abbreviated, **ED90**)

Lethal Concentration for **10%** of a population (abbreviated, **LC10**)

Lethal Concentration for **90%** of a population, etc. (abbreviated, **LC90**)

Parameters Relating to Specific Chemicals or Substances

Upper & Lower Explosive Limits

The Upper and Lower Explosive Limits (abbreviated **UEL** and **LEL**) refer to the upper and lower vapor concentration boundaries, for some specific compound or material of interest, within which the vapor-air mixture will propagate a flame [i.e., explode] if ignited.

Explosive Range

The **Explosive Range** for any chemical or compound is the range of concentrations that exist between its Upper and Lower Explosive Limits. For example, the **LEL** and **UEL** for n-hexane are, respectively, 1.2 & 7.7%; thus this chemical's **Explosive Range** is 1.2 to 7.7% in air. When considering the **Explosive Range** for any specific chemical, it is important to note that for any concentration of that chemical that falls outside of this range, the vapor-air mixture possessing that concentration will not propagate a flame when a source of ignition is introduced. For chemical concentrations lower than the **LEL**, there is an insufficient amount of the combustible chemical to permit flame propagation, and for concentrations greater than the **UEL**, there is insufficient ambient oxygen to permit flame propagation.

Flash Point

The Flash Point of any compound is that temperature to which it must be heated before its vapors can be ignited by a free flame in the presence of air. It is a measure of the flammability of any material, and as such it is a reasonably good criterion for this characteristic. The lower the Flash Point, the more flammable a material is. This value is affected by the relative volatility <u>and</u> the chemical composition of the material in question. Thus, ranked in the order of decreasing flammability, we would find the following to hold true (for a particular Flash Point designated as "**F**"):

$$\mathbf{F}_{\text{pure hydrocarbons}} < \mathbf{F}_{\text{oxygenated hydrocarbons}} < \mathbf{F}_{\text{partially halogenated hydrocarbons}}, \text{ etc.}$$

As typical examples, $\mathbf{F}_{gasoline}$ = –45°C, $\mathbf{F}_{isopropanol}$ = 12°C, & $\mathbf{F}_{lubricating\ oil}$ = 232°C, etc. Certain materials (i.e., carbon tetrachloride, CCl_4) do not have a Flash Point, since there is <u>no</u> temperature at which their vapors can be ignited.

Ambient Measurements of Concentration

"Volume-Based" Concentrations

Ambient concentrations are most frequently expressed in, and understood on, a basis that is understood to be:

a unit volume - per - multiple volumes

OR, more precisely

a unit molecule - per - multiple molecules basis.

Among the most commonly used volume-based concentration units are each of the following:

Concentration Unit	Form in which these concentration units are commonly expressed	Volume (or Molecular) Definition
parts per million	**ppm(vol)** OR simply **ppm**	Volumes (or molecules) of a specific material of interest PER million volumes (or PER million molecules) present in the total matrix being considered.
parts per billion	**ppb(vol)** OR simply **ppb**	Volumes (or molecules) of a specific material of interest PER billion volumes (or PER billion molecules) present in the total matrix being considered.
percent	%	Volumes (or molecules) of a specific material of interest PER hundred volumes (or PER hundred molecules) present in the total matrix being considered.

While there are other volume-based concentration units [i.e., parts per hundred thousand, etc.] they are far less common than any of the foregoing three.

Two very important factors and/or relationships must be understood about any volume-based concentration unit. These are:

1. The volume-based concentration unit can only be used to express the ambient concentration of a gas or vapor; it can never be used to express the ambient concentration of any form of particulate, dust, or aerosol.
2. Virtually all ambient gas analyzers provide concentration outputs in the form of "parts per million by volume" [ppm(vol)]. With very rare exceptions, the calibrated output of these analyzers will always have been referenced to the readout that would have been obtained for measurements made at sea level. If a gas analyzer is used at an altitude that is significantly different than sea level, its readout will be incorrect in proportion to the difference in the actual barometric pressures (1) at the location where the analyzer has been used, and (2) at sea level — see **Appendix A**, which goes into more detail on the atmosphere and its effects on a wide variety of measurements.

"Mass-Based" Concentrations

Ambient concentrations are less frequently expressed in, and understood on, the basis of the mass of material of interest-per-unit volume in the ambient matrix being considered. Virtually without exception, the two principal mass-based units of concentration, as applied to the ambient air, are:

milligrams of the material of interest - per - cubic meter of matrix **mg/m^3**

AND

micrograms of the material of interest - per - liter of matrix **$\mu g/l$**

For any situation that is being evaluated, the magnitude of the numeric value that would be obtained for either of these two mass-based ambient concentrations would be the same — i.e., if it is known that a workplace has an ambient silica dust concentration of 0.13 mg/m^3, it would turn out that the process of converting this concentration to the form of micrograms - per - liter would produce the numerically identical value, namely, 0.13 $\mu g/l$.

From the perspective of the potential adverse impacts that might be caused by some material that exists as a component of the ambient air, the mass-based concentration of this material is a somewhat more "absolute" parameter than its volume-based counterpart. To determine the ambient concentration of any particulate, dust, or aerosol in the ambient air, the mass-based concentration parameter is the only form that is used.

Any analyzer that provides ambient mass-based concentration outputs [for gases, vapors, and/or particulates] will provide correct values irrespective of the altitude where the measurements are made.

Components & Measurement Parameters of the Ambient Air

Gas

A **Gas** is a substance that is in the gaseous state at **NTP**.

Vapor

A **Vapor** is the gaseous state of any material that would, under **NTP**, exist principally as a solid or a liquid.

Aerodynamic Diameter

The **Aerodynamic Diameter** is the diameter of a unit density sphere (i.e., density = 1.00 gms/cm^3) that would have the same settling velocity as the particle or aerosol in question.

Aerosol

An **Aerosol** is a suspension of liquid or solid particles in the air. Actual physical particle diameters usually fall in the range:

$$0.01\mu \leq [\text{particle diameter}] \leq 100\mu.$$

Dust

Dust is any particulate material, usually generated by a mechanical process, such as crushing, grinding, etc. Typical dust particles have aerodynamic diameters in the range:

$$0.5\ \mu m \leq [\text{aerodynamic diameter}] \leq 50\ \mu m.$$

Mist

A **Mist** is an aerosol suspension of liquid particles in the air, usually formed either by condensation directly from the vapor phase or by some mechanical process. Typical mist droplets have aerodynamic diameters in the range:

$$40\mu m \leq [\text{aerodynamic diameter}] \leq 400\ \mu m.$$

Smoke

Smoke is an aerosol suspension, usually of solid particulates, and usually formed by either the combustion of organic materials or the sublimation of some material. Typical smoke particulates have aerodynamic diameters in the range:

$$0.01\mu m \leq [\text{aerodynamic diameter}] \leq 0.5\ \mu m.$$

Fume

A **Fume** is an aerosol made up of solid particulates formed by condensation directly from the vapor state. Typical fume particles have aerodynamic diameters in the range:

$$0.001\mu m \leq [\text{aerodynamic diameter}] \leq 0.2\ \mu m.$$

Aspect Ratio

The **Aspect Ratio** for any particle is the ratio of its longest or greatest dimension to its shortest or smallest dimension.

Fiber

A **Fiber** is any particle having an aspect ratio greater than 3.

RELEVANT FORMULAE & RELATIONSHIPS

Calculations Involving Time Weighted Averages [TWAs]

Equation #3-1:

Equation **#3-1** is used to obtain a **Time Weighted Average** for any sort of time related exposure or measurement; it applies to any Time Weighted Average determination and can be utilized for any total base time period or time interval. Typically this calculation will be made in order to develop a TWA exposure value for use in comparison against any of the currently published exposure limit standards [i.e., the TLV, the PEL, the REL, and/or the MAK]. As an example, any Short Term Exposure Limit exposure determination will use this Equation. The denominator for any such calculation will be determined by the specified STEL Time Interval that applies to the material of interest in the determination [i.e., if the material being evaluated — from the perspective of referencing against its established PEL-STEL — were to have been isopropanol, then the STEL Time Interval would have been 15 minutes]:

$$\mathbf{TWA} = \frac{\sum_{i=1}^{n} T_iC_i}{\sum_{i=1}^{n} T_i} = \frac{T_1C_1 + T_2C_2 + \ldots + T_nC_n}{T_1 + T_2 + \ldots + T_n}$$

Where:

TWA = The **T**ime **W**eighted **A**verage Concentration that existed during the Time Intervals given by the sum of the individual $\mathbf{T_i}$s used in this calculation, i.e., the $\sum_{i=1}^{n} T_i$.

$\mathbf{T_i}$ = the **i**th Time Interval from the overall time period, which for a TWA (in contrast to a STEL) would total up to a full 8 hours, i.e., $\sum_{i=1}^{n} T_i = 8$ hours.

$\mathbf{C_i}$ = the **i**th Concentration Value of the single component of interest — i.e., the ambient gas or particulate concentration — that existed during the specific **i**th Time Interval.

Equation #3-2:

Equation **#3-2** is used to determine the **Effective Percent Exposure Level** for any of the established parameters [i.e., the TLV, the PEL, the REL, or the MAK], resulting from the combined effects of <u>all</u> the potentially irritating, toxic, or hazardous components, whether gas or particulate, in any ambient air system that is being evaluated. To repeat, it is effective for any ambient air matrix, whether there is only a single volatile or aerosolized component in it, or many such components.

In applying Equation **#3-2**, the individual performing the calculation must know the specific value of the Standard being applied [i.e., for the case shown, the REL] that has been established for each of the various components that are contained in the ambient air that is being evaluated. If, for example, the evaluation were to involve an 8-hour Recommended Exposure Limit – Time Weighted Average determination for an air mass containing three different refrigerants: R-12, R-22, & R-112, then the REL-TWAs for these three compounds would have to be known. For reference in this example, these REL-TWA Values are as follows: for R-12, the $\mathbf{REL\text{-}TWA_{R\text{-}12}}$ = 1,000 ppm(vol); for R-22, the $\mathbf{REL\text{-}TWA_{R\text{-}22}}$ = 1,000 ppm(vol); and finally for R-112, the $\mathbf{REL\text{-}TWA_{R\text{-}112}}$ = 500 ppm(vol). The effective Time Weighted Average Concentration Values [i.e., the TWA_is] that are shown in the numerator of Equation **#3-2** would be determined by applying Equation **#3-1**, from Page 3-8, to the available information or data.

$$\% \, \mathbf{REL} = 100\left[\sum_{i=1}^{n} \frac{\mathbf{TWA_i}}{\mathbf{REL_i}}\right] = 100\left[\frac{\mathbf{TWA_1}}{\mathbf{REL_1}} + \frac{\mathbf{TWA_2}}{\mathbf{REL_2}} + \cdots + \frac{\mathbf{TWA_n}}{\mathbf{REL_n}}\right]$$

Where: **% REL** = the **Effective Percent Exposure Level** from the perspective of the Recommended Exposure Limit Standards that was achieved for the mixture being evaluated, expressed always as a percentage.

$\mathbf{TWA_i}$ = the **T**ime **W**eighted **A**verage Concentration of the **i**th component in the mixture being evaluated.

$\mathbf{REL_i}$ = the listed **R**ecommended **E**xposure **L**imit (or, in this case, the **REL-TWA**) of the **i**th component.

Note: If the **% REL** is equal to or less than 100%, then it can be inferred that the **Effective REL** for the mixture has not been exceeded; if this **% REL** is greater than 100%, then the inference is that the **Effective REL** for the mixture has been exceeded.

Calculations Involving Exposure Limits

Equation #3-3:

Equation **#3-3** is used to determine the **Effective Exposure Limit** that exists for any equilibrium vapor phase that is in contact with any well-defined liquid mixture containing two or more different volatile components. It can be applied to any of the established parameters [i.e., the TLV, the PEL, the REL, or the MAK]; the value of the individual Exposure Limit Standard for each of the components in the liquid mixture must be known. If this **Effective Exposure Limit** is determined EITHER by using different Exposure Limit Standards [i.e., TLV-TWAs coupled with PEL-TWAs], OR by using the same Exposure Limit Standard, with different time interval bases [i.e., the $\text{REL-STEL}_{\text{hydrazine}}$ which must be evaluated over a 120-minute period, and the $\text{REL-STEL}_{\text{1,2-dioxane}}$ which must be evaluated over a 30-minute period], then the resultant calculated **Effective Exposure Limit** would be considered to be NEITHER valid NOR accurate.

Equation **#3-3** assumes that the overall composition of the vapor phase existing above a volatile liquid mixture will be identical to the mass composition of the liquid mixture, and although this would rarely — if ever — be true, the Equation is considered to be a useful tool for determining "Order-of-Magnitude-Approximations" for the concentrations of the components in the vapor phase. This Equation was proposed for use by the ACGIH; therefore, it is used most commonly for the determination of this organization's Exposure Limit

Standard — namely, the TLV; it can, however, be applied to any of the commonly used Exposure Limit Standards.

If the user wishes a more precise treatment of this situation, then what is required would be the sequential application of several of the more basic laws and relationships, each of which has been documented in Chapter 2, which covered the basic laws of physics and the basic gas laws. For reference, these would include Dalton's Law of Partial Pressures [Equation **#s 1-15** & **1-16**, from Pages 1-22 & 1-23], and Raoult's Law [Equation **#s 1-17** & **1-18**, from Pages 1-23 & 1-24].

$$\mathbf{TLV_{effective}} = \left[\frac{1}{\sum_{i=1}^{n}\frac{f_i}{TLV_i}}\right] = \left[\frac{1}{\frac{f_1}{TLV_1} + \frac{f_2}{TLV_2} + \ldots + \frac{f_n}{TLV_n}}\right]$$

Where: $\mathbf{TLV_{effective}}$ = the **Effective Exposure Limit** — in this case a Threshold Limit Value. This parameter can be evaluated for <u>any</u> of the established Exposure Limits — i.e., for a PEL, REL, or MAK, etc. Remember, however, that this relationship applies <u>only</u> to the vapor phase existing above a mixture of volatile liquids. The user must understand that the result of this calculation is an "Order-of-Magnitude-Approximation" <u>only</u>. Also the calculated $\mathbf{TLV_{effective}}$ in this case will always be expressed in mass concentration units, mg/m^3, NEVER the more common volumetric unit, ppm(vol).

$\mathbf{f_i}$ = the Weight Fraction of the **i**th component in the mixture of volatile liquid components being considered, i.e., the $\left[\frac{\text{mass of the component of interest}}{\text{total mass of the liquid mixture}}\right]$;

$\mathbf{TLV_i}$ = the Threshold Limit Value [TLV] — this could be any established Exposure Limit — of the **i**th component, must always be expressed in mass-based units of concentration, mg/m^3, NEVER the more common volume-based unit, ppm(vol).

Calculations Involving the Conversion of Concentration Units

<u>Equation #s 3-4, 3-5, & 3-6</u>:

The following three Equations, **#s 3-4**, **3-5**, and **3-6**, are used to convert sets of **Mass-Based Concentration** units to their **Volume-Based Concentration** equivalents [e.g., converting from concentrations expressed in units such as mg/m^3 to those in one of the common volumetric-based sets of units, such as ppm(vol)]. The first Equation, **#3-4**, would be used to affect this conversion under conditions of Normal Temperature and Pressure [NTP]; the second, **#3-5**, would be used for conversions at Standard Temperature and

Pressure [STP] conditions, while the third, **#3-6**, applies to any mass-to-volume-based concentration conversion, regardless of the conditions of ambient temperature and pressure.

Equation #3-4:

$$C_{vol} = \frac{24.45}{MW_i}[C_{mass}] \qquad @\ NTP$$

Where:

C_{vol} = the Volume-Based Concentration of the component of interest, namely, the **i**th component, measured in ppm(vol);

C_{mass} = the Mass-Based Concentration of the same component, namely the **i**th component, in mg/m^3; &

MW_i = the Molecular Weight of the same **i**th component.

Equation #3-5:

$$C_{vol} = \frac{22.41}{MW_i}[C_{mass}] \qquad @\ STP$$

Where:

C_{vol} = the Volume-Based Concentration of the component of interest, namely, the **i**th component, measured in ppm(vol);

C_{mass} = the Mass-Based Concentration of the same component, namely the **i**th component, in mg/m^3; &

MW_i = the Molecular Weight of the same **i**th component.

Equation #3-6:

$$C_{vol} = \frac{RT}{P[MW_i]}[C_{mass}] \quad \text{under any conditions of ambient Temperature \& Pressure}$$

Where:

C_{vol} = the Volume-Based Concentration as defined immediately above for Equation #s **3-4** & **3-5**, in ppm(vol);

C_{mass} = the Mass-Based Concentration also as defined immediately above for Equation #s **3-4** & **3-5**, in mg/m^3;

MW_i = the Molecular Weight of the **i**th component, again as defined above;

T = the ambient Absolute Temperature, in some suitable units of absolute temperature, most probably, K;

P = the ambient Barometric Pressure, in some suitable units of pressure, such as mm Hg; &

R = the Universal Gas Constant in units consistent with those used for the prevailing ambient barometric pressure and temperature conditions — for example, any of the following values of this Universal Gas Constant could be used.

$$= 0.0821 \frac{(\text{liter})(\text{atmospheres})}{(\text{K})(\text{mole})};$$

$$= 62.36 \frac{(\text{liter})(\text{mm Hg})}{(\text{K})(\text{mole})};$$

$$= 1.631 \frac{(\text{feet})^3(\text{millibars})}{(^\circ\text{R})(\text{mole})}.$$

Equation #s 3-7, 3-8, & 3-9:

The following three Equations, **#s 3-7**, **3-8**, & **3-9**, are used to convert sets of Volume-Based Concentration units to their Mass-Based Concentration equivalents [e.g., converting from concentrations expressed in ppm(vol) to those expressed in mg/m^3]. The first Equation, **#3-7**, would be used to affect this conversion under Normal Temperature and Pressure [NTP] conditions; the second, **#3-8**, would be used under Standard Temperature and Pressure [STP] conditions; while the third, **#3-9**, applies in general to any mass-to-volume-based concentration conversion, regardless of the conditions of ambient temperature and pressure.

Equation #3-7:

$$C_{mass} = \frac{MW_i}{24.45}[C_{vol}] \qquad @\ NTP$$

Where:

C_{vol} = the Volume-Based Concentration of the component of interest, namely, the **i**th component, measured in ppm(vol);

C_{mass} = the Mass-Based Concentration of the same component, namely the **i**th component, in mg/m^3; &

MW_i = the Molecular Weight of the same **i**th component.

Equation #3-8:

$$C_{mass} = \frac{MW_i}{22.41}[C_{vol}] \qquad @\ STP$$

Where:

C_{vol} = the Volume-Based Concentration of the component of interest, namely, the **i**th component, measured in ppm(vol);

C_{mass} = the Mass-Based Concentration of the same component, namely the **i**th component, in mg/m³; &

MW_i = the Molecular Weight of the same **i**th component.

Equation #3-9:

$$C_{mass} = \frac{P\left[MW_i\right]}{R\,T}\left[C_{vol}\right]$$ under <u>any</u> conditions of ambient Temperature & Pressure

Where:

C_{vol} = the Volume-Based Concentration as defined for Equation **#s 3-7** & **3-8**, in ppm(vol), on Pages 3-11 & 3-12;

C_{mass} = the Mass-Based Concentration also as defined immediately above for Equation **#s 3-7** & **3-8**, in mg/m³, on Pages 3-11 & 3-12;

MW_i = the Molecular Weight of the **i**th component, again as defined for Equation **#s 3-7** & **3-8**, in ppm(vol), on Pages 3-11 & 3-12;

T = the ambient Absolute Temperature, in some suitable units of absolute temperature, most probably, K;

P = the ambient Barometric Pressure, in some suitable units of pressure, such as mm Hg; &

R = the Universal Gas Constant in units consistent with those used for the prevailing ambient barometric pressure and temperature conditions — for example, any of the following values of this Universal Gas Constant could be used.

$$= 0.0821\ \frac{(\text{liter})(\text{atmospheres})}{(\text{K})(\text{mole})};$$

$$= 62.36\ \frac{(\text{liter})(\text{mm Hg})}{(\text{K})(\text{mole})};$$

$$= 1.631\ \frac{(\text{feet})^3(\text{millibars})}{(^\circ\text{R})(\text{mole})}.$$

Calculations Involving TLV Exposure Limits for Free Silica Dusts in Ambient Air

Equation #s 3-10, 3-11, & 3-12:

The following three Equations, **#s 3-10**, **3-11**, & **3-12**, are now used only very infrequently, and only when one must develop a rough approximation of the respective required TLVs. These Equations were developed and promulgated by the ACGIH, and are used, therefore, only to determine Threshold Limit Value Exposure Limits. In order, these three Equations are as follows: **#3-10** — For Respirable Quartz Dusts; **#3-11** — For Total Dusts; and **#3-12** — For Mixtures of the Three Most Common Types of Silica Dusts. These three equations are shown here.

Equation #3-10:

$$\mathbf{TLV_{quartz}} = \left[\frac{\mathbf{10mg/m^3}}{\mathbf{\%RQ + 2}}\right]$$

Where: $\mathbf{TLV_{quartz}}$ = the calculated 8-hour TWA Threshold Limit Value for respirable quartz dust, in mg/m^3; &

$\mathbf{\%RQ}$ = the Mass or Weight Fraction of Respirable Quartz Dusts in the sample being evaluated, expressed as a percentage.

Equation #3-11:

$$\mathbf{TLV_{dust}} = \left[\frac{\mathbf{30mg/m^3}}{\mathbf{\%Q + 2}}\right]$$

Where: $\mathbf{TLV_{dust}}$ = the calculated 8-hour TWA Threshold Limit Value for total dusts, in mg/m^3; &

$\mathbf{\%Q}$ = the Mass or Weight Fraction of Quartz Dusts in the sample being evaluated, expressed as a percentage.

Equation #3-12:

$$\mathbf{TLV_{mix}} = \left[\frac{\mathbf{10mg/m^3}}{\mathbf{\%Q + 2(\%C) + 2(\%T) + 2}}\right]$$

Where: $\mathbf{TLV_{mix}}$ = the calculated 8-hour TWA Threshold Limit Value for the complex mixture of silica dusts, in mg/m^3;

$\mathbf{\%Q}$ = the Mass or Weight Fraction of Quartz Dusts in the sample being evaluated, expressed as a percentage;

%C	=	the Mass or Weight Fraction of Cristobalite Dusts in the sample being evaluated, also expressed as a percentage; &
%T	=	the Mass or Weight Fraction of Tridymite Dusts in the sample being evaluated, this one, also, expressed as a percentage.

WORKPLACE AMBIENT AIR PROBLEM SET

Problem #3.1:

A worker was exposed, over his work shift, to various levels of n-pentane, and for varying periods of time, as shown in the tabulated exposure history below. The appropriate ACGIH Standards that apply to n-pentane are: the TLV-TWA = 600 ppm(vol), and the TLV-STEL = 750 ppm(vol). What was this employee's 8-hour TWA exposure to n-pentane?

2.0 hours	451 ppm(vol)
3.0 hours	728 ppm(vol)
1.5 hours	619 ppm(vol)
1.5 hours	501 ppm(vol)

Applicable Definitions:	Threshold Limit Values	Page 3-1
	Time Weighted Averages	Page 3-2
	Short Term Exposure Limits	Page 3-2
Applicable Formula:	Equation **#3-1**	Page 3-8
Solution to this Problem:	Page 3-47	

Problem Workspace

Problem #3.2:

During the concluding 15 minutes of the final 1.5-hour segment of his workday, what would this worker's exposure to n-pentane have to have been so that his overall 8-hour TWA would have been in violation of the established ACGIH Threshold Limit Value?

Applicable Definitions:	Threshold Limit Values	Page 3-1
	Time Weighted Averages	Page 3-2
Applicable Formula:	Equation **#3-1**	Page 3-8
Solution to this Problem:	Pages 3-47 & 3-48	

Workspace for Problem **#3.2**

Problem #3.3:

During the concluding 15 minutes of the final 1.5-hour segment of his workday, what would his exposure have to have been so that his work experience for that entire day would have involved a TLV-STEL violation? Would this 15-minute exposure also have produced a TLV-TWA violation?

Applicable Definitions:	Threshold Limit Values	Page 3-1
	Time Weighted Averages	Page 3-2
	Short Term Exposure Limits	Page 3-2
Applicable Formula:	Equation **#3-1**	Page 3-8
Solution to this Problem:	Page 3-48	

Problem Workspace

Problem #3.4:

An Office Machine Repair Technician was exposed to a relatively high concentration level of ozone from a malfunctioning copier he had been called upon to repair. He observed the operation of the machine for 30 minutes in order to diagnose its problems. He then fixed the copier by replacing a defective part. Finally, he observed the operation of the then properly functioning machine for an additional 30 minutes to ensure that it was, in fact, working properly. His ozone exposures for the day on which he fixed this machine were as follows:

Task Description	Exposure Time	Ozone Exposures
Diagnostic Effort	30 min	289 ppb(vol)
Repair Effort	60 min	42 ppb(vol)
Repair Observing Time	30 min	93 ppb(vol)
Balance of the Work Day	6 hours	8 ppb(vol)

The PEL-TWA for ozone is 0.1 ppm(vol), and the PEL-STEL for this material is 0.3 ppm(vol). What was this worker's 8-hour TWA to ozone? Did this worker experience a PEL-STEL violation?

Applicable Definitions:	Permissible Exposure Limits	Page 3-1
	Time Weighted Averages	Page 3-2
	Short Term Exposure Limits	Page 3-2
Applicable Formula:	Equation **#3-1**	Page 3-8
Solution to this Problem:	Pages 3-48 & 3-49	

Problem Workspace

Problem #3.5:

The same Office Machine Repair Technician from the previous problem, Problem **#3.4**, worked in a geographic region that was characterized by a very bad "smog" condition, under which the exterior ambient air — on the particular day when this individual made the copier repairs described in Problem **#3.4** — had an ozone concentration of 222 ppb(vol). If this technician spent 2.5 of the final 6 hours of that work day in his repair vehicle driving from one job to the next, and if all his other repairs on that day were to typewriters where he experienced only the previously mentioned indoor 8-ppb(vol) ozone background level, what would his new 8-hour TWA to ozone have been? For reference, his timed ozone exposures were broken down as follows:

Task Description	Exposure Time	Ozone Exposures
Initial Copier Repairs from Problem **#3.4**	120 min	Determine from Problem **#3.4**
Repair Effort on Typewriters	210 min	8 ppb(vol)
Driving Time between Jobs	150 min	222 ppb(vol)

Applicable Definitions:	Time Weighted Averages	Page 3-2
Data from:	Problem **#3.4**	Page 3-18
Applicable Formula:	Equation **#3-1**	Page 3-8
Solution to this Problem:	Page 3-49	

Problem Workspace

Problem #3.6:

An Engineering Technician for a large metropolitan hospital spent one full 8-hour workday servicing the gas sterilizer in the hospital's Central Supply Department. Her ethylene oxide exposures were measured continuously by an Industrial Hygienist using an infrared spectrophotometer [the minimum detectable ethylene oxide level — the MDL — for this analyzer was 0.4 ppm(vol)]. Ethylene oxide TWA measurements were made for various periods of time, and were reported by this IH as follows:

2.2 hours	0.7 ppm(vol)
1.4 hours	1.2 ppm(vol)
1.5 hours	1.6 ppm(vol)
2.9 hours	< 0.4 ppm(vol)

Recognizing that the 8-hour PEL-TWA for ethylene oxide is 1.0 ppm(vol), and its 8-hour TWA Action Level is 0.5 ppm(vol), would this employee's TWA exposure on the date in question have to be reported as a violation? What was her exposure?

Applicable Definitions:	Permissible Exposure Limits	Page 3-1
	Time Weighted Averages	Page 3-2
	Action Levels	Page 3-3
Applicable Formula:	Equation **#3-1**	Page 3-8
Solution to this Problem:	Pages 3-49 & 3-50	

Problem Workspace

Problem Workspace Continued on the Following Page

Continuation of Workspace for Problem **#3.6**

Problem #3.7:

The production employees in a custom fiberglass fabrications shop are routinely exposed to ambient styrene vapors [for styrene, the PEL-TWA = 50 ppm(vol), and the 15-minute PEL-STEL = 100 ppm(vol)]. After one particularly busy day, an employee's styrene dosimeter indicated an 8-hour TWA styrene exposure of 48 ppm(vol). If her 8-hour workday included two 5-minute coffee breaks and a 30-minute lunch period, all three of which were spent in a room where the ambient styrene concentration level never exceeded 0.1 ppm(vol) with the balance of her workday being spent in the production area, what must this employee's average styrene exposure have been during those time periods when she was actually involved in productive work? Did this exposure involve a PEL-STEL violation?

Applicable Definitions:	Permissible Exposure Limits	Page 3-1
	Time Weighted Averages	Page 3-2
	Short Term Exposure Limits	Page 3-2
Applicable Formula:	Equation **#3-1**	Page 3-8
Solution to this Problem:	Pages 3-50 through 3-52	

Problem Workspace

Problem Workspace Continued on the Following Page

Continuation of Workspace for Problem **#3.7**

Problem #3.8:

The Safety Manager of a large Cold Storage Plant, designed for holding apples in a fresh condition over extended periods of time, determined that the workers in his company's cold rooms have been experiencing an overall 84% 8-hour TWA exposure to carbon dioxide. If: (1) each worker must spend 6 hours/day in these Cold Rooms; if (2) the PEL-TWA for CO_2 is 10,000 ppm(vol); and if (3) the outside air, to which each worker is exposed for the remainder of the work day, contains an average of 425 ppm(vol) of CO_2, what must the average concentration of CO_2 be in the Cold Rooms?

Applicable Definitions:	Permissible Exposure Limits	Page 3-1
	Time Weighted Averages	Page 3-2
Applicable Formula:	Equation **#3-1**	Page 3-8
Solution to this Problem:	Page 3-52	

Problem Workspace

Problem #3.9:

If the Safety Manager of the foregoing Cold Storage Plant [Problem **#3.8**] has determined: (1) that his workers should never experience more than a 50% PEL exposure to CO_2, and (2) that the minimum CO_2 concentration he can permit in the Cold Rooms [in order to provide for proper apple storage] is 1.5% by volume, what then will be the maximum time period he will be able to allow his company's employees to work in the Cold Rooms each day from now on?

Applicable Definitions:	Permissible Exposure Limits	Page 3-1
	Time Weighted Averages	Page 3-2
Data from:	Problem **#3.8**	Page 3-23
Applicable Formula:	Equation **#3-1**	Page 3-8
Solution to this Problem:	Pages 3-52 & 3-53	

Problem Workspace

Problem #3.10:

John Smith, an Industrial Hygienist, working for the ABC Company determined the following Time Weighted Average employee exposures to the following list of four solvent vapors, for those workers who operated the company's paint spray booth:

No.	Solvent	TWA	TLV-TWA
1.	MIBK	12 ppm(vol)	50 ppm(vol)
2.	toluene	17 ppm(vol)	100 ppm(vol)
3.	methanol	55 ppm(vol)	200 ppm(vol)
4.	IPA	91 ppm(vol)	400 ppm(vol)

What was the % TLV exposure of these workers?

Applicable Definitions:	Threshold Limit Values	Page 3-1
	Time Weighted Averages	Page 3-2
Applicable Formula:	Equation **#3-2**	Pages 3-8 & 3-9
Solution to this Problem:	Page 3-53	

Problem Workspace

Problem #3.11:

Kraft pulp mill employees involved in bleaching kraft paper are potentially exposed to both chlorine and chlorine dioxide. For these two chemicals, the published PEL-TWAs and PEL-STELs are as follows:

Bleaching Gas	PEL-TWA	PEL-STEL
Chlorine	500 ppb(vol)	1,000 ppb(vol)
Chlorine Dioxide	100 ppb(vol)	300 ppb(vol)

On a heavy bleaching day, one employee's dosimeters indicated that his 8-hour TWA exposures to these two materials had been 0.42 ppm(vol) for chlorine, and 0.08 ppm(vol) for chlorine dioxide. What was this employee's overall % PEL exposure to these hazardous vapors? Was his employer in violation of any OSHA Standards?

Applicable Definitions:	Permissible Exposure Limits	Page 3-1
	Time Weighted Averages	Page 3-2
	Short Term Exposure Limits	Page 3-2
Applicable Formula:	Equation **#3-2**	Pages 3-8 & 3-9
Solution to this Problem:	Pages 3-53 & 3-54	

Problem Workspace

Problem Workspace Continued on the Following Page

Continuation of Workspace for Problem **#3.11**

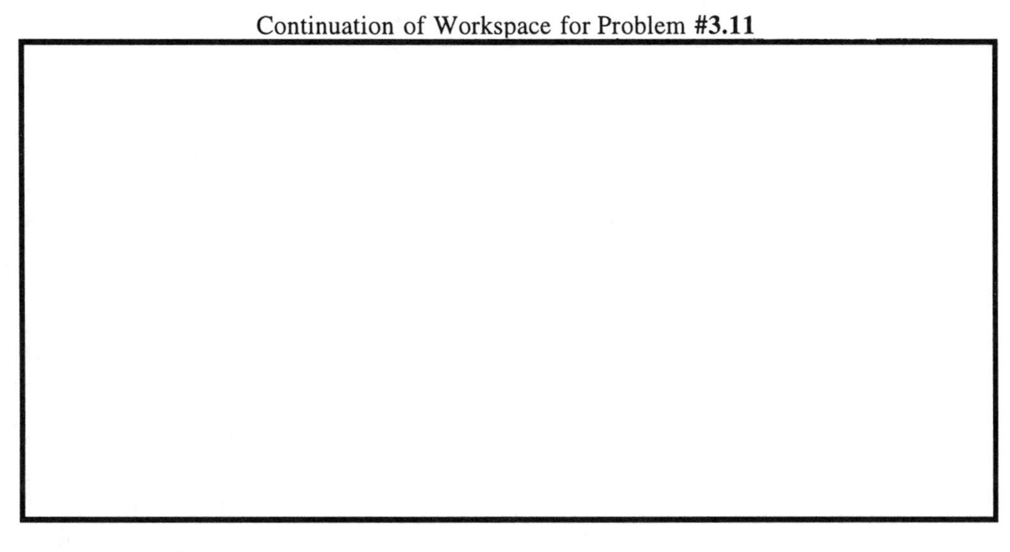

Problem #3.12:

A vapor degreasing solvent is made up of the following four chemicals, according to the following proportions by weight:

Component Number	Component	Weight % in Solvent	TLV-TWA	Molecular Weight
1.	Freon 11	25 %	1,000 ppm(vol)	137.38
2.	Freon 113	55 %	1,000 ppm(vol)	187.38
3.	Methyl Chloroform	15 %	350 ppm(vol)	133.41
4.	Methylene Chloride	5 %	50 ppm(vol)	84.93

Assuming that the composition of the degreasing vapor is equal to that of the liquid mixture [a very unlikely condition], what would be the "Order-of-Magnitude-Approximation" for the Effective TLV Exposure Limit [in both mg/m^3 & ppm(vol)] for the vapor space in this vapor degreaser? You may assume that this machine is operated at STP.

What would the corresponding "Order-of-Magnitude-Approximations" for the vapor space concentrations of each of these four components be [again, in both mg/m^3 & ppm(vol)] on the assumption that these four concentrations, in combination, actually resulted in this vapor space being operated at the calculated overall Effective TLV Exposure Limit?

It should be noted that only very rarely would the vapor space in any sort of a degreaser ever be a "Workplace". On occasion, however, particularly during periods when facilities of this type are being serviced, a worker might well have to place his or her head into this vapor space; thus, the calculation would have significance for those situations. As you will see from the calculations that follow, placing one's head in the vapor space of a degreaser when it is in operation would be about as hazardous an action that one might contemplate.

Applicable Definitions:	Threshold Limit Values	Page 3-1
	Time Weighted Averages	Page 3-2
	Standard Temperature & Pressure	Page 1-14
Applicable Formulae:	Equation **#3-3**	Pages 3-9 & 3-10
	Equation **#3-5**	Pages 3-10 & 3-11
	Equation **#3-8**	Pages 3-12 & 3-13
Solution to this Problem:	Pages 3-54 through 3-57	

Workspace for Problem **#3.12**

Problem Workspace Continued on the Following Page

Continuation of Workspace for Problem **#3.12**

Problem #3.13:

For the vapor degreaser of the previous problem, **#3.12**, please recalculate the equilibrium vapor space concentrations of the four solvents [again, in both mg/m^3 & ppm(vol)], but for this case, please use the more basic laws of physical chemistry as listed in the first chapter of this volume. The results of these calculations will be more accurate and precise than the approximations you developed for Problem **#3.12**. You may find the following information to be of value for making this determination.

Component Number	Component	Weight % in Solvent	Molecular Weight	Component Vapor Pressure at STP
1.	Freon 11	25 %	137.38	302.8 mm Hg
2.	Freon 113	55 %	187.38	111.8 mm Hg
3.	Methyl Chloroform	15 %	133.41	36.0 mm Hg
4.	Methylene Chloride	5 %	84.93	144.0 mm Hg

Applicable Definition:	Standard Temperature & Pressure	Page 1-14
Data from:	Problem **#3.12**	Page 3-27
Applicable Formulae:	Equation **#1-10**	Pages 1-19 & 1-20
	Equation **#1-16**	Pages 1-22 & 1-23
	Equation **#1-17**	Page 1-23
	Equation **#3-8**	Pages 3-12 & 3-13
Solution to this Problem:	Pages 3-57 through 3-61	

Problem Workspace

Problem Workspace Continued on the Following Page

Continuation of Workspace for Problem **#3.13**

Problem #3.14:

A solution containing two moderately volatile solvents is used widely in the wafer fabrications area of a large semiconductor manufacturer. This solution is made up, by weight, of 65% cellosolve [2-ethoxyethanol] and 35% t-butyl alcohol. Assuming that the composition of the vapor is equal to the composition of the liquid solvent mixture, what is the Effective TLV [in both mg/m^3 & ppm(vol)] for this mixture? You may also assume that the wafer fabrications area is operated at NTP. What will the ambient air concentrations for each of the two components of this mixture be [in both mg/m^3 & ppm(vol)] when the total vapor concentration in the wafer fabrications area is equal to this Effective TLV? The following data may be useful for you:

Component	Structural Formulae	TLV-TWA	Mol. Weight
cellosolve	$HO\text{-}CH_2\text{-}CH_2\text{-}O\text{-}CH_2\text{-}CH_3$	200 ppm(vol)	90.12 amu
t-butyl alcohol	$[CH_3]_3\text{-}C\text{-}OH$	100 ppm(vol)	74.12 amu

Applicable Definitions:	Threshold Limit Values	Page 3-1
	Time Weighted Averages	Page 3-2
	Short Term Exposure Limits	Page 3-2
Applicable Formulae:	Equation **#3-3**	Pages 3-9 & 3-10
	Equation **#3-4**	Pages 3-10 & 3-11
	Equation **#3-7**	Page 3-12
Solution to this Problem:	Pages 3-61 & 3-62	

Problem Workspace

Problem Workspace Continued on the Next Page

Continuation of Workspace for Problem **#3.14**

Problem #3.15:

For the volatile solvent mixture of the previous problem, **#3.14**, please recalculate the equilibrium vapor space concentrations of the two solvents [again, in both mg/m^3 & ppm(vol)], but for this case, please again use the more basic laws of physical chemistry as listed in the first chapter of this volume. Again the results of these calculations will be considerably more accurate and precise than the approximations you developed for Problem **#3.14**. You may find the following information to be of value for making this determination.

Component	Weight % in Solvent	Molecular Weight	Component Vapor Pressure at NTP
cellosolve	65%	90.12 amu	5.3 mm Hg
t-butyl alcohol	35%	74.12 amu	17.4 mm Hg

Applicable Definition:	Standard Temperature & Pressure	Page 1-14
Data from:	Problem **#3.14**	Page 3-32
Applicable Formulae:	Equation **#1-10**	Pages 1-19 & 1-20
	Equation **#1-16**	Pages 1-22 & 1-23
	Equation **#1-17**	Page 1-23
	Equation **#3-7**	Page 3-12
Solution to this Problem:	Pages 3-62 through 3-65	

Problem Workspace

Problem Workspace Continued on the Next Page

Continuation of Workspace for Problem **#3.15**

Problem #3.16:

A U.S. Navy shipboard refrigerant consists of a well-defined mixture of the following three chlorofluorocarbons, made up according to the following proportions:

Chemical	Wt. % in Refrigerant	Molecular Weight	TLV-TWA
Freon 12	40 %	120.92	1,000 ppm(vol)
Freon 21	20 %	102.92	10 ppm(vol)
Freon 112	40 %	170.92	500 ppm(vol)

Obviously any leak of this pressurized liquid refrigerant will produce a vapor whose composition will be the same as that of the liquid material. What will be the Effective TLV for this mixture [in both mg/m^3 & ppm(vol)]? U.S. Navy ships operate their refrigeration spaces at NTP.

Applicable Definitions:	Threshold Limit Values	Page 3-1
	Time Weighted Averages	Page 3-2
Applicable Formulae:	Equation **#3-3**	Pages 3-9 & 3-10
	Equation **#3-4**	Pages 3-10 & 3-11
	Equation **#3-7**	Page 3-12
Solution to this Problem:	Pages 3-65 through 3-67	

Problem Workspace

Problem Workspace Continued on the Next Page

Continuation of Workspace for Problem **#3.16**

Problem #3.17:

The U.S. Navy uses a bulkhead mounted, digital readout, highly specific Freon 21 monitor (i.e., it does not respond to any vapor other than Freon 21) in each of the shipboard refrigeration spaces that employ the special refrigerant described in Problem **#3.16**. The response range of this gas monitor is 0 to 20.0 ppm(vol) [actually, 0.0 to 19.9 ppm(vol)]. The readout is in the form, "XX.X". In this range the analyzer's accuracy is ± 0.05 ppm(vol). What concentration must the two bi-level audible alarm settings be on this instrument? Typically, a bi-level Alert Alarm will be set at 60% of the effective TLV-TWA, and a second, higher level Evacuate Alarm, at 90% of this same effective TLV-TWA. As stated in Problem **#3.16**, Navy ships usually operate their refrigeration spaces at NTP.

Applicable Definitions:	Threshold Limit Values	Page 3-1
	Time Weighted Averages	Page 3-2
Data from:	Problem **#3.16**	Page 3-36
Applicable Formula:	Equation **#3-1**	Page 3-8
Solution to this Problem:	Page 3-67	

Problem Workspace

Problem #3.18:

An Industrial Hygienist who works for a large automobile manufacturer has initiated a program at her company designed to quantify the ambient levels of airborne dusts [principally alumina dust] in the facility's main assembly area. She has decided to monitor 16 assembly area workers, and has chosen to fit each of these employees with an active dust sampling system that consists of a personal sampling pump connected to a particulate filter cassette that uses a pre-weighed, 37-mm diameter Teflon filter on which the dust is to be collected, and ultimately quantified by weighing. The sampling pumps she has chosen provide a very stable 1.75 liters per minute flow rate over an entire 8-hour work shift; each pump works for a continuous, uninterrupted 8-hour period. The actual time each employee spends in the assembly area is less than a full 8 hours, since the company provides all employees with a 30-minute lunch break, and two 10-minute coffee breaks per full 8-hour work shift. Workers always take their lunch and coffee breaks in an area that is completely dust free, although their personal sampling pumps continue to operate during these periods. The net weight gains of the 16 filters in this evaluation are given below:

Empl. No.	Cassette Net Wt.	Empl. No.	Cassette Net Wt.	Empl. No.	Cassette Net Wt.	Empl. No.	Cassette Net Wt.
1	4.95 mg	5	7.90 mg	9	7.44 mg	13	2.26 mg
2	6.19 mg	6	1.08 mg	10	5.08 mg	14	8.11 mg
3	3.85 mg	7	4.33 mg	11	1.52 mg	15	6.87 mg
4	2.63 mg	8	6.81 mg	12	7.82 mg	16	4.90 mg

Assuming these data are fully representative of the situation in the main assembly area of this plant, what was the average airborne dust concentration, expressed in mg/m^3? If the PEL-TWA for alumina is 10 mg/m^3, should this Industrial Hygienist be concerned about the dust levels in the assembly area?

Applicable Definitions:	Permissible Exposure Limits	Page 3-1
	Time Weighted Averages	Page 3-2
Applicable Formulae:	Equation **#2-2**	Page 2-4
	Equation **#3-1**	Page 3-8
Solution to this Problem:	Pages 3-68 & 3-69	

Problem Workspace

Problem Workspace Continued on the Next Page

Continuation of Workspace for Problem **#3.18**

Problem #3.19:

Glutaraldehyde is the "sterilant" used as the active ingredient in numerous proprietary Cold Sterilizing solutions that are, in turn, used by many hospitals and other health care facilities. The German MAK-TWA for this material is 0.8 mg/m^3. Express this mass-based concentration in its volume-based equivalent, assuming NTP. The molecular weight of glutaraldehyde is 100.12 amu.

Applicable Definitions:	Max. Concentration in the Workplace	Page 3-2
	Normal Temperature & Pressure	Page 1-14
Applicable Formula:	Equation **#3-4**	Pages 3-10 & 3-11
Solution to this Problem:	Page 3-70	

Problem Workspace

Problem #3.20:

The OSHA PEL-STEL for tetrahydrofuran is 735 mg/m^3. The PEL-TWA is 590 mg/m^3. Please express these concentrations in ppm(vol) units, assuming STP conditions. The formula weight of tetrahydrofuran is 721 amu.

Applicable Definitions:	Permissible Exposure Limits	Page 3-1
	Time Weighted Averages	Page 3-2
	Short Term Exposure Limits	Page 3-2
Applicable Formula:	Equation **#3-5**	Pages 3-10 & 3-11
Solution to this Problem:	Page 3-70	

Problem Workspace

Problem Workspace Continued on the Next Page

Continuation of Workspace for Problem **#3.20**

Problem #3.21:

What would be the volume-based concentration of nitric oxide in the interior of an unpressurized aircraft flying at an altitude of 10,000 feet and having an interior temperature of 18°C, if the mass-based concentration of this material was found to be at the PEL-TWA level of 30 mg/m^3? For your information: (1) the normal pressure in the interior of this type of aircraft flying at this altitude would be 0.70 atmospheres, and (2) the molecular weight of nitric oxide is 30 amu.

Applicable Definitions:	Permissible Exposure Limits	Page 3-1
	Time Weighted Averages	Page 3-2
Applicable Formulae:	Equation **#1-1**	Page 1-16
	Equation **#3-6**	Pages 3-10 & 3-11
Solution to this Problem:	Pages 3-70 & 3-71	

Problem Workspace

Problem #3.22:

The 8-hour PEL-TWA and the Action Level for ethylene oxide are, respectively, 1.0 ppm(vol) and 0.5 ppm(vol). Express these concentrations in mg/m^3, assuming NTP. The molecular weight of ethylene oxide is 44.05.

Applicable Definitions:	Permissible Exposure Limits	Page 3-1
	Time Weighted Averages	Page 3-2
	Action Level	Page 3-3
Applicable Formula:	Equation **#3-7**	Page 3-12
Solution to this Problem:	Page 3-71	

Problem Workspace

Problem #3.23:

Benzene is a relatively unique volatile organic in that it has three published OSHA Permissible Exposure Limits — its PEL-TWA = 1.0 ppm(vol), its PEL-STEL = 5.0 ppm(vol), and its PEL-C = 25.0 ppm(vol). Express these concentrations in mg/m^3, assuming STP conditions. Benzene's molecular weight is 78.11.

Applicable Definitions:	Permissible Exposure Limits	Page 3-1
	Time Weighted Averages	Page 3-2
	Short Term Exposure Limits	Page 3-2
	Ceiling Values	Page 3-2
Applicable Formula:	Equation **#3-8**	Pages 3-12 & 3-13
Solution to this Problem:	Page 3-72	

Workspace for Problem **#3.23**

Problem #3.24:

The earth background concentration level of carbon dioxide is approximately 350 ppm(vol). This concentration level holds and is valid at any altitude of 20,000 feet or less. What would be the concentration of carbon dioxide in the mile-high city of Denver, Colorado, on a day when the air temperature was 32°C? The normal barometric pressure in Denver is 626 mm Hg, and the molecular weight of carbon dioxide is 44.01 amu.

Applicable Formulae:	Equation **#1-1**	Page 1-16
	Equation **#3-9**	Pages 3-12 & 3-13
Solution to this Problem:	Page 3-72 & 3-73	

Problem Workspace

Problem Workspace Continued on the Next Page

Continuation of Workspace for Problem **#3.24**

Problem #3.25:

The Safety Manager for a Building Materials Supply Company has determined — for the respirable quartz fraction in the river gravel that is currently being crushed in his plant — that the 8-hour Time Weighted Average $TLV_{quartz} = 0.22\ mg/m^3$. What is the approximate percentage of respirable quartz in the crushed gravel?

Applicable Definitions:	Threshold Limit Values	Page 3-1
	Time Weighted Averages	Page 3-2
Applicable Formula:	Equation **#3-10**	Page 3-14
Solution to this Problem:	Page 3-73	

Problem Workspace

Problem #3.26:

What is the 8-hour Time Weighted Average TLV_{mix} for the mixed quartz species in a sample that contains 80% total silica, if the three categories of silica that are present in the sample are in the following ratio:

[quartz] : [cristobalite] : [tridymite] = 18 : 13 : 9?

Applicable Definitions:	Threshold Limit Values	Page 3-1
	Time Weighted Averages	Page 3-2
Applicable Formula:	Equation **#3-12**	Pages 3-14 & 3-15
Solution to this Problem:	Pages 3-73 & 3-74	

Problem Workspace

SOLUTIONS TO THE WORKPLACE AMBIENT AIR PROBLEM SET

Problem #3.1:

To solve this problem, we must apply Equation **#3-1**, from Page 3-8:

$$TWA = \frac{\sum_{i=1}^{n} T_i C_i}{\sum_{i=1}^{n} T_i} = \frac{T_1C_1 + T_2C_2 + \ldots + T_nC_n}{T_1 + T_2 + \ldots + T_n} \qquad \text{[Eqn. \#3-1]}$$

$$TWA = \frac{(2.0)(451) + (3.0)(728) + (1.5)(619) + (1.5)(501)}{2.0 + 3.0 + 1.5 + 1.5}$$

$$TWA = \frac{902.0 + 2,184.0 + 928.5 + 751.5}{8.0} = \frac{4,766.0}{8.0} = 595.75$$

∴ The Time Weighted Average ~ 596 ppm(vol).

Problem #3.2:

To solve this problem, we must again apply Equation **#3-1**, from Page 3-8. To generate a solution, we must assume that this individual's resultant full 8-hour work shift Time Weighted Average exposure to n-hexane must have been equal to or greater than 600 ppm(vol). From this starting point, we must recognize that the concentration level to which he or she was exposed over the final 15 minutes (= 0.25 hours) of his or her work shift must have been the source of the violation of the ACGIH's TLV-TWA. We must consider the final 1.5-hour period as having been made up of two separate and distinct time periods, the first one being 1.25 hours (= 75 minutes) — during which this worker was exposed to the average 501 ppm(vol) level of n-pentane, and the second being 0.25 hours (= 15 minutes) — during which he or she was exposed to an unknown average n-pentane concentration, to be designated as "C". It is this concentration level of n-pentane that will have to be determined; thus:

$$TWA = \frac{\sum_{i=1}^{n} T_i C_i}{\sum_{i=1}^{n} T_i} = \frac{T_1C_1 + T_2C_2 + \ldots + T_nC_n}{T_1 + T_2 + \ldots + T_n} \qquad \text{[Eqn. \#3-1]}$$

$$600 = \frac{(2.0)(451) + (3.0)(728) + (1.5)(619) + (1.25)(501) + 0.25C}{2.0 + 3.0 + 1.5 + 1.5}$$

$$600 = \frac{902.0 + 2,184.0 + 928.5 + 626.25 + 0.25C}{8.0}$$

$$600 = \frac{4,640.75 + 0.25C}{8.0} = \frac{4,640.75}{8.0} + \frac{0.25C}{8.0} = 580.09 + \frac{C}{32}$$

$$600 = 580.09 + \frac{C}{32}$$

Now since we are seeking an exposure level that would produce a violation, we can assume that we are looking at a "greater than or equal to" type situation.

$$\frac{C}{32} \geq 600 - 580.09 \geq 19.91$$

$$C \geq (19.91)(32) \geq 637.0$$

∴ The final 15-minute concentration level of n-pentane ≥ 637 ppm(vol).

Problem #3.3:

To solve this problem, we must consider the Definitions of the TLV, the TWA, and the STEL, from Pages 3-1 & 3-2.

Since the TLV-STEL [a 15-minute Time Weighted Average] for n-pentane is 750 ppm(vol), and since we are dealing here only with the final 15-minute segment of this worker's shift, which up until these final 15 minutes has had no TLV-STEL violation, it can be concluded that these final 15 minutes must have been the source of the TLV-STEL violation; and, therefore, it must have been characterized by a n-pentane exposure level of 750 ppm(vol) — or possibly even more than this level.

∴ The concentration during the final 15-minutes of this work shift was 750+ ppm(vol).

∴ Obviously, this n-pentane concentration level would also have produced a TLV-TWA violation — since, from Problem **#3.2**, we can see that a final 15-minute exposure to only 637 ppm(vol) produced a TLV-TWA violation. Therefore, this employee's exposure to an even higher level [namely, 750+ ppm(vol)] would certainly also have produced such a violation.

Problem #3.4:

To solve this problem, we must again apply Equation **#3-1**, from Page 3-8.

For this problem, we shall consider concentrations expressed in parts per billion by volume [ppb(vol)], since that is the format in which the concentrations have been provided. Because of this, we must convert the PELs from the part per million [ppm(vol)] units in which they have been provided in the problem statement, to their corresponding part per billion [ppb(vol)] units. Remembering that 1,000 ppb(vol) = 1 ppm(vol), we can convert the established ozone Permissible Exposure Limits, as follows:

$$\text{PEL-TWA}_{\text{ozone}} = 100 \text{ ppb(vol)}, \&$$

$$\text{PEL-STEL}_{\text{ozone}} = 300 \text{ ppb(vol)}.$$

We proceed; thus:

$$TWA = \frac{\sum_{i=1}^{n} T_i C_i}{\sum_{i=1}^{n} T_i} = \frac{T_1C_1 + T_2C_2 + \ldots + T_nC_n}{T_1 + T_2 + \ldots + T_n} \qquad \text{[Eqn. \#3-1]}$$

$$TWA = \frac{(0.5)(289) + (1.0)(42) + (0.5)(93) + (6.0)(8)}{0.5 + 1.0 + 0.5 + 6.0}$$

$$TWA = \frac{144.5 + 42.0 + 46.5 + 48.0}{8.0} = \frac{281.0}{8.0} = 35.13$$

∴ The TWA = 35 ppb(vol).

∴ This Technician probably did not experience a PEL-STEL violation; however, his 30-minute diagnostic effort, for which the average exposure was 289 ppb(vol), could — at least in principle — have involved one or more 15-minute periods [as subsets of the longer 30-minute period] for which the 15-minute Time Weighted Average exposures could have exceeded the 300 ppb(vol) PEL-STEL.

Problem #3.5:

To solve this additional twist on Problem **#3.4**, we must again apply Equation **#3-1**, from Page 3-8. In this case, we must reexamine the final 6 hours of this Technician's day, considering it to be made up of two time periods, as follows:

Time Period #1:	2.5 hours @	222 ppb(vol), &
Time Period #2:	3.5 hours @	8 ppb(vol).

Again applying Equation **#3-1**, thus:

$$\text{TWA} = \frac{\sum_{i=1}^{n} T_i C_i}{\sum_{i=1}^{n} T_i} = \frac{T_1C_1 + T_2C_2 + \ldots + T_nC_n}{T_1 + T_2 + \ldots + T_n} \qquad \text{[Eqn. \#3-1]}$$

$$\text{TWA} = \frac{(0.5)(289) + (1.0)(42) + (0.5)(93) + (3.5)(8) + (2.5)(222)}{0.5 + 1.0 + 0.5 + 3.5 + 2.5}$$

$$\text{TWA} = \frac{144.5 + 42.0 + 46.5 + 28.0 + 555.0}{8.0} = \frac{816.0}{8.0} = 102.0$$

∴ The TWA = 102 ppb(vol) > 0.1 ppm(vol).

∴ Clearly, this Technician had experienced an 8-hour TWA ozone exposure that exceeded the established OSHA PEL-TWA of 0.1 ppm(vol) — 102 ppb(vol) > 0.1 ppm(vol).

Problem #3.6:

To solve this Problem, we must again apply Equation **#3-1**, from Page 3-8. The analysis of this problem is comparable to those for the previous two; however, since one of the timed average ethylene oxide exposures listed in this problem statement has been provided in a "less than" format, we must think carefully about the implications of this. Specifically, what this means is that the ethylene oxide concentration, during that specific 2.9-hour time period, would be more accurately and usefully presented in the following format:

$$0.0 \text{ ppm(vol)} \leq \text{ethylene oxide concentration} < 0.4 \text{ ppm(vol)}$$

As a consequence of this, the final result will, of necessity, also have to be in a slightly different format than was the case for any of the previous problems. To develop the solution, we will have to consider two different scenarios: (1) a maximum possible TWA value, and (2) a minimum possible TWA value.

The maximum TWA value can be calculated by assuming that this employee's 2.9-hour time segment actually involved an exposure at a concentration of just barely less than 0.4 ppm(vol) — or for mathematical purposes, we shall assume that this exposure was equal to 0.4 ppm(vol). Clearly, a calculation based on this assumption will result in a more adverse exposure determination for this worker than what would have been the actual "worst possible" case. The number that results from this calculation will identify the employee's TWA exposure at a level that will be higher than what was the true value for the situation.

The minimum TWA value, on the other hand, will be developed by assuming that this same 2.9-hour exposure time segment was at an ethylene oxide concentration level of 0.0 ppm(vol). This concentration level clearly is less than 0.4 ppm(vol); it is also equal to, or less than, the actual "indeterminable" concentration level ["indeterminable", because the Minimum Detection Limit [MDL] for the analyzer that was used was 0.4 ppm(vol)] for this segment; thus:

$$\text{TWA} = \frac{\sum_{i=1}^{n} T_i C_i}{\sum_{i=1}^{n} T_i} = \frac{T_1C_1 + T_2C_2 + \ldots + T_nC_n}{T_1 + T_2 + \ldots + T_n} \qquad \text{[Eqn. \#3-1]}$$

Consider first the "worst possible case" situation, i.e., where the calculated TWA will be at its maximum value, namely, when the ethylene oxide exposure was at 0.4 ppm(vol):

$$\text{TWA}_{\text{maximum}} = \frac{(2.2)(0.7) + (1.4)(1.2) + (1.5)(1.6) + (2.9)(0.4)}{2.2 + 1.4 + 1.5 + 2.9}$$

$$\text{TWA}_{\text{maximum}} = \frac{1.54 + 1.68 + 2.40 + 1.16}{8.0} = \frac{6.78}{8.0} = 0.85$$

Next, consider the alternative scenario, i.e., where the TWA was 0.0 ppm(vol) for the period:

$$\text{TWA}_{\text{minimum}} = \frac{(2.2)(0.7) + (1.4)(1.2) + (1.5)(1.6) + (2.9)(0.0)}{2.2 + 1.4 + 1.5 + 2.9}$$

$$\text{TWA}_{\text{minimum}} = \frac{1.54 + 1.68 + 2.40 + 0.0}{8.0} = \frac{5.62}{8.0} = 0.70$$

∴ The requested result is 0.7 ppm(vol) $\leq$ TWA $<$ 0.9 ppm(vol).

∴ Clearly, the exposure was less than the 8-hour PEL-TWA of 1.0 ppm(vol); however, it must be noted that it was greater than the 0.5 ppm(vol) Action Level. As such, it indicates the existence of a potentially bad situation that a prudent professional should examine — from a preventative perspective — in order to prevent any future PEL violations. It would have to be reported as an exposure that exceeded the Action Level for the worker involved.

Problem #3.7:

To solve this problem, we must again apply Equation **#3-1**, from Page 3-8. There are aspects to this problem that are identical to those in Problem **#3.6** — specifically, in the area of having certain timed average exposures to styrene, reported in a "less than" format. The approach to this problem, therefore, will be directly analogous to the one employed for Problem **#3.6**. In this case, however, since we have been given the overall 8-hour TWA, we shall have to evaluate the situation by considering two different time intervals — namely, Time Interval #1, which was the 7.33-hour segment [or 7 hours, 20 minutes] in the Production Area; and Time Interval #2, which was the 0.67-hour [or 40-minute] segment in the "Low Level Room". In addition, we will have to perform the analysis from two different perspectives, which are as follows:

1. In order to have developed the given overall 48 ppm(vol) TWA styrene exposure, it is clear that the highest concentration level of styrene vapors that could have existed in the Production Area must have corresponded to the lowest styrene concentration level that could have existed in the "Low Level Room". These two first perspective concentrations will be designated as $C_{maximum}$ in the Production Area, and 0.0 ppm(vol) in the "Low Level Room".
2. Conversely, in order to have produced the given overall 48 ppm(vol) TWA styrene exposure, it is also clear the lowest concentration level of styrene vapors that could have existed in the Production Area must correspond to the highest styrene concentration level that could have existed in the "Low Level Room". These two second perspective concentrations will be designated as $C_{minimum}$ in the Production Area, and 0.1 ppm(vol) in the "Low Level Room".

Thus we view this situation, as follows:

Time Interval #	Duration	Location	"1st Perspective" Concentrations	"2nd Perspective" Concentrations
1	7.33 hours	Production Area	$C_{maximum}$	$C_{minimum}$
2	0.67 hours	"Low Level Room"	0.0 ppm(vol)	0.1 ppm(vol)
Total	8.00 hours			

With this approach, we can proceed, as follows:

$$\text{TWA} = \frac{\sum_{i=1}^{n} T_i C_i}{\sum_{i=1}^{n} T_i} = \frac{T_1 C_1 + T_2 C_2 + \ldots + T_n C_n}{T_1 + T_2 + \ldots + T_n} \qquad \text{[Eqn. \textbf{\#3-1}]}$$

Solving for the "First Perspective" $C_{maximum}$ concentration, we see:

$$48 = \frac{(0.67)(0.0) + (7.33)(C_{maximum})}{0.67 + 7.33} = \frac{0.00 + (7.33)(C_{maximum})}{8.00}$$

$$48 = \frac{0.00}{8.00} + \frac{(7.33)(C_{maximum})}{8.00} = 0.00 + \frac{(7.33)(C_{maximum})}{8.00} = (0.917)(C_{maximum}) \quad \&$$

$$C_{maximum} = \frac{48}{0.917} = 52.364$$

Solving next for the "Second Perspective" $C_{minimum}$ concentration, we see:

$$48 = \frac{(0.67)(0.1) + (7.33)(C_{minimum})}{0.67 + 7.33} = \frac{0.067 + (7.33)(C_{minimum})}{8.00}$$

$$48 = \frac{0.067}{8.00} + \frac{(7.33)(C_{minimum})}{8.00} = 8.38 \times 10^{-3} + (0.916)(C_{minimum})$$

$$(0.916)(C_{minimum}) = 48 - 8.38 \times 10^{-3} = 47.992$$

$$C_{minimum} = \frac{47.992}{0.917} = 52.355$$

As you can see, these two calculations have been carried out to a precision of 0.001 ppm(vol). At this level, there is a mathematical difference between the two styrene exposure levels of only 0.009 ppm(vol), or 9 ppb(vol) — i.e., the difference between the $C_{maximum}$ which equals 52.3**64** ppm(vol), and the $C_{minimum}$ which equals 52.3**55** ppm(vol).

Such a difference is both unjustified by the precision of the data that was provided in this problem, and, in addition, completely inconsequential; thus, it should and will be ignored.

> ∴ Production Area styrene ambient concentration ~ 52 ppm(vol).

> ∴ Again, in this case, the competent professional would probably focus some of his or her attention on this situation, since this scenario is quite close to being a PEL violation, with dangerously high styrene concentration levels in the Production Area.

Problem #3.8:

To solve this problem, we must again apply Equation **#3-1**, from Page 3-8. Clearly, the average worker's 8-hour TWA exposure to CO_2 is 8,400 ppm(vol) [84% of the PEL-TWA, which is currently 10,000 ppm(vol)]. To determine the Cold Room concentration of CO_2, which we shall designate as $C_{\text{Cold Room}}$, we must proceed thus:

$$TWA = \frac{\sum_{i=1}^{n} T_i C_i}{\sum_{i=1}^{n} T_i} = \frac{T_1C_1 + T_2C_2 + \ldots + T_nC_n}{T_1 + T_2 + \ldots + T_n} \qquad \text{[Eqn. \textbf{\#3-1}]}$$

$$8,400 = \frac{(6.0)(C_{\text{Cold Room}}) + (2.0)(425)}{6.0 + 2.0} = \frac{(6.0)(C_{\text{Cold Room}}) + 850}{8.0}$$

$$8,400 = \frac{(6.0)(C_{\text{Cold Room}})}{8.0} + \frac{850}{8.0} = (0.75)(C_{\text{Cold Room}}) + 106.25$$

$$(0.75)(C_{\text{Cold Room}}) = 8,400 - 106.25 = 8,293.75$$

$$C_{\text{Cold Room}} = \frac{8,293.75}{0.75} = 11,058.3$$

> ∴ Cold Room average CO_2 concentration ~ 11,058 ppm(vol).

Problem #3.9:

Again, to solve this problem, we must apply Equation **#3-1**, from Page 3-8. This problem is both straightforward and analogous to all of the preceding problems, except that in this case, we have been asked for, and will focus on, two unknown time intervals, the first of which shall be referred to as t_{maximum}, which will represent the maximum number of hours in each workday that the Cold Storage Plant workers will be permitted to work in the Cold Rooms [which will have a minimum background CO_2 level of 15,000 ppm(vol)]. Under the new CO_2 exposure requirements — as they have been defined by the Safety Manager — the second time interval, which makes up the balance of the 8-hour workday, will be identified as $(8 - t_{\text{minimum}})$:

$$TWA = \frac{\sum_{i=1}^{n} T_i C_i}{\sum_{i=1}^{n} T_i} = \frac{T_1C_1 + T_2C_2 + \ldots + T_nC_n}{T_1 + T_2 + \ldots + T_n} \qquad \text{[Eqn. \textbf{\#3-1}]}$$

$$5,000 = \frac{(15,000)(t_{maximum}) + (425)(8 - t_{maximum})}{t_{maximum} + (8 - t_{maximum})}$$

$$5,000 = \frac{(15,000 - 425)t_{maximum} + (425)(8)}{8 + t_{maximum} - t_{maximum}}$$

$$5,000 = \frac{(14,575)(t_{maximum}) + 3,400}{8} = \frac{(14,575)(t_{maximum})}{8} + \frac{3,400}{8}$$

$$5,000 = (1,821.9)(t_{maximum}) + 425$$

$$(1,821.9)(t_{maximum}) = 5,000 - 425 = 4,575$$

$$t_{maximum} = \frac{4,575}{1,821.9} = 2.51$$

∴ The maximum time period out of each workday that any Worker can be permitted to be in the Cold Room would be 2.51 hours, or 2 hours, 30 minutes, & 36 seconds, or approximately 2 hours, 30 minutes = 2.5 hours.

Problem #3.10:

To solve this problem, we must apply Equation **#3-2**, from Pages 3-8 & 3-9. This problem is very straightforward and can be solved directly:

$$\% \text{TLV} = 100\left[\sum_{i=1}^{n} \frac{\text{TWA}_i}{\text{TLV}_i}\right] = 100\left[\frac{\text{TWA}_1}{\text{TLV}_1} + \frac{\text{TWA}_2}{\text{TLV}_2} + \ldots + \frac{\text{TWA}_n}{\text{TLV}_n}\right] \quad \text{[Eqn. \#3-2]}$$

$$\% \text{TLV} = (100)\left(\frac{12}{50} + \frac{17}{100} + \frac{55}{200} + \frac{91}{400}\right)$$

$$\% \text{TLV} = (100)(0.240 + 0.170 + 0.275 + 0.228)$$

$$\% \text{TLV} = (100)(0.9125) = 91.25$$

These Workers experience just over 91% of the established "composite" TLV for these four volatile materials.

∴ % TLV ~ 91.3%

Problem #3.11:

To solve this problem, we must again apply Equation **#3-2**, from Pages 3-8 & 3-9. For this problem, we will perform two different sets of calculations, only one of which will have any validity. In order, the first of these two sets of calculations will be for a %PEL-TWA, which will be the completely valid determination, while the second will be for a %PEL-STEL, which will be invalid and only useful as an approximation of potential problems — the stated dosages are 8-hour TWAs, whereas the %PEL-STEL Standard is a 15-minute TWA; thus, the two are completely mutually inconsistent. We must also again deal with differences in the listed concentration units. The average exposures are given in parts per million by volume, while the OSHA Standards are all listed in parts per billion by volume. It will be easiest to convert everything into parts per billion by volume; therefore, we have the Employee's average 8-hour TWA chlorine exposure = 420 ppb(vol), and his 8-hour

TWA chlorine dioxide counterpart = 80 ppb(vol). Let us start with the %PEL-TWA determination, using Equation **#3-2**, from Pages 3-8 & 3-9; thus:

$$\% \text{TLV} = 100\left[\sum_{i=1}^{n} \frac{\text{TWA}_i}{\text{TLV}_i}\right] = 100\left[\frac{\text{TWA}_1}{\text{TLV}_1} + \frac{\text{TWA}_2}{\text{TLV}_2} + \ldots + \frac{\text{TWA}_n}{\text{TLV}_n}\right] \quad \text{[Eqn. \#3-2]}$$

$$\% \text{PEL-TWA} = (100)\left(\frac{420}{500} + \frac{80}{100}\right) = (100)(0.84 + 0.80)$$

$$\% \text{PEL-TWA} = (100)(1.64) = 164.00$$

This is clearly a bad situation; it represents the case where the employee has experienced approximately a 164% combined exposure to these two vapors.

Now let us develop the "invalid" %PEL-STEL, that we may possibly be able to use as a measure of how good or bad things might be from the perspective of the Short Term Exposure Limit, again using Equation **#3-2** from Pages 3-8 & 3-9; thus:

$$\% \text{TLV} = 100\left[\sum_{i=1}^{n} \frac{\text{TWA}_i}{\text{TLV}_i}\right] = 100\left[\frac{\text{TWA}_1}{\text{TLV}_1} + \frac{\text{TWA}_2}{\text{TLV}_2} + \ldots + \frac{\text{TWA}_n}{\text{TLV}_n}\right] \quad \text{[Eqn. \#3-2]}$$

$$\% \text{PEL-STEL} = (100)\left(\frac{420}{1,000} + \frac{80}{300}\right) = (100)(0.420 + 0.267)$$

$$\% \text{PEL-STEL} = (100)(0.687) = 68.7$$

To put both of the foregoing results into a proper perspective, the following can be said — clearly, there is a problem with this kraft pulp bleaching mill's ambient environment, particularly on heavy bleaching days. The employee identified in this problem received an exceptionally dangerous exposure to these two very irritating chemicals. The result of calculating the fictitious %PEL-STEL of ~ 69% appears to be all right; however, the different time bases — as stated earlier — make this value only marginally useful, and then just as a general indication.

∴ This Worker's %PEL (based on the 8-hour PEL-TWA Standard) = 164%.

∴ Clearly this worker's employer, the company operating the kraft pulp mill, is in violation — severely so, as a matter of fact — of the established PEL-TWA Standard, as it applies to an environment that contains, simultaneously, both of these two very hazardous, toxic, and irritating chemicals. It is also interesting to note that neither of these two chemicals' individual PEL-TWAs has been exceeded — at least as the exposures have been defined in this problem — when each is considered by itself.

Problem #3.12:

To solve this problem, we must apply a total of three of the listed relationships, and do so in the following sequence: Equation **#3-5**, from Pages 3-10 & 3-11; then Equation **#3-3**, from Page 3-9 & 3-10; and then finally Equation **#3-8**, from Pages 3-12 & 3-13.

Again, note that the relationship listed in Equation **#3-3** is an approximation, one that relies upon the assumption that the overall composition of the vapor space above a liquid mixture of volatile solvents has the same composition as does the solution itself. Such a situation would virtually never exist in the real world; however, it is the basis for the approximation developed in this relationship. The first thing that must be done in obtaining a solution to this problem is to convert the four volume-based TLV-TWAs that have been provided in the problem statement into their mass-based equivalents, using Equation **#3-5**, from Pages 3-10 & 3-11:

$$C_{mass_i} = \frac{MW_i}{22.41}\left[C_{vol_i}\right] \quad @ \text{ STP Conditions} \qquad \text{[Eqn. \#3-5]}$$

1. For Freon 11:

$$C_{mass_{Freon\ 11}} = \frac{137.38}{22.41}(1,000) = \frac{137,380}{22.41} = 6,174.9$$

$$\text{TLV-TWA}_{Freon\ 11} = 6,175 \text{ mg/m}^3$$

2. For Freon 113:

$$C_{mass_{Freon\ 113}} = \frac{187.38}{22.41}(1,000) = \frac{187,380}{22.41} = 8,361.4$$

$$\text{TLV-TWA}_{Freon\ 113} = 8,361 \text{ mg/m}^3$$

3. For Methyl Chloroform:

$$C_{mass_{Methy\ Chloroform}} = \frac{133.41}{22.41}(350) = \frac{46,693.50}{22.41} = 2,083.6$$

$$\text{TLV-TWA}_{Methyl\ Chloroform} = 2,084 \text{ mg/m}^3$$

4. For Methylene Chloride:

$$C_{mass_{Methylene\ Chloride}} = \frac{84.93}{22.41}(50) = \frac{4,246.50}{22.41} = 189.5$$

$$\text{TLV-TWA}_{Methylene\ Chloride} = 190 \text{ mg/m}^3$$

Next, we must apply Equation **#3-3**, from Pages 3-9 & 3-10, since we have now converted all the TLV concentrations from their volumetric bases [i.e., ppm(vol)] to their corresponding mass bases (i.e., mg/m^3), as required for incorporation into this Equation. We can now determine the overall ambient air mass concentrations of this vaporous mix of four degreasing solvents (in milligrams of solvent vapor per cubic meter of air) as they would hypothetically exist in equilibrium with the air. Remember that this result is an approximation that relies upon the unlikely situation that the vapor and the liquid compositions are identical; thus:

$$TLV_{effective} = \left[\frac{1}{\sum_{i=1}^{n}\frac{f_i}{TLV_i}}\right] = \left[\frac{1}{\frac{f_1}{TLV_1} + \frac{f_2}{TLV_2} + \cdots + \frac{f_n}{TLV_n}}\right] \qquad \text{[Eqn. \#3-3]}$$

$$TLV_{effective} = \left(\frac{1}{\frac{0.25}{6,175} + \frac{0.55}{8,361} + \frac{0.15}{2,084} + \frac{0.05}{190}}\right)$$

$$TLV_{effective} = \left(\frac{1}{4.05\times10^{-5} + 6.58\times10^{-5} + 7.20\times10^{-5} + 2.64\times10^{-4}}\right)$$

$$TLV_{effective} = \frac{1}{4.42\times10^{-4}} = 2,261.9$$

∴ The effective overall TLV for this mixture of solvent vapors = 2,262 mg/m^3.

Now that we have determined — for this overall 4-solvent mixture — an overall mass-based conce tration value [expressed in mg/m^3], we can now determine the concentration of each of the individu solvent components [also expressed in mg/m^3]. We do this simply by applying the solution weig ratios for each of the four components of this mixture, all of which were given in the initial proble statement, to this just determined total vapor concentration.

1. For Freon 11:

$$C_{mass_{Freon\ 11}} = (2,261.9)(0.25) = 565.47$$

$$C_{Freon\ 11} \sim 565 \text{ mg/m}^3$$

2. For Freon 113:

$$C_{mass_{Freon\ 113}} = (2,261.9)(0.55) = 1,244.04$$

$$C_{Freon\ 113} \sim 1,244 \text{ mg/m}^3$$

3. For Methyl Chloroform:

$$C_{mass_{Methyl\ Chloroform}} = (2,261.9)(0.15) = 339.28$$

$$C_{Methyl\ Chloroform} \sim 339 \text{ mg/m}^3$$

4. For Methylene Chloride:

$$C_{mass_{Methylene\ Chloride}} = (2,261.9)(0.05) = 113.10$$

$$C_{Methylene\ Chloride} \sim 113 \text{ mg/m}^3$$

With these four individual mass-based concentration values, we can finally apply the last necessa relationship, namely, Equation **#3-8**, from Pages 3-12 & 3-13, thereby converting these mass-bas values [expressed in mg/m^3] to their equivalent volume-based concentrations [expressed in ppm(vol thus:

$$C_{vol_i} = \frac{22.41}{MW_i}\left[C_{mass_i}\right] \quad @ \text{ STP Conditions} \qquad \text{[Eqn. \#3-8]}$$

1. For Freon 11:

$$C_{vol_{Freon\ 11}} = \frac{22.41}{137.38}(565.5) = \frac{12,672.18}{137.38} = 92.24$$

$$C_{Freon\ 11} \sim 92 \text{ ppm(vol)}$$

2. For Freon 113:

$$C_{vol_{Freon\ 113}} = \frac{22.41}{187.38}(1,244.04) = \frac{27,878.94}{187.38} = 148.78$$

$$C_{Freon\ 113} \sim 149 \text{ ppm(vol)}$$

3. For Methyl Chloroform:

$$C_{vol_{Methyl\ Chloroform}} = \frac{22.41}{133.41}(339.28) = \frac{7,603.27}{133.41} = 56.99$$

$$C_{Methyl\ Chloroform} \sim 57 \text{ ppm(vol)}$$

4. For Methylene Chloride:

$$C_{vol\ \text{Methylene Chloride}} = \frac{22.41}{84.93}(113.10) = \frac{2,534.57}{84.93} = 29.84$$

$$C_{\text{Methylene Chloride}} \sim 30 \text{ ppm(vol)}$$

Finally, summarizing these results, which are the data set that was asked for in the problem statement, we have the following tabulation of the ambient solvent concentrations.

∴	Freon 11 Concentration	=	565 mg/m^3
		=	92 ppm(vol)
	Freon 113 Concentration	=	1,244 mg/m^3
		=	149 ppm(vol)
	Methyl Chloroform Concentration	=	339 mg/m^3
		=	57 ppm(vol)
	Methylene Chloride Concentration	=	113 mg/m^3
		=	30 ppm(vol)

Problem #3.13:

To solve this problem we must employ a large number of relationships: we start with Equation **#1-10**, from Pages 1-19 & 1-20 — this relationship permits us to determine the number of moles of each component in a specific volume of the liquid vapor degreasing solvent mixture; we next determine the mole fractions of each of the components in this liquid mixture; we then apply Raoult's Law, Equation **#1-17**, from Page 1-23 — this relationship will give us the partial vapor pressure of each component of this mixture in the vapor space above the liquid phase; next we apply Dalton's Law of Partial Pressures, Equation **#1-16**, from Pages 1-22 & 1-23 — this relationship will give us the volume-based concentrations of each of the four components in the vapor space above this liquid mixture; and finally Equation **#3-8**, from Pages 3-12 & 3-13 — to provide the equivalent mass-based concentrations of each of these four component vapors.

As stated above, to make the initial determination of the number of moles of each component in this mixture, we will assume that we are working with a total of 1,000 grams of this liquid mixture, and then apply Equation **#1-10**, from Pages 1-19 & 1-20:

$$n_i = \frac{m_i}{MW_i} \qquad \text{[Eqn. \#1-10]}$$

1. Freon 11:

$$n_{\text{Freon 11}} = \frac{(1,000)(0.25)}{137.38} = \frac{250}{137.38} = 1.82$$

$$n_{\text{Freon 11}} = 1.82 \text{ moles}$$

2. Freon 113:

$$n_{\text{Freon 113}} = \frac{(1,000)(.55)}{187.38} = \frac{550}{187.38} = 2.94$$

$$n_{\text{Freon 113}} = 2.94 \text{ moles}$$

3. Methyl Chloroform:

$$n_{\text{Methyl Chloroform}} = \frac{(1,000)(.15)}{133.41} = \frac{150}{133.41} = 1.12$$

$$n_{\text{Methyl Chloroform}} = 1.12 \text{ moles}$$

4. Methylene Chloride:

$$n_{\text{Methylene Chloride}} = \frac{(1,000)(.05)}{84.93} = \frac{50}{84.93} = 0.59$$

$$n_{\text{Methylene Chloride}} = 0.59 \text{ moles}$$

Next, we must determine the mole fractions of each component in this mixture of solvents; this parameter is simply the ratio of the number of moles of the component of interest to the total number of moles in the entire mixture, thus:

$$\text{Mole Fraction of the ith component in a mixture} = m_i = \frac{n_i}{\sum_{i=1}^{p} n_i}$$

$$\text{For reference, } \sum_{i=1}^{p} n_i = 1.82 + 2.94 + 1.12 + 0.59 = 6.47 \text{ total moles}$$

The total number of moles of the four solvents in 1,000 grams of this mixture = 6.47 moles

1. Freon 11:

$$m_{\text{Freon 11}} = \frac{1.82}{6.47} = 0.281$$

$$m_{\text{Freon 11}} = 0.281$$

2. Freon 113:

$$m_{\text{Freon 113}} = \frac{2.94}{6.47} = 0.455$$

$$m_{\text{Freon 113}} = 0.455$$

3. Methyl Chloroform:

$$m_{\text{Methyl Chloroform}} = \frac{1.12}{6.47} = 0.173$$

$$m_{\text{Methyl Chloroform}} = 0.173$$

4. Methylene Chloride:

$$m_{\text{Methylene Chloride}} = \frac{0.59}{6.47} = 0.091$$

$$m_{\text{Methylene Chloride}} = 0.091$$

We can now apply Raoult's Law, Equation **#1-17**, from Page 1-23 to determine the partial pressures of each of the components in the vapor space above the liquid mixture, thus:

$$PP_i = m_i VP_i \qquad \text{[Eqn. \textbf{\#1-17}]}$$

1. Freon 11:

$$PP_{\text{Freon 11}} = m_{\text{Freon 11}} VP_{\text{Freon 11}} = (0.281)(302.3) = 84.95$$

$$PP_{\text{Freon 11}} = 84.95 \text{ mm Hg}$$

2. Freon 113:

$$PP_{Freon\ 113} = m_{Freon\ 113}VP_{Freon\ 113} = (0.455)(111.8) = 50.87$$

$$PP_{Freon\ 113} = 50.87 \text{ mm Hg}$$

3. Methyl Chloroform:

$$PP_{Methyl\ Chloroform} = m_{Methyl\ Chloroform}VP_{Methyl\ Chloroform} = (0.173)(36.0) = 6.23$$

$$PP_{Methyl\ Chloroform} = 6.23 \text{ mm Hg}$$

4. Methylene Chloride:

$$PP_{Methylene\ Chloride} = m_{Methylene\ Chloride}VP_{Methylene\ Chloride} = (0.091)(144.0) = 13.10$$

$$PP_{Methylene\ Chloride} = 13.10 \text{ mm Hg}$$

Next we apply Dalton's Law of Partial Pressures, Equation **#1-16**, from Pages 1-22 & 1-23 to convert these partial pressures into their volume-based concentration counterparts. Remember that we are making these computations at STP where the total ambient barometric pressure is 1.0 atmosphere, or 760 mm Hg.

$$C_i = \frac{[1,000,000][PP_i]}{P_{total}} \qquad \text{[Eqn. \#1-16]}$$

1. Freon 11:

$$C_{Freon\ 11} = \frac{(1,000,000)(PP_{Freon\ 11})}{P_{total}}$$

$$C_{Freon\ 11} = \frac{(1,000,000)(84.95)}{760} = 111,771$$

$$C_{Freon\ 11} = 111,771 \text{ ppm(vol)}$$

2. Freon 113:

$$C_{Freon\ 113} = \frac{(1,000,000)(PP_{Freon\ 113})}{P_{total}}$$

$$C_{Freon\ 113} = \frac{(1,000,000)(50.87)}{760} = 66,934$$

$$C_{Freon\ 113} = 66,934 \text{ ppm(vol)}$$

3. Methyl Chloroform:

$$C_{Methyl\ Chloroform} = \frac{(1,000,000)(PP_{Methyl\ Chloroform})}{P_{total}}$$

$$C_{Methyl\ Chloroform} = \frac{(1,000,000)(6.23)}{760} = 8,534$$

$$C_{Methyl\ Chloroform} = 8,534 \text{ ppm(vol)}$$

4. Methylene Chloride:

$$C_{Methylene\ Chloride} = \frac{(1,000,000)(PP_{Methylene\ Chloride})}{P_{total}}$$

$$C_{Methylene\ Chloride} = \frac{(1,000,000)(13.1)}{760} = 17,237$$

$$C_{\text{Methylene Chloride}} = 17{,}237 \text{ ppm(vol)}$$

Finally, we can convert these volume-based concentrations for each component in the vapor space to the mass-based counterparts for each by using Equation **#3-8**, from Pages 3-12 & 3-13:

$$C_{\text{mass}_i} = \frac{MW_i}{22.41}\left[C_{\text{vol}_i}\right] \quad @ \text{ STP Conditions} \qquad \text{[Eqn. \textbf{\#3-8}]}$$

1. Freon 11:

$$C_{\text{mass}_{\text{Freon 11}}} = \frac{MW_{\text{Freon 11}}}{22.41}\left[C_{\text{vol}_{\text{Freon 11}}}\right]$$

$$C_{\text{mass}_{\text{Freon 11}}} = \frac{(137.38)(111{,}771)}{22.41} = 685{,}190$$

$$C_{\text{Freon 11}} = 685{,}190 \text{ mg/m}^3 = 685.2 \text{ gms/m}^3$$

2. Freon 113:

$$C_{\text{mass}_{\text{Freon 113}}} = \frac{MW_{\text{Freon 113}}}{22.41}\left[C_{\text{vol}_{\text{Freon 113}}}\right]$$

$$C_{\text{mass}_{\text{Freon 113}}} = \frac{(187.38)(66{,}934)}{22.41} = 559{,}665$$

$$C_{\text{Freon 113}} = 559{,}665 \text{ mg/m}^3 = 559.7 \text{ gms/m}^3$$

3. Methyl Chloroform:

$$C_{\text{mass}_{\text{Methyl Chloroform}}} = \frac{MW_{\text{Methyl Chloroform}}}{22.41}\left[C_{\text{vol}_{\text{Methyl Chloroform}}}\right]$$

$$C_{\text{mass}_{\text{Methyl Chloroform}}} = \frac{(133.41)(8{,}534)}{22.41} = 50{,}804$$

$$C_{\text{Methyl Chloroform}} = 50{,}804 \text{ mg/m}^3 = 50.8 \text{ gms/m}^3$$

4. Methylene Chloride:

$$C_{\text{mass}_{\text{Methylene Chloride}}} = \frac{MW_{\text{Methylene Chloride}}}{22.41}\left[C_{\text{vol}_{\text{Methylene Chloride}}}\right]$$

$$C_{\text{mass}_{\text{Methylene Chloride}}} = \frac{(84.93)(17{,}237)}{22.41} = 65{,}325$$

$$C_{\text{Methylene Chloride}} = 65{,}325 \text{ mg/m}^3 = 65.3 \text{ gms/m}^3$$

It should now be very clear that the "Order-of-Magnitude-Approximations" that were obtained in developing answers for Problem **#3.12** probably should not even be characterized as "Order-of-Magnitude-Approximations", since they are incorrect in the extreme when compared to the more precise numbers obtained by relying on accepted theory and the application of the basic laws of physical chemistry. It is also easy to understand why Equation **#3-3** enjoys such great relative popularity for making this determination. It is clear that using the alternate, more accurate approach, and applying accepted theoretical considerations and relationships simply requires 2 to 3 times as much time and calculating effort. Suffice it to say — from the perspective of having an accurate characterization of the situation — it would probably be slightly less dangerous to place one's head in a vapor space of this type of vapor degreaser — i.e., a space that could be characterized by the previously calculated vapor concentrations — than it would be to kayak over Niagara Falls; however, the survival probabilities are very poor in both cases!

∴	Freon 11 Concentration	=	685.2 gms/m^3
		=	111,771 ppm(vol)
	Freon 113 Concentration	=	559.7 gms/m^3
		=	66,934 ppm(vol)
	Methyl Chloroform Concentration	=	50.8 gms/m^3
		=	8,534 ppm(vol)
	Methylene Chloride Concentration	=	65.3 gms/m^3
		=	17,237 ppm(vol)

Problem #3.14:

This problem is very similar to Problem **#3.12**, except that this one involves only two, rather than four, volatile chemicals. Like its predecessor, it will require the use basically of the same three listed relationships [this time, however, using the relationships that apply to NTP, rather than STP conditions]. We will again follow the same sequence, namely: Equation **#3-4**, from Pages 3-10 & 3-11; then Equation **#3-3**, from Pages 3-9 & 3-10; and then finally Equation **#3-7**, from Page 3-12. Again, we must first make the conversion of the two volume-based TLV-TWAs [expressed in ppm(vol)], to their mass-based equivalents [expressed in mg/m^3], using Equation **#3-4**, from Pages 3-10 & 3-11:

$$C_{mass_i} = \frac{MW_i}{24.45}\left[C_{vol_i}\right] \quad @ \text{ NTP Conditions} \qquad [\text{Eqn. } \mathbf{\#3\text{-}4}]$$

1. For cellosolve:

$$C_{mass_{cellosolve}} = \frac{90.12}{24.45}(200) = \frac{18,024}{24.45} = 737.18$$

$$\text{TLV-TWA}_{cellosolve} \sim 737 \text{ mg/m}^3$$

2. For t-butyl alcohol:

$$C_{mass_{t\text{-butyl alcohol}}} = \frac{74.12}{24.45}(100) = \frac{7,412}{24.45} = 303.15$$

$$\text{TLV-TWA}_{t\text{-butyl alcohol}} \sim 303 \text{ mg/m}^3$$

Next, we must use these mass-based values to determine the Effective TLV for the mixture of solvent vapors that might exist above the solution, again — as with Problem **#3.12** — using Equation **#3-3**, from Pages 3-9 & 3-10, thus:

$$TLV_{effective} = \left[\frac{1}{\sum_{i=1}^{n}\frac{f_i}{TLV_i}}\right] = \left[\frac{1}{\frac{f_1}{TLV_1} + \frac{f_2}{TLV_2} + \cdots + \frac{f_n}{TLV_n}}\right] \qquad [\text{Eqn. } \mathbf{\#3\text{-}3}]$$

$$TLV_{effective} = \left(\frac{1}{\frac{0.65}{737.18} + \frac{0.35}{303.15}}\right) = \left(\frac{1}{8.82\times10^{-4} + 1.15\times10^{-3}}\right)$$

$$TLV_{effective} = \frac{1}{2.04\times10^{-3}} = 491.09$$

∴ The effective overall TLV for this mixture of solvent vapors ~ 491 mg/m^3.

It will now be quite easy to develop the mass-based concentrations for the two solvents in this solution, simply by multiplying the overall concentration value (the one that was just determined above) by the solution weight fractions that were provided in the problem statement; thus:

1. For cellosolve:

$$C_{mass_{cellosolve}} = (491.09)(0.65) = 319.21$$

$$C_{cellosolve} \sim 319 \text{ mg/m}^3$$

2. For t-butyl alcohol:

$$C_{mass_{t\text{-butyl alcohol}}} = (491.09)(0.35) = 171.88$$

$$C_{t\text{-butyl alcohol}} \sim 172 \text{ mg/m}^3$$

With these two mass-based concentrations, we can now apply Equation **#3-4**, from Pages 3-10 & 3-11, to obtain the final pieces of information requested in the problem statement; thus:

$$C_{vol_i} = \frac{24.45}{MW_i}\left[C_{mass_i}\right] \quad @ \text{ NTP Conditions} \qquad \text{[Eqn. \#3-4]}$$

1. For cellosolve:

$$C_{vol_{cellosolve}} = \frac{24.45}{90.12}(319.21) = \frac{7,804.64}{90.12} = 86.60$$

$$C_{cellosolve} \sim 87 \text{ ppm(vol)}$$

2. For t-butyl alcohol:

$$C_{vol_{t\text{-butyl alcohol}}} = \frac{24.45}{74.12}(171.88) = \frac{4,202.50}{74.12} = 56.70$$

$$C_{t\text{-butyl alcohol}} \sim 57 \text{ ppm(vol)}$$

The final set of data that was requested in the statement of the problem is a summary of the concentrations of these two volatile solvents, as they might exist in the vapor space above the solution.

∴	cellosolve concentrations	=	319 mg/m^3
		=	87 ppm(vol)
	t-butyl alcohol concentrations	=	172 mg/m^3
		=	57 ppm(vol)

Problem #3.15:

To solve this problem we must again employ a large number of relationships starting with: Equation **#1-10**, from Pages 1-19 & 1-20 — this relationship permits us to determine the number of moles of each component in a specific volume of the volatile solvent mixture; we next determine the mole fractions of each of the two components in this liquid mixture; we then apply Raoult's Law, Equation **#1-17**, from Page 1-23 — this relationship will give us the partial vapor pressure of each component of this mixture in the vapor space above the liquid phase; next we apply Dalton's Law of Partial Pressures, Equation **#1-16**, from Pages 1-22 & 1-23 — this relationship will give us the volume-based concentrations of each of the two components in the vapor space above this liquid mixture; and finally, we apply Equation **#3-7**, from Page 3-12 — to provide the equivalent mass-based concentrations of each of these two component vapors.

As stated above, to make the initial determination of the number of moles of each component in this mixture, we will assume that we are working with a total of 1,000 grams of this liquid mixture, and then apply Equation **#1-10**, from Pages 1-19 & 1-20:

$$n_i = \frac{m_i}{MW_i} \qquad \text{[Eqn. \#1-10]}$$

1. cellosolve:

$$n_{cellosolve} = \frac{(1,000)(0.65)}{90.12} = \frac{650}{90.12} = 7.21$$

$$n_{cellosolve} = 7.21 \text{ moles}$$

2. t-butyl alcohol:

$$n_{t\text{-butyl alcohol}} = \frac{(1,000)(.35)}{74.12} = \frac{350}{74.12} = 4.72$$

$$n_{t\text{-butyl alcohol}} = 4.72 \text{ moles}$$

Next, we must determine the mole fractions of each component in this mixture of solvents; this parameter is simply the ratio of the number of moles of the component of interest to the total number of moles in the entire mixture, thus:

$$\text{Mole Fraction of the ith Component in a Mixture} = m_i = \frac{n_i}{\sum_{i=1}^{p} n_i}$$

$$\text{For reference, } \sum_{i=1}^{p} n_i = 7.21 + 4.72 = 11.93 \text{ total moles}$$

The total number of moles of the two solvents in 1,000 grams of this mixture = 11.93 moles. From the relationship listed above, the mole fractions of each of these two solvents are then given by the following:

1. cellosolve:

$$m_{cellosolve} = \frac{7.21}{11.93} = 0.604$$

$$m_{cellosolve} = 0.604$$

2. t-butyl alcohol:

$$m_{t\text{-butyl alcohol}} = \frac{4.72}{11.93} = 0.396$$

$$m_{t\text{-butyl alcohol}} = 0.396$$

We can now apply Raoult's Law, Equation **#1-17**, from Page 1-23, to determine the partial pressures of each of the components in the vapor space above the liquid mixture, thus:

$$PP_i = m_i VP_i \qquad \text{[Eqn. \#1-17]}$$

1. cellosolve:

$$PP_{cellosolve} = m_{cellosolve} VP_{cellosolve} = (0.604)(5.3) = 3.20$$

$$PP_{Freon\ 11} = 3.20 \text{ mm Hg}$$

2. t-butyl alcohol:

$$PP_{t\text{-butyl alcohol}} = m_{t\text{-butyl alcohol}} VP_{t\text{-butyl alcohol}} = (0.396)(17.4) = 6.88$$

$$PP_{t\text{-butyl alcohol}} = 6.88 \text{ mm Hg}$$

Next we apply Dalton's Law of Partial Pressures, Equation **#1-16**, from Pages 1-22 & 1-23, to convert these partial pressures into their volume-based concentration counterparts. Remember that we are making these computations at NTP where the total ambient barometric pressure is 1.0 atmosphere, or 760 mm Hg.

$$C_i = \frac{[1{,}000{,}000][PP_i]}{P_{total}} \qquad \text{[Eqn. \#1-16]}$$

1. cellosolve:

$$C_{cellosolve} = \frac{(1{,}000{,}000)(PP_{cellosolve})}{P_{total}}$$

$$C_{cellosolve} = \frac{(1{,}000{,}000)(3.20)}{760} = 4{,}210.5$$

$$C_{cellosolve} = 4{,}210.5 \text{ ppm(vol)}$$

2. t-butyl alcohol:

$$C_{t\text{-butyl alcohol}} = \frac{(1{,}000{,}000)(PP_{t\text{-butyl alcohol}})}{P_{total}}$$

$$C_{t\text{-butyl alcohol}} = \frac{(1{,}000{,}000)(6.88)}{760} = 9{,}052.6$$

$$C_{t\text{-butyl alcohol}} = 9{,}052.6 \text{ ppm(vol)}$$

Finally, we can convert these volume-based concentrations for each component in the vapor space to the mass-based counterparts for each by using Equation **#3-7**, from Page 3-12, thus:

$$C_{mass_i} = \frac{MW_i}{24.45}\left[C_{vol_i}\right] \quad @ \text{ NTP Conditions} \qquad \text{[Eqn. \#3-7]}$$

1. cellosolve:

$$C_{mass_{cellosolve}} = \frac{MW_{cellosolve}}{24.45}\left[C_{vol_{cellosolve}}\right]$$

$$C_{mass_{cellosolve}} = \frac{(90.12)(4{,}210.5)}{24.45} = 15{,}519.4$$

$$C_{cellosolve} = 15{,}519.4 \text{ mg/m}^3 = 15.5 \text{ gms/m}^3$$

2. t-butyl alcohol:

$$C_{mass_{t\text{-butyl alcohol}}} = \frac{MW_{t\text{-butyl alcohol}}}{24.45}\left[C_{vol_{t\text{-butyl alcohol}}}\right]$$

$$C_{mass_{t\text{-butyl alcohol}}} = \frac{(74.12)(9{,}052.6)}{24.45} = 27{,}442.9$$

$$C_{t\text{-butyl alcohol}} = 27{,}442.9 \text{ mg/m}^3 = 27.4 \text{ gms/m}^3$$

It should now be very clear for a second time that the "Order-of-Magnitude-Approximations" that were obtained in developing answers for Problem **#3.14** probably should not be characterized as "Order-of-Magnitude-Approximations", since they are very substantially incorrect when compared to the more precise numbers obtained in the more fundamental approach for this solution to the problem, which solution relies on accepted theory and the application of the basic laws of physical chemistry. It is also easy to understand why Equation **#3-3** enjoys such great relative popularity for making this determination. It is clear that using the alternate approach and applying accepted theoretical considerations and relationships simply requires 2 to 3 times as much time and calculating effort. Suffice it to say — from the perspective of having an accurate characterization of the situation — it is better to expend the additional time in calculating.

∴	cellosolve concentration	=	15.5 gms/m³
		=	4,210.5 ppm(vol)
	t-butyl alcohol concentration	=	27.4 gms/m³
		=	9,052.6 ppm(vol)

Problem #3.16:

This problem is also very similar to Problem **#s 3.12 & 3.14**, except that this one involves three, rather than four or two, different volatile chemicals. Like these two previous problems, we must apply a total of three of the listed relationships, and do so in the following sequence: Equation **#3-4**, from Pages 3-10 & 3-11; then Equation **#3-3**, from Pages 3-9 & 3-10; and then finally Equation **#3-7**, from Page 3-12.

As has been noted in these earlier problems, Equation **#3-3** is an approximation, one that relies upon the assumption that the overall equilibrium composition of the vapor space above a liquid mixture of volatile solvents will have the same composition as does the liquid mixture itself. For this problem, however, we are not dealing with an equilibrium vaporization situation; instead we are working with a leak of the three refrigerants. Because of this, it is very clear that the vapor composition resulting from a leak will, in fact, have a composition that is identical to that of the liquid refrigerant mixture. To start the solution to this problem, we must first make the conversion of the volume-based TLV-TWAs [expressed in ppm(vol)] for each of these three refrigerants to their mass-based equivalents [expressed in mg/m³], using Equation **#3-4** from Pages 3-10 & 3-11:

$$C_{mass_i} = \frac{MW_i}{24.45}\left[C_{vol_i}\right] \quad \text{@ NTP Conditions} \qquad \text{[Eqn. \#3-4]}$$

1. For Freon 12:

$$C_{mass_{Freon\ 12}} = \frac{120.92}{24.45}(1,000) = \frac{120,920}{24.45} = 4,945.60$$

$$\text{TLV-TWA}_{Freon\ 12} \sim 4,946 \text{ mg/m}^3$$

2. For Freon 21:

$$C_{mass_{Freon\ 21}} = \frac{102.92}{24.45}(10) = \frac{1,029.20}{24.45} = 42.09$$

$$\text{TLV-TWA}_{Freon\ 21} \sim 42 \text{ mg/m}^3$$

3. For Freon 112:

$$C_{mass_{Freon\ 112}} = \frac{170.92}{24.45}(500) = \frac{85,460.00}{24.45} = 3,495.30$$

$$\text{TLV-TWA}_{\text{Freon 112}} \sim 3{,}495 \text{ mg/m}^3$$

Next, these three mass-based TLV-TWA values can be readily used to determine the Effective TLV for the mixture, using Equation **#3-3** from Pages 3-9 & 3-10; thus:

$$\text{TLV}_{\text{effective}} = \left[\frac{1}{\sum_{i=1}^{n}\frac{f_i}{\text{TLV}_i}}\right] = \left[\frac{1}{\frac{f_1}{\text{TLV}_1}+\frac{f_2}{\text{TLV}_2}+\cdots+\frac{f_n}{\text{TLV}_n}}\right] \qquad \text{[Eqn. \textbf{\#3-3}]}$$

$$\text{TLV}_{\text{effective}} = \left[\frac{1}{\frac{0.40}{4{,}945.60}+\frac{0.20}{42.09}+\frac{0.40}{3{,}495.30}}\right]$$

$$\text{TLV}_{\text{effective}} = \left[\frac{1}{8.09\times10^{-5} + 4.75\times10^{-3} + 1.14\times10^{-4}}\right]$$

$$\text{TLV}_{\text{effective}} = \frac{1}{4.95\times10^{-3}} = 202.16 \text{ mg/m}^3$$

This is the initial piece of information that was asked for in the problem statement.

∴ The Effective TLV-TWA for this mixture of three different refrigerant vapors ~ 202 mg/m³.

We must now calculate the three mass-based concentrations that would make up an overall TLV-TWA level concentration of these three refrigerants. As with the previous two problems, this is easily accomplished simply by multiplying the overall mixture TLV-TWA obtained above by the weight percentages of each of the three refrigerants, as they were provided in the problem statement; thus:

1. For Freon 12:

$$C_{\text{mass}_{\text{Freon 12}}} = (202.16)(0.40) = 80.86$$
$$C_{\text{Freon 12}} \sim 80.9 \text{ mg/m}^3$$

2. For Freon 21:

$$C_{\text{mass}_{\text{Freon 21}}} = (202.16)(0.20) = 40.43$$
$$C_{\text{Freon 21}} \sim 40.4 \text{ mg/m}^3$$

3. For Freon 112:

$$C_{\text{mass}_{\text{Freon 112}}} = (202.16)(0.40) = 80.86$$
$$C_{\text{Freon 112}} \sim 80.9 \text{ mg/m}^3$$

With these three mass-based concentrations, we can now apply Equation **#3-7**, from Page 3-12, to obtain the final pieces of information requested in the problem statement; thus:

$$C_{\text{vol}_i} = \frac{24.45}{\text{MW}_i}\left[C_{\text{mass}_i}\right] \quad \text{@ NTP Conditions} \qquad \text{[Eqn. \textbf{\#3-7}]}$$

1. For Freon 12:

$$C_{\text{vol}_{\text{Freon 12}}} = \frac{24.45}{120.92}(80.86) = \frac{1{,}977.12}{120.92} = 16.35$$
$$C_{\text{Freon 12}} \sim 16.4 \text{ ppm(vol)}$$

2. For Freon 21:

$$C_{vol_{Freon\ 21}} = \frac{24.45}{102.92}(40.43) = \frac{988.56}{102.92} = 9.61$$

$$C_{Freon\ 21} \sim 9.6 \text{ ppm(vol)}$$

3. For Freon 112:

$$C_{vol_{Freon\ 112}} = \frac{24.45}{170.92}(80.86) = \frac{1,977.12}{170.92} = 11.57$$

$$C_{Freon\ 112} \sim 11.6 \text{ ppm(vol)}$$

We now can tabulate the final set of data asked for in the problem statement.

∴	Freon 12 concentration	~	80.9 mg/m^3
		~	16.4 ppm(vol)
	Freon 21 concentration	~	40.4 mg/m^3
		~	9.6 ppm(vol)
	Freon 112 concentration	~	80.9 mg/m^3
		~	11.6 ppm(vol)

Problem #3.17:

The solution to this problem relies in part on the results obtained in Problem **#3.16**. From the results of the determinations made for Problem **#3.16**, we see that the concentration of Freon 21 [i.e., the material to which the U.S. Navy's highly specific, bulkhead mounted ambient air analyzer responds] must be 9.6 ppm(vol) whenever the calculated overall Effective TLV-TWA for this mixture has been achieved. This then must be the concentration upon which the Alarm Settings for this analyzer will have to be based. Remembering, too, that the precision of this analyzer is ± 0.05 ppm(vol), we can proceed, thus:

1. For the Alert Alarm: $C_{Alert\ Alarm} = (0.60)(9.61) = 5.76$

   Alert Alarm Setting must be 5.7 ppm(vol)

2. For the Evacuate Alarm: $C_{Evacuate\ Alarm} = (0.90)(9.61) = 8.64$

   Evacuate Alarm Setting must be 8.6 ppm(vol)

With these settings, either alarm will be energized — considering the precision of the analyzer in question — at an ambient concentration that is either *just slightly below* the target concentration for that alarm [i.e., the Alert Alarm], or *most likely just slightly below* BUT *possibly just barely above* the target concentration for that alarm [i.e., the Evacuate Alarm, where a concentration 10 ppb(vol) higher than the desired concentration — 8.65 ppm(vol) vs. 8.64 ppm(vol) — could be achieved before the alarm sounded]. For confirmation of this situation, please see the tabulation that follows:

Alarm Type	Alarm Setting	[Concentration] Range at which an Alarm will Sound	Target Concentration
Alert	5.7 ppm(vol)	5.65 ppm ≤ [Freon 21] ≤ 5.75 ppm	5.76 ppm(vol)
Evacuate	8.6 ppm(vol)	8.55 ppm ≤ [Freon 21] ≤ 8.65 ppm	8.64 ppm(vol)

∴ The Alert Alarm Setting = 5.7 ppm(vol) & the Evacuate Alarm Setting = 8.6 ppm(vol).

Problem #3.18:

The solution to this problem requires first the use of Equation **#2-2**, from Page 2-4. In this case we must consider the Total Volume, as pumped over the entire 8-hour period by the 16 personal sampling pumps, as being made up of two separate and distinct sub-volumes, namely:

1. The Non-Dusty 50-minute period [30 + 10 + 10 = 50] during which each worker was in a dust free area, either on a coffee break or eating lunch; and
2. The Dusty 430-minute period [480 – 50 = 430] — i.e., the balance of the 8-hour work day — during which time period each worker was involved in productive activities in the dusty main assembly area.

Of these two sub-volumes, we are interested primarily in the 7.167-hour [430 minutes] "Dusty" period during which each worker was in an area where he or she could have been exposed to airborne particulate matter. It is clear that the accumulation of mass on any employee's filter cassette must have occurred during this "Dusty" period. We must, however, calculate the volumes of air that passed through each filter cassette during both the 50-minute "Non-Dusty", and the 430-minute "Dusty" periods for each employee. We must have both pieces of data to determine the overall Time Weighted Average.

$$\text{Total Volume} = [\text{Flow Rate}][\text{Time Interval}] \qquad \text{[Eqn. \#2-2]}$$

$$\text{Total Volume}_{\text{Non-Dusty}} = (1.75)(50) = 87.5 \text{ liters}$$
$$\text{Total Volume}_{\text{Non-Dusty}} = (87.5)(0.001) = 0.0875 \text{ m}^3$$

$$\text{Total Volume}_{\text{Dusty}} = (1.75)(430) = 752.5 \text{ liters}$$
$$\text{Total Volume}_{\text{Dusty}} = (752.5)(0.001) = 0.7525 \text{ m}^3$$

With these two total sub-volume numbers we can use the tabulated mass accumulations for each of the 16 filter cassettes to calculate average ambient dust concentrations characteristic of each of the two distinctly different time intervals. Obviously, the Non-Dusty average dust concentrations were simply 0.00 mg/m^3, since there was no dust in those areas where the employees took their coffee breaks and/or ate their lunches. Only in the main production area was there a dust problem. To calculate the dust concentrations in these areas, we simply must apply the following very basic relationship, thus:

$$\text{Dust Concentration} = \frac{\text{Cassette Net Weight Gain}}{\text{Total Volume of Air Passing Through the Cassette}}$$

In terms of just the 430-minute duration "Dusty" period, this relationship would be:

$$[\text{Average Dust Concentration}]_{\text{"Dusty" Time Period}} = \frac{\text{Cassette Net Weight Gain}}{\text{"Dusty" Period Volume}}$$

The results of performing this calculation — i.e., dividing by 0.7525 m^3 — for each of the 16 employees is shown in the following tabulation:

Average Dust Concentration During the Sum of the Daily "Dusty" Time Periods

Empl. No.	Dust Conc. in mg/m^3	Empl. No.	Dust Conc. in mg/m^3	Empl. No.	Dust Conc. in mg/m^3	Empl. No.	Dust Conc. in mg/m^3
1	6.58	5	10.50	9	9.89	13	3.00
2	8.23	6	1.44	10	6.75	14	10.78
3	5.12	7	5.75	11	1.97	15	9.13
4	3.50	8	9.05	12	10.39	16	6.51

Now to calculate the overall Time Weighted Average Exposures for these 16 employees, we must apply the following formula.

$$\text{TWA} = \frac{\sum_{i=1}^{n} T_i C_i}{\sum_{i=1}^{n} T_i} = \frac{T_1 C_1 + \ldots + T_n C_n}{T_1 + \ldots + T_n} \qquad \text{[Eqn. \textbf{\#3-1}]}$$

Clearly, the second, or "Non-Dusty", factor will always be 0.0 (mg)(min)/m^3, since there was no dust exposure [i.e., the Average Dust Concentration = 0.0 mg/m^3] for any employee during those periods away from the main production floor. Because of this, the second term in the numerator of this expression — namely, the T_2C_2 term — will always be equal to zero since C_2 = 0 mg/m^3. We actually could have made the overall TWA calculation by applying the same basic relationship used on the previous page, but using the full 8-hour or 480-minute time period Total Volume of Air as the denominator. Remembering that the Total Volume of air that passed through each cassette would simply be the sum of the "Dusty" and the "Non-Dusty" Volumes [i.e., 0.0875 + 0.7525 = 0.84 m^3], we can now develop the required answer, thus:

$$[\text{Average Dust Concentration}]_{\text{Full Shift}} = \frac{\text{Cassette Net Weight Gain}}{\text{Full Shift Volume}}$$

$$[\text{Average Dust Concentration}]_{\text{Full Shift}} = \frac{\text{Cassette Net Weight Gain}}{0.84 \text{ m}^3}$$

The calculated 8-hour, Full Shift TWAs for each of these 16 employees are as follows:

Employee 8-hour, Full Shift TWA Dust Exposures

Empl. No.	Dust Conc. in mg/m^3	Empl. No.	Dust Conc. in mg/m^3	Empl. No.	Dust Conc. in mg/m^3	Empl. No.	Dust Conc. in mg/m^3
1	5.89	5	9.40	9	8.86	13	2.69
2	7.37	6	1.29	10	6.05	14	9.65
3	4.58	7	5.15	11	1.81	15	8.18
4	3.13	8	8.11	12	9.31	16	5.83

The set of average ambient dust concentrations during the "Dusty" period, as tabulated on the previous page, should be useful to any Industrial Hygienist as a measure of the potential for dust exposure problems as these potential problems relate either to some specific employee or work location. From this perspective, there would probably be some concern about the exposures to Employee #s 2, 5, 8, 9, 12, 14, & 15 — all of whom appear to have had "Dusty" time period exposures > 8.00 mg/m^3, which is 80% of the PEL-TWA standard of 10.0 mg/m^3 for alumina dust. The greatest potential concern should be for Employee #s 5, 12, & 14, whose "Dusty" period exposures exceeded this 10.0 mg/m^3 standard. Clearly, the time period involved for each of these employees was less than, but close to, 8 full hours; however, their exposures during this almost 8-hour period were at levels that might possibly indicate a future problem.

∴ Based on the calculated Full Shift TWA Dust Exposures of the 16 employees, the Industrial Hygienist in this case should not anticipate any problems from alumina dust exposures; however, since some of these exposures are very close to the borderline, this IH might logically choose to remain very vigilant in this area.

Problem #3.19:

To solve this problem, we must apply Equation **#3-4**, from Pages 3-10 & 3-11:

$$C_{vol_i} = \frac{24.45}{MW_i}\left[C_{mass_i}\right] \quad @ \text{ NTP Conditions} \qquad [\text{Eqn. } \#\mathbf{3\text{-}4}]$$

For the MAK-TWA:

$$C_{vol} = \frac{24.45}{100.12}(0.8) = \frac{19.56}{100.12} = 0.20$$

$$\text{MAK-TWA}_{glutaraldehyde} = 0.20 \text{ ppm(vol)}$$

$$\therefore \quad \text{The MAK-TWA}_{glutaraldehyde} = 0.2 \text{ ppm(vol).}$$

Problem #3.20:

To solve this problem, we must apply Equation **#3-5**, from Pages 3-10 & 3-11:

$$C_{vol_i} = \frac{22.41}{MW}\left[C_{mass_i}\right] \quad @ \text{ STP Conditions} \qquad [\text{Eqn. } \#\mathbf{3\text{-}5}]$$

1. For the PEL-STEL:

$$C_{vol} = \frac{22.41}{72.11}(735) = \frac{16,471.35}{72.11} = 228.42$$

$$\text{PEL-STEL}_{tetrahydrofuran} \sim 228 \text{ ppm(vol)}$$

2. For the PEL-TWA:

$$C_{vol} = \frac{22.41}{72.11}(590) = \frac{13,221.90}{72.11} = 183.36$$

$$\text{PEL-STEL}_{tetrahydrofuran} \sim 183 \text{ ppm(vol)}$$

$$\therefore \quad \text{PEL-STEL}_{tetrahydrofuran} \sim 228 \text{ ppm(vol)}$$
$$\text{PEL-STEL}_{tetrahydrofuran} \sim 183 \text{ ppm(vol)}$$

Problem #3-21:

To solve this problem, we must eventually apply Equation **#3-6**, from Pages 3-10 & 3-11, but we must first decide on the set of units and/or dimensions we will use to develop the solution. In particular, we are concerned with the units of temperature and pressure since these will determine the dimensions and value of the Universal Gas Constant, "R". Because we have been given the barometric pressure in "atmospheres" and the temperature in "°C", which we can readily convert to "K", we will select, from Page 1-19, as a value for R:

$$R = 0.0821 \frac{(\text{liter})(\text{atmospheres})}{(\text{K})(\text{mole})}$$

Let us first convert the relative Metric temperature [°C] to its absolute equivalent [K] using Equation **#1-1**, from Page 1-16:

$$t_{Metric} + 273.16 = T_{Metric} \qquad [\text{Eqn. } \#\mathbf{1\text{-}1}]$$

$$18 + 273.16 = T_{Metric}$$

$$T_{Metric} = 291.16 \text{ K}$$

Now using Equation **#3-6**, from Pages 3-10 & 3-11, we can develop the result asked for in the problem statement, thus:

$$C_{vol} = \frac{R\,T}{P\,[MW_i]}[C_{mass}] \qquad \text{[Eqn. \#3-6]}$$

$$C_{vol} = \frac{(0.0821)(291.16)}{(0.70)(30)}(30) = \frac{23.90}{21}(30) = 34.15$$

$$C_{vol} \sim 34.2 \text{ ppm(vol)}$$

It is interesting to note that the concentration of nitric oxide, when expressed in mass-based units, is precisely at the PEL-TWA specified concentration level of 30 mg/m^3; however, when we convert this value to its volume-based equivalent, we observe that this "equivalent" is greater than the published volume-based PEL-TWA [i.e., 32.4 ppm(vol) > 24.5 ppm(vol)]!!! This situation provides two important pieces of information:

1. Mass-Based concentration units are always more "absolute" than their volume-based "equivalents"; &
2. The basic assumption for all published volume-based PEL concentrations is that these measurements are made under conditions of Normal Temperature & Pressure [NTP]. This is true also for all TLVs, RELs, and MAKs.

> ∴ The nitric oxide concentration in the interior of this aircraft under the stated conditions of pressure and temperature would be 34.2 ppm(vol).

Problem #3.22:

The solution to this problem will require the use of Equation **#3-7**, from Page 3-12, thus:

$$C_{mass_i} = \frac{MW_i}{24.45}[C_{vol_i}] \quad @ \text{ NTP Conditions} \qquad \text{[Eqn. \#3-7]}$$

1. For the PEL-TWA:

$$C_{mass} = \frac{44.05}{24.45}(1.0) = \frac{44.05}{24.45} = 1.80$$

$$\text{PEL-TWA}_{\text{ethylene oxide}} = 1.80 \text{ mg/m}^3$$

2. For the Action Level:

$$C_{mass} = \frac{44.05}{24.45}(0.5) = \frac{22.03}{24.45} = 0.90$$

$$[\text{Action Level}]_{\text{ethylene oxide}} = 0.90 \text{ mg/m}^3$$

> ∴ $\text{PEL-TWA}_{\text{ethylene oxide}} = 1.80 \text{ mg/m}^3$
> $[\text{Action Level}]_{\text{ethylene oxide}} = 0.90 \text{ mg/m}^3$

Problem #3.23:

To solve this problem, we must apply Equation **#3-8**, from Page 3-12, thus:

$$C_{mass_i} = \frac{MW_i}{22.41}\left[C_{vol_i}\right] \quad @ \text{ STP Conditions} \qquad \text{[Eqn. \#3-8]}$$

1. For the PEL-TWA:

$$C_{mass} = \frac{78.11}{22.41}(1.0) = \frac{78.11}{22.41} = 3.49$$

$$\text{PEL-TWA}_{benzene} \sim 3.5 \text{ mg/m}^3$$

2. For the PEL-STEL:

$$C_{mass} = \frac{78.11}{22.41}(5.0) = \frac{390.55}{22.41} = 17.43$$

$$\text{PEL-STEL}_{benzene} \sim 17.4 \text{ mg/m}^3$$

3. For the PEL-C:

$$C_{mass} = \frac{78.11}{22.41}(25.0) = \frac{1,952.75}{22.41} = 87.14$$

$$\text{PEL-C}_{benzene} \sim 87.1 \text{ mg/m}^3$$

$$\therefore \quad \text{PEL-TWA}_{benzene} \sim 3.5 \text{ mg/m}^3$$
$$\text{PEL-STEL}_{benzene} \sim 17.4 \text{ mg/m}^3$$
$$\text{PEL-C}_{benzene} \sim 87.1 \text{ mg/m}^3$$

Problem #3.24:

To solve this problem, we will eventually apply Equation **#3-9**, from Pages 3-12 & 3-13; however, we must first decide on the set of units and/or dimensions that will be required to develop the solution. Because the problem statement has provided the barometric pressure in units of "mm Hg" and the temperature in "°C", which we can readily convert to "K", we will select for R:

$$R = 62.36 \frac{(\text{liter})(\text{mm Hg})}{(\text{K})(\text{mole})}$$

Let us first convert the relative Metric temperature [°C] to its absolute equivalent [K] using Equation **#1-1**, from Page 1-16:

$$t_{Metric} + 273.16 = T_{Metric} \qquad \text{[Eqn. \#1-1]}$$

$$32 + 273.16 = T_{Metric}$$

$$T_{Metric} = 305.16 \text{ K}$$

Now using Equation **#3-9**, from Pages 3-12 & 3-13, we can develop the result asked for in the problem statement:

$$C_{mass_i} = \frac{P\left[MW_i\right]}{RT}\left[C_{vol_i}\right] \qquad \text{[Eqn. \#3-9]}$$

$$C_{mass} = \frac{(626)(44.01)}{(62.36)(305.16)}(350) = \frac{27,550.26}{19,029.78}(350) = 506.71$$

$$C_{mass} = 506.7 \text{ mg/m}^3$$

∴	The mass-based concentration of carbon dioxide in Denver would be 506.7 mg/m³.

Problem # 3.25:

The solution to this problem requires the application of Equation **#3-10**, from Page 3-14:

$$TLV_{quartz} = \frac{10\text{mg}/\text{m}^3}{\%RQ + 2} \qquad \text{[Eqn. \#3-10]}$$

We must transpose this equation so as to let us solve for the % of respirable quartz. In its transposed form, the Equation would be:

$$\%RQ = \frac{10 \text{ mg/m}^3}{TLV_{quartz}} - 2$$

$$\%RQ = \frac{10}{0.22} - 2 = 45.45 - 2 = 43.45\%$$

∴	There appears to be approximately 43.5% respirable quartz in the crushed gravel.

Problem #3.26:

This problem clearly will require the eventual use of Equation **#3-12**, from Pages 3-14 & 3-15. In order to apply this formula, however, we must first determine the percentages of each of the three types of silica that make up the material mix being handled. We have been provided with a characteristic "mix ratio" for this material — one that identifies the proportions of each of the three basic silica components in the mix being evaluated; thus, we can see:

$$18 + 13 + 9 = 40$$

We have been further advised of the fact that this mix contains 80% silica [i.e., this 80% consists of the three basic types of silica — quartz, cristobalite, and tridymite]; thus, for the total material mix being considered, we can say:

1. For the percent quartz:

$$\%Q = \frac{18}{40}(80) = 36\%$$

2. For the percent cristobalite:

$$\%C = \frac{13}{40}(80) = 26\%$$

3. For the percent tridymite:

$$\%T = \frac{9}{40}(80) = 18\%$$

We can now apply Equation **#3-12**, from Pages 3-14 & 3-15:

$$TLV_{mix} = \frac{10 mg/m^3}{\%Q + [2][\%C] + [2][\%T] + 2} \quad \text{[Eqn. \#3-12]}$$

$$TLV_{mix} = \frac{10}{36 + (2)(26) + (2)(18) + 2}$$

$$TLV_{mix} = \frac{10}{36 + 52 + 36 + 2} = \frac{10}{126} = 0.079$$

∴ The TLV_{mix} ~ 0.08 mg/m^3.

Chapter 4
Ventilation

This chapter will discuss the parameters, measurements, mathematical relationships, and the various pieces of hardware, equipment, etc. that constitute the general area of ventilation.

RELEVANT DEFINITIONS

Ventilation Equipment & Basic Parameters

Ventilation is an operation whereby air is moved to or from various locations within a facility by and/or through a variety of different pieces of equipment or hardware. Selecting from the wide variety of available ventilation equipment will always be a function of the task that is to be accomplished. Tasks such as providing fresh air to a office, removing vapors or fumes produced by some manufacturing process, etc. are all recognizable objectives that can be readily achieved by the proper choice and operation of appropriate ventilation equipment. The various items of hardware, and the factors and parameters that characterize each, will be defined and described below.

Hood

A **Hood** is a shaped inlet designed and positioned so as to capture air that contains some sort of contaminant, and then conduct this contaminated air into the ducts of the exhaust system. Hoods may be plain or flanged depending upon a variety of design factors.

Flange

A **Flange** is generally a flat piece of metal or rigid plastic that surrounds the shaped inlet of a hood and serves the purpose of minimizing the drawing of air from nearby zones where no contaminant is known or thought to exist.

Capture Velocity

The **Capture Velocity** is the air velocity at any point in front of a hood, or hood opening, that is necessary to overcome the opposing air currents and other factors, and thereby "capture" the contaminated air at that point by causing it to flow or be drawn into the hood.

Slot Velocity

The **Slot Velocity** is the actual linear velocity of air as it enters a hood through its slot opening.

Face Velocity

The **Face Velocity** is the linear velocity of air, averaged over the entire opening, as it enters a hood.

Coefficient of Entry

The **Coefficient of Entry** is the actual rate of flow produced by a given hood static pressure, when compared to the theoretical flow that would result if the static pressure in the hood could be converted, with 100% efficiency, into the velocity pressure of the fluid being drawn into the hood. The **Coefficient of Entry** is the ratio of the actual flow rate to the theoretical maximum flow rate.

Entry Loss

Entry Loss is the loss or decrease in pressure experienced by air as it flows into or enters a duct or hood. **Entry Losses** are usually expressed in "inches of water - gauge".

Duct

A **Duct** is a hollow, cylindrical [circular, rectangular, etc.] connected section of tubing that serves as the main conduit for the flow of air in any ventilation system.

Blast Gate or Damper

A **Blast Gate** or **Damper** is a sliding, swinging, or pivoting "door" that can shut off or decrease the cross sectional area of a duct, thereby limiting the flow of air through it. The principal purpose of a **Blast Gate** or **Damper** is, obviously, to permit the easy adjustment of the volume of air that is flowing through the duct. Adjustments of this type are usually required in order to improve or balance the general flow of ventilation air to all locations.

Plenum

A **Plenum** is a chamber within an overall ventilation system that serves the purpose of equalizing potentially different pressures that might have arisen from various different streams of air as they merge or separate in an overall system of ducting.

Flow Rate or Volumetric Flow Rate

Flow Rate or **Volumetric Flow Rate** is a measure of the volume of a gas (usually air) that is moving through a fan or a duct. Its most commonly used dimensions are "volume per unit time" in units such as cubic feet per minute [cfm], etc.

Velocity or Duct Velocity

Velocity or **Duct Velocity** is a measure of the speed with which a gas (again, usually air) is moving through a duct. This velocity is always considered to be in a direction that is perpendicular to the cross section of the duct through which the gas is flowing. Its most commonly used dimensions are "distance per unit time" in units such as feet per minute [fpm].

Air Handler or Blower or Fan

An **Air Handler**, **Blower**, or **Fan** is the basic or prime motive force that is integral to every ventilation system. The movement of air in any such system will always have been caused by such a prime mover.

Air Horsepower

The **Air Horsepower** is the theoretical horsepower required to operate a fan, blower, or air handler on the assumption that this item of hardware can be operated without any internal losses [i.e., the horsepower that would be required to operate a 100% efficient, totally friction free unit].

Brake Horsepower

The **Brake Horsepower** is the horsepower actually required to operate a fan. Brake Horsepower involves all the internal losses in the fan and can only be measured by conducting an actual test on the fan.

Pressure Terms

To understand the area of ventilation, one must have an in-depth knowledge of the various types of pressure measurements that characterize this general subject. The various pressure terms will be defined and discussed in detail.

Static Pressure

Static Pressure is the pressure exerted in all directions by a fluid at rest. For any fluid in motion, this parameter is measured in a direction normal to the direction of fluid flow. **Static Pressure** is usually expressed in "inches of water - gauge" when dealing with air flowing in a duct; this parameter can be either negative or positive.

Velocity Pressure

Velocity Pressure is the kinetic pressure that is necessary in order to cause a fluid to flow at a particular velocity; it is measured in the direction of fluid flow. **Velocity Pressure** is usually expressed in "inches of water - gauge" when dealing with air flowing in a duct; this parameter is always positive. It is measured in a direction parallel to the direction of fluid flow, looking "upstream", or directly into the flowing fluid.

Total Pressure

Total Pressure is the algebraic sum of the Static and the Velocity Pressures. When dealing with air flowing in a duct, **Total Pressure**, like the other ventilation related pressure terms, will usually be expressed in "inches of water - gauge".

Manometer

A **Manometer** is an instrument that is used for measuring the pressure in a stationary or flowing fluid or gas. A **Manometer** is very often a U-tube filled with water, light oil, or mercury. This implement is usually constructed so that the observed displacement of liquid in it will indicate the pressure being exerted on it by the fluid being monitored. In a Static Pressure measuring situation, the plane of the opening of the Manometer tube in the fluid being monitored will be parallel to the direction of flow of that fluid [i.e., the Manometer tube, itself, will be situated perpendicular to the flow of that fluid]. In a Velocity Pressure measuring situation, the plane of the opening of the Manometer tube in the fluid being monitored will be perpendicular to the direction of flow of that fluid, and will be directed so that the opening at the end of the tube faces directly into that flow [i.e., the Manometer tube, itself, will be situated parallel to the flow of that fluid, with the tube opening facing "upstream"].

Vapor Pressure

Vapor Pressure is the static pressure exerted by a vapor. If the vapor phase of some material of interest is in equilibrium with its liquid phase, then the **Vapor Pressure** that exists is referred to as the Saturated Vapor Pressure for that material. The saturated vapor pressure for any material is solely dependent on its temperature. Frequently vapor pressure and saturated vapor pressure are used synonymously.

Humidity Factors

Humidity, or the presence of water vapor in the air in a structure or flowing in a duct, is always a source of concern to individuals who must work with ventilation systems. Several measures of Humidity will be discussed in this section.

Absolute Humidity

Absolute Humidity is the weight of the vaporous water in the air measured per unit volume of that air, usually expressed in lbs/ft^3 or $grams/cm^3$.

Relative Humidity

Relative Humidity is the ratio of the actual or measured vapor pressure of water in an air mass to the saturated vapor pressure of pure water at the same temperature as that of the air mass. **Relative Humidity** can also be thought of as the percentage that the actual measured absolute concentration of water in the air under any ambient conditions is to the saturation concentration of water in air under the same ambient conditions.

Dew Point

The **Dew Point** of an air mass is also a measure of relative humidity. It is the temperature at which the water vapor present in an air mass would start to condense — or in the context of the term, **Dew Point**, it is the temperature at which that water vapor would turn to "dew". It can be determined experimentally simply by cooling a mass of air until "dew" or condensate forms; the temperature at which this event occurs is defined to be the **Dew Point** of that air mass.

Ventilation Standards

Ventilation Standard Air

Standard Air, as Ventilation Engineers view it, differs somewhat from air at STP or NTP [as defined earlier]. **Standard Air** is air possessing the following characteristics:

Temperature:	70°F 21.1°C 529.67°R 294.26°K
Barometric Pressure:	1.00 atmosphere 760 mm Hg 29.92 in. of Hg 14.70 psia 0.00 psig 1,013.25 millibars 760 Torr
Humidity Level:	Relative Humidity = 0.0%, Absolute Humidity = 0.0 grams/cm^3 Water Vapor Concentration = 0.0 ppm(vol) = 0.0 mg/m^3
Specific Gravity:	Density = 0.075 lbs/ft^3
Heat Capacity:	Specific Heat = 0.24 BTU/lb/°F.

RELEVANT FORMULAE & RELATIONSHIPS

Calculations Involving Gases Moving in Ducts

Most of the important calculations in the area of ventilation involve some gas, usually air, that is being conducted through ducting in order to achieve one or more specific purposes. The relationships that follow permit the engineer, whose responsibility it is to operate a ventilation system so as to accomplish the stated goals or tasks, to determine exactly how his system can be operated in order to achieve the designed purposes or goals.

Equation #4-1:

Equation **#4-1** is probably the most basic relationship in the general area of ventilation. It relates the **Volumetric Flow Rate** in a duct to the **Cross Sectional Area** of that duct, and to the **Velocity** (the **Duct Velocity**) of the gas flowing in it.

$$Q = AV$$

Where:

Q = the **Volumetric Flow Rate** in the duct, measured in cubic feet/minute = ft^3/min [cfm];

A = the **Cross Sectional Area** of the duct under consideration, in square feet [ft^2]; &

V = the **Velocity**, or **Duct Velocity**, of the gases moving in the duct, in feet/minute = ft/min [fpm].

Equation #4-2:

Equation **#4-2** is the first of two basic relationships involving the **Velocity** of the gases that are flowing in a duct. This expression relates this parameter to the **Velocity Pressure** and the **Density** of the gases that are flowing in the duct.

$$V = 1,096\sqrt{\frac{VP}{\rho}}$$

Where:

V = the **Velocity** of the gases moving in the duct, or the **Duct Velocity**, measured in feet per minute [fpm];

VP = the **Velocity Pressure** of the gases moving in the duct, measured in inches of water; &

ρ = the **Density** of the gases that are flowing in the duct, measured in lbs per cubic foot [lbs/ft^3] — for reference, the density of Ventilation Standard Air = 0.075 lbs/ft^3.

Equation #4-3:

Equation **#4-3** is the second of the two basic relationships involving the **Duct Velocity** of the gases that are flowing in a duct. This expression, which simply relates this parameter to the **Velocity Pressure** of the gases that are flowing in the duct, is shown below.

$$\mathbf{V = 4{,}005\sqrt{VP}}$$

Where:

V = the **Velocity** of the gases moving in the duct, or the **Duct Velocity**, measured in feet per minute [fpm];

VP = the **Velocity Pressure**, and is as defined for Equation **#4-2** on the previous page, namely, Page 4-6.

The following three Equations, namely, **#s 4-4**, **4-5**, & **4-6**, are presented as a group — they are known as the *General Rules of Ventilation* — because: (1) they constitute three relationships that will always apply to gases — usually air — that happen to be flowing in a duct, and (2) they are closely interrelated.

Equation #4-4:

$$\mathbf{TP_1 = TP_2 + losses}$$

Where:

$\mathbf{TP_1}$ = the **Total Pressure** at Point #1 in the duct, measured in inches of water;

$\mathbf{TP_2}$ = the **Total Pressure** at Point #2 — which is downstream of Point #1 — in the duct, also measured in inches of water; &

losses = **Pressure Losses** from various duct internal factors [i.e., friction] that affect the flow of gases passing through the duct — these total **losses** are almost always measured in inches of water. In general, the factors that can be used to quantify these **losses** will be provided by the duct's manufacturer, usually in the form of a "Specific Pressure Loss or Pressure Drop per Unit Length of Duct". These design factors will be expressed in units of "inches of water loss per lineal foot of duct traversed".

Equation #4-5:

$$\mathbf{TP = SP + VP}$$

Where:

TP = the **Total Pressure** in the duct, and is as defined above on this page, namely, Page

4-7, for Equation **#4-4**, where it was listed either as TP_1 or TP_2, and measured in inches of water;

SP = the **Static Pressure** in the duct, usually measured in inches of water; &

VP = the **Velocity Pressure** in the duct, and is exactly as defined for Equation **#4-2** on Page 4-6.

Equation #4-6:

$$SP_1 + VP_1 = SP_2 + VP_2 + \text{losses}$$

Where: SP_i = the **Static Pressure** at the **i**th point in the duct, and is as defined for Equation **#4-5** on Pages 4-7 & 4-8, where it is listed simply as **SP**;

VP_i = the **Velocity Pressure** at the **i**th point in the duct, and is as defined for Equation **#4-5** on Pages 4-7 & 4-8, where it is listed simply as **VP**; &

losses = the **Pressure Losses** and are exactly as defined for Equation **#4-4** on the previous page, namely Page 4-7.

Calculations Involving Hoods

The following two Equations, namely, **#s 4-7** & **4-8**, are the relationships that provide the **Capture Velocity** — i.e., the hood opening centerline velocity — for either a square or a round hood opening having a known cross sectional area — for hoods either without Flanges [Equation **#4-7**], or those with Flanges [Equation **#4-8**] — with the evaluation of this velocity being made at some specified distance directly in front of the plane of the hood opening being considered and evaluated.

Equation #4-7: [Hoods WITHOUT Flanges]:

$$V = \frac{Q}{10x^2 + A}$$

Where: **V** = the **Capture Velocity** — i.e., the center-line velocity of the air entering the hood under consideration, at a point "**x**" feet directly in front of the face of the hood. This **Capture Velocity** is usually measured in feet per minute [fpm];

Q = the **Volumetric Flow Rate** of the hood, measured in cubic feet per minute [cfm];

A = the **Cross Sectional Area** of the hood opening, measured in square feet [ft^2]; &

x = the **Distance** from the plane of the hood opening to the point directly in front of it where the **Capture Velocity** is to be determined, measured in feet [ft].

Equation #4-8: [Hoods WITH Flanges]:

$$V = \frac{Q}{0.75\left[10x^2 + A\right]} = \frac{4Q}{3\left[10x^2 + A\right]}$$

Where: **V**, **Q**, **A**, & **x** are all exactly as defined for Equation **#4-7**, both above on this page, and on the previous page, namely, Pages 4-8 & 4-9.

Equation #4-9:

Equation #4-9 is usually referred to as the *Simple Hood Formula.* It is widely used to determine the **Hood Static Pressure**, which is one of the most important operating factors applying to the operation of any hood.

$$SP_h = VP_d + h_e$$

Where: SP_h = the **Hood Static Pressure**, measured usually in inches of water;

VP_d = the **Velocity Pressure** in the hood duct, also measured in inches of water; &

h_e = the **Hood Entry Losses**, also measured in inches of water.

Equation #4-10:

Equation **#4-10** is used to calculate the **Coefficient of Entry** for any hood. This dimensionless parameter serves, functionally at least, as a Hood Efficiency Rating.

$$C_e = \sqrt{\frac{VP_d}{SP_h}} = \frac{\text{Actual Flow}}{\text{Theoretical Flow}}$$

Where: C_e = the **Coefficient of Entry** for the hood under investigation, this is a dimensionless parameter;

VP_d = the **Velocity Pressure** in the hood duct, and is as defined above on this page for Equation **#4-9**; &

SP_h = the **Hood Static Pressure**, and is also as defined on the previous page, namely, Page 4-9, for Equation **#4-9**.

Equation #4-11:

The following two forms of **Equation #4-11** are known, collectively, as the *Hood Throat-Suction Equations*. They, too, are widely used to determine the volumetric flow rates of hoods.

$$Q = 4{,}005A\sqrt{VP_d} \quad \text{— Form \#1}$$

$$Q = 4{,}005AC_e\sqrt{SP_h} \quad \text{— Form \#2}$$

Where:

Q = the **Volumetric Flow Rate** and is as defined for Equation #s **4-7** & **4-8**, on Pages 4-8 & 4-9;

A = the **Cross Sectional Area** of the hood opening, and is also as defined for Equation #s **4-7** & **4-8**, on Pages 4-8 & 4-9;

VP_d = the **Velocity Pressure** in the hood duct, and is as defined for Equation **#4-9**, on the previous page, namely, Page 4-9;

C_e = is the **Coefficient of Entry**, and is also as defined for Equation **#4-10**, on Pages 4-9 & 4-10; &

SP_h = is the **Hood Static Pressure**, and is also as defined for Equation **#4-10**, on Pages 4-9 & 4-10.

Equation #4-12:

This equation is known as the *Hood Entry Loss Equation*, and simply uses most of the previously defined parameters to develop a different but very useful relationship.

$$h_e = \left[\frac{1 - C_e^2}{C_e^2}\right]VP_h$$

Where:

h_e = the **Hood Entry Losses**, and is as defined for Equation **#4-9**, on the previous page, namely, Page 4-9;

C_e = the **Coefficient of Entry** for the hood under consideration, and is also as defined for Equation **#4-10**, on Pages 4-9 & 4-10; &

VP_h = the **Velocity Pressure** in the hood, which is measured in inches of water.

Equation #4-13:

The following Equation **#4-13** is known as the *Hood Entry Loss Factor Equation*, and it, too, is widely used.

$$F_h = \frac{h_e}{VP_h}$$

Where:

F_h = the **Hood Entry Loss Factor**, which is a dimensionless parameter;

h_e = the **Hood Entry Losses**, and is as defined for Equation **#4-9**, on Page 4-9;

VP_h = is the **Velocity Pressure** in the hood, and is as defined for Equation **#4-12**, on the previous page, namely, Page 4-10.

Equation #4-14:

The following two variations of **Equation #4-14**, identified as Equation **#s 4-14A & 4-14B**, are known jointly as the *Compound Hood Equations*. They provide a vehicle for calculating the **Hood Static Pressure** for all conditions in which there are quantitative differences in the **Duct Velocity** and the **Hood Slot Velocity**. There are two specific circumstances that will be considered:

Condition #1: the **Duct Velocity** is GREATER than the **Hood Slot Velocity** [Equation **#4-14A**]; &

Condition #2: the **Hood Slot Velocity** is GREATER than the **Duct Velocity** [Equation **#4-14B**].

In addition to these two *Compound Hood Equations*, there is an additional pair of relationships — actually approximations — that provide useful estimates as to the magnitudes of the **Entry Losses** in either the **Hood Slots** or the **Hood Ducts** of any compound hood. These two approximations have been designated as Equation **#s 4-14C & 4-14D**. These two Entry Losses are be included as factors in both Equation **#s 4-14A & #4-14B**.

Equation #4-14A — [Condition #1]:

$$SP_h = h_{ES} + h_{ED} + VP_d \qquad \text{whenever: } V_d > V_s$$

Where:

SP_h = the **Hood Static Pressure**, and is as defined for Equation **#4-10**, on Pages 4-9 & 4-10;

h_{ES} = the **Hood Slot Entry Loss**, measured in inches of water;

h_{ED} = the **Hood Duct Entry Loss**, measured in inches of water;

VP_d = the **Duct Velocity Pressure**, is as defined for Equation **#4-9**, on Page 4-9, and is measured in inches of water;

$\mathbf{V_s}$ = the **Slot Velocity**, and is measured in velocity units, such as feet per minute [fpm]; &

$\mathbf{V_d}$ = the **Duct Velocity**, which is measured in the same velocity units as $\mathbf{V_s}$.

Equation #4-14B — [Condition #2]:

$$SP_h = h_{ES} + h_{ED} + VP_s \qquad \text{whenever: } V_s > V_d$$

Where: $\mathbf{SP_h}$, $\mathbf{h_{ES}}$, $\mathbf{h_{ED}}$, $\mathbf{V_s}$, & $\mathbf{V_d}$ are all exactly as defined for Equation **#4-14A**, both above on this page, and on the previous page, namely, Pages 4-11 & 4-12; &

$\mathbf{VP_s}$ = the **Slot Velocity Pressure**, and is measured in inches of water.

Note: for virtually any hood and/or duct situation, the following two approximations will always hold:

Equation #4-14C:

$$h_{ES} = 1.78\left[VP_s\right]$$

Where: $\mathbf{h_{ES}}$ & $\mathbf{VP_s}$ are exactly as defined for Equation **#s 4-14A & 4-14B**, above on this page, and on the previous page, namely, Pages 4-11 & 4-12.

Equation #4-14D:

$$h_{ED} = 0.25\left[VP_d\right]$$

Where: $\mathbf{h_{ED}}$ & $\mathbf{VP_d}$ are exactly as defined for Equation **#s 4-14A & 4-14B**, above on this page, and on the previous page, namely, Pages 4-11 & 4-12.

Calculations Involving the Rotational Speeds of Fans

Equation #4-15:

Equation **#4-15** relates the **Air Discharge Volume** being provided by a fan to its **Rotational Speed**. This is very important whenever one is trying to decide on the relative sizes of the pulleys that will be used [i.e., a pulley on the fan itself, and a second one on the motor that is to serve as the motive force for the fan]. This relationship, as well as each of the five that will follow, assumes that the fan being evaluated may well be used under **different** operating circumstances, but only while handling fluids or gases of the **same** density.

$$\frac{CFM_1}{CFM_2} = \frac{RPM_1}{RPM_2}$$

Where: CFM_i = the **Air Discharge Volume** of the fan when it is operating at the **i**th set of conditions, in cubic feet per minute [cfm]; &

RPM_i = the **i**th operating **Rotational Speed** of the fan, in revolutions per minute [rpm].

Equation #4-16:

Equation **#4-16** relates the **Static Discharge Pressure** of a fan to its **Rotational Speed**. As with Equation **#4-15** above, this relationship assumes that the fan will be evaluated only for gases or fluids of the same density.

$$\frac{SP_{Fan_1}}{SP_{Fan_2}} = \left[\frac{RPM_1}{RPM_2}\right]^2$$

Where: SP_{Fan_i} = the **Static Discharge Pressure** of a fan when it is operating at the **i**th set of conditions; &

RPM_i = is the **i**th operating **Rotational Speed** of the fan, as defined for Equation **#4-15**, above on this page.

Equation #4-17:

Equation **#4-17** relates the motive **Brake Horsepower** required to operate a fan at any of its possible **Rotational Speeds**. As with the two previous equations, this relationship assumes that the fan will be evaluated only for different **Rotational Speed** applications that involve gases and fluids having the same density.

$$\frac{BHP_1}{BHP_2} = \left[\frac{RPM_1}{RPM_2}\right]^3$$

Where: **BHP_i** = the **Brake Horsepower** required for a fan to be operated at the **i**th set of conditions, measured in brake horsepower units [bhp]; &

RPM_i = is the **i**th operating **Rotational Speed** of the fan, as defined for Equation **#4-15**, on the previous page, namely, Page 4-13.

Calculations Involving the Diameters of Fans

Equation #4-18:

Equation **#4-18** relates the **Air Discharge Volume** being produced by a fan to its **Diameter**. This relationship, like the previous three, as well as each of the two that will follow, assumes that the fan being examined will only be evaluated for **different** diameters, while handling gases or fluids of the **same** density.

$$\frac{CFM_1}{CFM_2} = \left[\frac{d_1}{d_2}\right]^3$$

Where: **CFM_i** = the **Air Discharge Volume** of the fan at its **i**th **Diameter**, this **Air Discharge Volume** is measured in cubic feet per minute [cfm]; &

d_i = the **Diameter** of the **i**th Fan, measured in inches.

Equation #4-19:

Equation **#4-19** relates the **Static Discharge Pressure** being developed by a fan to its **Diameter**. This relationship, like the previous four, assumes that the fan being evaluated will only be operated at **different** diameters while handling gases and fluids of the **same** density.

$$\frac{SP_{Fan_1}}{SP_{Fan_2}} = \left[\frac{d_1}{d_2}\right]^2$$

Where: **SP_{Fan_i}** = the **Static Discharge Pressure** of a fan of the **i**th **Diameter**, this **Static Discharge Pressure** is measured in inches of water; &

d_i = the **Diameter** of the **i**th Fan, as defined for Equation **#4-18**, immediately above on this page.

Equation #4-20:

Equation **#4-20** relates the **Brake Horsepower** required to operate Fans of various **Diameters**. This relationship — as was true with the previous five — assumes that the fan under evaluation will only be operated at **different** diameters while handling gases or fluids of the **same** density, and at the same **Fan Discharge Volume**.

$$\frac{BHP_1}{BHP_2} = \left[\frac{d_1}{d_2}\right]^5$$

Where:

BHP_i = the **Brake Horsepower** required for a fan of the ith **Diameter** to deliver some specified, and for this application, identical **Air Discharge Volume**. The **Brake Horsepower** term is always measured in brake horsepower units [bhp]; &

d_i = the **Diameter** of the ith fan, as defined for Equation **#4-18**, on the previous page, namely, Page 4-14.

Calculations Involving Various Other Fan-Related Factors

Equation #4-21:

Equation **#4-21** is commonly known as the *Fan Brake Horsepower Equation.* It combines most of the factors that must be considered in the choice of a fan for any ventilation application.

$$BHP = \frac{1}{63.56}\left[\frac{(CFM)(TP)}{FME}\right] = 0.0157\left[\frac{(CFM)(TP)}{FME}\right]$$

Where:

BHP = the **Brake Horsepower** that will be required for the fan to be able to provide the required performance level, measured in brake horsepower units [bhp];

CFM = the required **Fan Discharge Volume**, measured in cubic feet per minute [cfm];

TP = the **Fan Total Pressure**, measured in inches of water; &

FME = the **Fan Mechanical Efficiency**, which is a dimensionless percentage $\leq$ 100%.

Equation #4-22:

Equation **#4-22** is known as the *Fan Total Pressure Equation.* It is a very simple, straightforward, and widely used relationship.

$$TP_{Fan} = TP_{out} - TP_{in}$$

Where:

TP_{Fan} = the **Fan Total Pressure**, measured in inches of water;

TP_{out} = the **Fan Total Output Pressure**, measured in inches of water; &

TP_{in} = the **Fan Total Input Pressure**, measured in inches of water.

Equation #4-23:

Equation **#4-23** is known as the *Fan Static Pressure Equation*, simply because it defines the overall **Fan Static Pressure**.

$$SP_{Fan} = SP_{out} - SP_{in} - VP_{in}$$

Where:

SP_{Fan} = the **Fan Static Discharge Pressure**, measured in inches of water;

SP_{out} = the **Fan Outlet Static Pressure**, measured in inches of water;

SP_{in} = the **Fan Inlet Static Pressure**, measured in inches of water; &

VP_{in} = the **Fan Inlet Velocity Pressure**, also measured in inches of water.

Calculations Involving Air Flow Balancing at a Duct Junction

Equation #4-24:

Equation **#4-24A**, and the family of relationships that follow it, define completely the process of balancing a duct junction. This set makes up the *Duct Junction Balancing System.* This system is logical in its development, and in the solution path it implies. Basically, this procedure considers the volumetric flow rate in each of two ducts that join to form a single larger duct.

Equation #4-24A:

$$R = \frac{SP_{greater}}{SP_{lesser}}$$

Where:

R = the **Junction Balance Ratio**, which is a dimensionless number;

$SP_{greater}$ = the **Duct Static Pressure** in the duct that carries the greater flow volume into the duct junction, measured in inches of water; &

SP_{lesser} = the **Duct Static Pressure** in the duct carrying the lesser flow volume into the duct junction, also measured in inches of water.

We must now consider Three Different Scenarios:

1. **$R < 1.05$** Consider the duct junction to be balanced.
2. **$R > 1.20$** The duct junction is so very badly imbalanced that the entire ventilation system must be completely redesigned from the ground up.
3. **$1.05 \leq R \leq 1.20$** Increase the volumetric flow rate in the lesser volume branch from Q_{former} to Q_{new}, according to the following relationship, identified below as Equation **#4-24B**:

Equation #4-24B:

$$Q_{new} = Q_{former}\sqrt{\frac{SP_{greater}}{SP_{lesser}}} = Q_{former}\sqrt{R}$$

Where:

Q_{new} = the INCREASED flow rate in the branch in which the lesser volume has been flowing, measured in cubic feet/minute [cfm];

$\mathbf{Q_{former}}$ = the STARTING flow rate in the branch in which the lesser volume has been flowing, measured in the same units as $\mathbf{Q_{new}}$.

$\mathbf{SP_{greater}}$ = the **Duct Static Pressure** in the duct that carries the greater flow volume into the duct junction, as defined on the previous page, namely, Page 4-17, for Equation **#4-24A**, measured in inches of water;

$\mathbf{SP_{lesser}}$ = the **Duct Static Pressure** in the duct carrying the lesser flow volume into the duct junction, also as defined on the previous page, namely, Page 4-17, for Equation **#4-24A**, and also measured in inches of water; &

R = the **Junction Balance Ratio**, which is a dimensionless number, as defined on the previous page, namely, Page 4-17, for Equation **#4-24A**, and also measured in inches of water.

Calculations Involving Dilution Ventilation

<u>Equation #4-25</u>:

Equation **#4-25** provides a single relationship that makes it possible to calculate the steady state, **Equilibrium Concentration** that would be produced in a room — or any enclosed space for which the overall volume can be determined — by the complete evaporation of some specific volume of any identifiable volatile solvent. This is probably the single most complicated equation with which any professional will ever have to work — its principal saving grace is that it works! This relationship assumes Normal Temperature and Pressure [NTP].

$$C = 6.24 \times 10^{7}\left[\frac{V_s \rho T}{P_{atm} V_{room} (MW_s)}\right]$$

Where:

C = the **Equilibrium Concentration** of the volatile solvent that would be produced in the room by the evaporation of the known volume of solvent, measured in ppm(vol);

$\mathbf{V_s}$ = the liquid **Volume** of the solvent that has evaporated, measured in milliliters [ml];

ρ = the **Density** of the solvent, measured in grams per cubic centimeter [gms/cm^3];

T = the absolute **Temperature** in the room, measured in degrees Kelvin [K];

$\mathbf{MW_s}$ = the **Molecular Weight** of the solvent;

P_{atm} = the **Ambient Barometric Pressure** that is prevailing in the room, measured in millimeters of mercury [mm Hg];

V_{room} = the **Volume** of the Room, measured in liters; &

6.24×10^7 = the **Proportionality Constant** that makes this Equation valid under NTP conditions.

Equation #4-26:

Equation **#4-26** is known as the *Basic Room Purge Equation.* It provides the necessary relationship for determining the time required to reduce a known initial high level concentration of any vapor — existing in a defined closed space or room — to a more acceptable ending lower level concentration. The relationship has been provided in two formats: one employing natural logarithms and the other, common logarithms.

$$D_t = \left[\frac{V}{Q}\right] ln\left[\frac{C_{initial}}{C_{ending}}\right] \quad \text{— Natural Log Format}$$

$$D_t = 2.3\left[\frac{V}{Q}\right] log\left[\frac{C_{initial}}{C_{ending}}\right] \quad \text{— Common Log Format}$$

Where:

D_t = the **Time Required** to reduce the vapor concentration in the closed space or room, as required, measured in minutes;

V = the **Volume** of the closed space or room, measured in cubic feet [ft^3];

Q = the **Ventilation Rate** at which the closed space or room will be purged by whatever air handling system is available for that purpose, measured in cubic feet per minute [cfm];

$C_{initial}$ = the **Initial High Level Concentration** of the vapor in the ambient air of the closed space or room, which concentration is to be reduced — by purging at **Q** cfm — to a more acceptable ending lower level concentration, measured in ppm(vol); &

C_{ending} = the desired **Ending Lower Level Concentration** of the vapor that is to result from the purging effort in the closed space or room, also measured in ppm(vol).

Equation #4-27:

Equation **#4-27** is known as the *Purge-Dilution Equation*. It is the most basic and fundamental relationship available for determining the various parameters associated with reducing the concentration of the vapor of any volatile material in a closed space or room.

$$C = C_0 e^{-\left[V_{removed}/V_{room}\right]}$$

Where:

C = the **Ending Concentration** of the vapor in the closed space or room, which concentration, measured in ppm(vol), resulted from the purging activities;

C_o = the **Initial Concentration** of the vapor in the closed space or room that is to be reduced by purging, also measured in ppm(vol);

$V_{removed}$ = the **Air Volume** that has been withdrawn from the closed space or room, measured in any suitable volumetric units, usually in cubic feet [ft^3]; &

V_{room} = the **Volume** of the room, measured in the same volumetric units as $V_{removed}$, which is usually in cubic feet [ft^3].

VENTILATION PROBLEM SET

Problem #4.1:

What is the volumetric flow rate, in cfm, that exists in a 10-inch diameter duct in which air has been determined to be flowing with a duct velocity of 2,500 fpm?

Applicable Definitions:	Duct	Page 4-2
	Volumetric Flow Rate	Page 4-2
	Duct Velocity	Page 4-2
Applicable Formula:	Equation **#4-1**	Page 4-6
Solution to this Problem:	Page 4-61	

Problem Workspace

Problem #4.2:

The supply duct in the HVAC System in a 40-story office building has a volumetric flow rate of 25,135 cfm. If air in this duct is flowing at a velocity of 8,000 fpm, what is the duct's diameter?

Applicable Definitions:	Duct	Page 4-2
	Volumetric Flow Rate	Page 4-2
	Duct Velocity	Page 4-2
Applicable Formula:	Equation **#4-1**	Page 4-6
Solution to this Problem:	Page 4-61	

Workspace for Problem **#4.2**

Problem #4.3:

Assuming that the density of the air flowing in the supply duct listed in Problem **#4.2** is 0.075 lbs/ft^3, what is the velocity pressure, in inches of water, that will exist in this duct?

Applicable Definitions:	Duct	Page 4-2
	Velocity Pressure	Page 4-3
Data from:	Problem **#4.2**	Page 4-21
Applicable Formula:	Equation **#4-2**	Page 4-6
Solution to this Problem:	Pages 4-61 & 4-62	

Problem Workspace

Problem #4.4:

What is the duct velocity of the air flowing under a velocity pressure of 2.32 inches of water?

Applicable Definitions:	Duct	Page 4-2
	Duct Velocity	Page 4-2
	Velocity Pressure	Page 4-3
Applicable Formula:	Equation **#4-3**	Page 4-7
Solution to this Problem:	Page 4-62	

Problem Workspace

Problem #4.5:

An Industrial Hygienist has measured the velocity pressure in an 8-inch duct, and found it to be 0.45 inches of water. What is the velocity, in fpm, of the gas that is flowing in that duct?

Applicable Definitions:	Duct	Page 4-2
	Duct Velocity	Page 4-2
	Velocity Pressure	Page 4-3
Applicable Formula:	Equation **#4-3**	Page 4-7
Solution to this Problem:	Page 4-62	

Problem Workspace

Problem #4.6:

A recently graduated Industrial Hygienist was assigned to measure the velocity pressure in a duct carrying air at a velocity of 25 fps. He reported a value of 0.23 inches of water. Comment on his skill at measuring velocity pressures in a duct.

Applicable Definitions:	Duct	Page 4-2
	Duct Velocity	Page 4-2
	Velocity Pressure	Page 4-3
Applicable Formula:	Equation **#4-3**	Page 4-7
Solution to this Problem:	Pages 4-62 & 4-63	

Problem Workspace

Problem #4.7:

The total pressure at a duct exhaust point was measured to be 0.41 inches of water. If the static pressure in this duct was 0.01 inches of water, what would you expect the velocity pressure to be?

Applicable Definitions:	Duct	Page 4-2
	Static Pressure	Page 4-3
	Velocity Pressure	Page 4-3
	Total Pressure	Page 4-3
Applicable Formula:	Equation **#4-5**	Pages 4-7 & 4-8
Solution to this Problem:	Page 4-63	

Workspace for Problem **#4.7**

Problem #4.8:

If a velocity pressure of 0.55 inches of water, and a total pressure of 0.24 inches of water were measured in a duct, at the inlet to the fan for that ventilation system, what would you expect the static pressure to be?

Applicable Definitions:	Duct	Page 4-2
	Fan	Page 4-2
	Static Pressure	Page 4-3
	Velocity Pressure	Page 4-3
	Total Pressure	Page 4-3
Applicable Formula:	Equation **#4-5**	Pages 4-7 & 4-8
Solution to this Problem:	Page 4-63	

Problem Workspace

Problem #4.9:

If the fan from Problem **#4.8** was located 220 feet downstream from its main hood inlet, into which all the air that ultimately passes through the fan is flowing, and if the frictional losses in this duct have been determined to be 0.02 inches of water per each 20 lineal feet of duct, and finally, if the velocity pressure, as well as the duct velocity, at the fan inlet and at the hood entry are identical, what would you expect the static pressure to be at the hood inlet?

Applicable Definitions:	Hood	Page 4-1
	Duct	Page 4-2
	Fan	Page 4-2
	Static Pressure	Page 4-3
	Velocity Pressure	Page 4-3
	Total Pressure	Page 4-3
Data from:	Problem **#4.8**	Page 4-25
Applicable Formula:	Equation **#4-6**	Page 4-8
Solution to this Problem:	Pages 4-63 & 4-64	

Problem Workspace

Problem #4.10:

An engineer measured the static pressure in a 10-inch diameter duct at two points separated from each other by 75 feet. He found the static pressure at the upstream point to be –0.35 inches of water, and at the downstream point, –0.41 inches of water. What are the frictional losses in this duct, expressed in inches of water per each 10 lineal feet of duct?

Applicable Definitions:	Duct	Page 4-2
	Static Pressure	Page 4-3
Applicable Formula:	Equation **#4-6**	Page 4-8
Solution to this Problem:	Page 4-64	

Problem Workspace

Problem #4.11:

The source of an irritating vapor is located 2.5 feet from the face of an unflanged rectangular hood opening that measures 12 inches by 36 inches. If a capture velocity of 100 fpm is necessary to completely remove this vapor from the room, what must the minimum volumetric flow rate, in cfm, be in order to be sure that vapor capture occurs?

Applicable Definitions:	Hood	Page 4-1
	Capture Velocity	Page 4-1
	Volumetric Flow Rate	Page 4-2
Applicable Formula:	Equation **#4-7**	Pages 4-8 & 4-9
Solution to this Problem:	Pages 4-64 & 4-65	

Workspace for Problem **#4.11**

Problem #4.12:

If the hood opening in Problem **#4.11** were to have been flanged, and if the same hood volumetric flow rate were maintained, how far from the hood opening could the irritating vapor source now be located and still guarantee that adequate vapor capture would continue to occur?

Applicable Definitions:	Hood	Page 4-1
	Flange	Page 4-1
	Capture Velocity	Page 4-1
	Volumetric Flow Rate	Page 4-2
Data from:	Problem **#4.11**	Page 4-27
Applicable Formula:	Equation **#4-8**	Page 4-9
Solution to this Problem:	Page 4-65	

Problem Workspace

Problem Workspace Continued on the Following Page

Continuation of Workspace for **Problem #4.12**

Problem #4.13:

A source of respirable dust particles must be mitigated by the use of an exhaust hood. The hood entry slot is to be 3 inches high by 12 inches wide. It is felt that the hood must be flanged, since the source of respirable particles can be no closer than 3.0 feet from the face of this hood slot. The air handler in this exhaust system will draw a total of 8,500 cfm. What will the centerline velocity of this hood be when it operates under these conditions? If the required capture velocity for the respirable particulate material is 2 fps, will this hood system be able to operate satisfactorily and successfully capture the respirable dust particles?

Applicable Definitions:	Hood	Page 4-1
	Flange	Page 4-1
	Capture Velocity	Page 4-1
	Volumetric Flow Rate	Page 4-2
	Air Handler	Page 4-2
Applicable Formula:	Equation **#4-8**	Page 4-9
Solution to this Problem:	Pages 4-65 & 4-66	

Problem Workspace

Problem #4.14:

The hood entry losses for a simple hood slot were found to be 8.1 inches of water. If the velocity pressure in the duct is 15.2 inches of water, what would you expect the hood static pressure, in inches of water, to be?

Applicable Definitions:	Hood	Page 4-1
	Static Pressure	Page 4-3
	Velocity Pressure	Page 4-3
Applicable Formula:	Equation **#4-9**	Page 4-9
Solution to this Problem:	Page 4-66	

Problem Workspace

Problem #4.15:

What will be the Coefficient of Entry for the hood in Problem **#4.14**?

Applicable Definitions:	Hood	Page 4-1
	Coefficient of Entry	Page 4-2
Data from:	Problem **#4.14**	Page 4-30
Applicable Formulae:	Equation **#4-10**	Pages 4-9 & 4-10
Solution to this Problem:	Page 4-67	

Problem Workspace

Problem #4.16:

The static pressure in a hood under investigation was measured to be 12.5 inches of water. The total pressure in the duct leading away from this hood, as well as the duct static pressure, were measured, respectively, to be 6.2 and –1.2 inches of water. What are the hood entry losses?

Applicable Definitions:	Hood	Page 4-1
	Duct	Page 4-2
	Entry Loss	Page 4-2
	Static Pressure	Page 4-3
	Total Pressure	Page 4-3
Applicable Formulae:	Equation **#4-5**	Pages 4-7 & 4-8
	Equation **#4-9**	Page 4-9
Solution to this Problem:	Page 4-67	

Problem Workspace

Problem #4.17:

What will be the Coefficient of Entry for the hood in Problem **#4.16**?

Applicable Definitions:	Hood	Page 4-1
	Coefficient of Entry	Page 4-2
	Entry Loss	Page 4-2
	Static Pressure	Page 4-3
	Total Pressure	Page 4-3
Data from:	Problem **#4.16**	Page 4-31
Applicable Formula:	Equation **#4-10**	Pages 4-9 & 4-10
Solution to this Problem:	Page 4-67	

Workspace for Problem **#4.17**

Problem #4.18:

An Engineer has measured a static pressure of –1.75 inches of water in a 4-inch diameter duct, 12 inches downstream from its unflanged opening. Page 4-12 of the 19th Edition of the ACGIH Volume, ***Industrial Ventilation***, identifies the Coefficient of Entry for such a hood opening to be 0.72. Calculate the Volumetric Flow Rate, in cfm, for this system.

Applicable Definitions:	Hood	Page 4-1
	Coefficient of Entry	Page 4-2
	Duct	Page 4-2
	Volumetric Flow Rate	Page 4-2
	Static Pressure	Page 4-3
Applicable Formula:	Equation **#4-11**	Page 4-10
Solution to this Problem:	Pages 4-67 & 4-68	

Problem Workspace

Problem #4.19:

For the hood system described in Problem **#4.18**, calculate the duct velocity, in fpm.

Applicable Definitions:	Hood	Page 4-1
	Duct	Page 4-2
	Duct Velocity	Page 4-2
Data from:	Problem **#4.18**	Page 4-32
Applicable Formula:	Equation **#4-1**	Page 4-6
Solution to this Problem:	Page 4-68	

Problem Workspace

Problem #4.20:

Finally, for the hood system described in Problem **#4.18**, what will be the average duct velocity pressure, in inches of water?

Applicable Definitions:	Hood	Page 4-1
	Velocity Pressure	Page 4-3
Data from:	Problem **#4.18**	Page 4-32
Applicable Formula:	Equation **#4-3**	Page 4-7
Solution to this Problem:	Pages 4-68 & 4-69	

Workspace for Problem **#4.20**

Problem #4.21:

What will be the total pressure in the exit duct of a hood that has the following operating characteristics?

Volumetric Flow Rate:	3,000 cfm
Hood Opening:	unflanged & square, measuring 6.0 inches on a side
Duct Static Pressure:	–4.5 inches of water

Applicable Definitions:	Hood	Page 4-1
	Duct	Page 4-2
	Volumetric Flow Rate	Page 4-2
	Static Pressure	Page 4-3
	Total Pressure	Page 4-3
Applicable Formulae:	Equation **#4-5**	Pages 4-7 & 4-8
	Equation **#4-11**	Page 4-10
Solution to this Problem:	Page 4-69	

Problem Workspace

Problem Workspace Continued on the Following Page

Continuation of Workspace for Problem **#4.21**

Problem #4.22:

The Coefficient of Entry for a particular hood was measured to be 0.85. If the velocity pressure in this hood is 21.0 inches of water, what will be the hood entry losses, in inches of water?

Applicable Definitions:	Hood	Page 4-1
	Coefficient of Entry	Page 4-2
	Entry Loss	Page 4-2
	Velocity Pressure	Page 4-3
Applicable Formula:	Equation **#4-12**	Page 4-10
Solution to this Problem:	Pages 4-69 & 4-70	

Problem Workspace

Problem #4.23:

Hood entry losses for a hood with a velocity pressure of 12.3 inches of water were found to be 6.5 inches of water. What is the Coefficient of Entry for this hood?

Applicable Definitions:	Hood	Page 4-1
	Coefficient of Entry	Page 4-2
	Entry Loss	Page 4-2
	Velocity Pressure	Page 4-3
Applicable Formula:	Equation **#4-12**	Page 4-10
Solution to this Problem:	Page 4-70	

Problem Workspace

Problem #4.24:

What is the hood entry loss factor for the hood in Problem **#4.23**?

Applicable Definitions:	Hood	Page 4-1
	Entry Loss	Page 4-2
	Velocity Pressure	Page 4-3
Data from:	Problem **#4.23**	Page 4-36
Applicable Formula:	Equation **#4-13**	Page 4-11
Solution to this Problem:	Page 4-70	

Workspace for Problem **#4.24**

Problem #4.25:

A compound hood was measured and found to have a slot velocity of 3,400 fpm, and a duct velocity of 5,600 fpm. What is the hood static pressure, in inches of water, for this compound hood?

Applicable Definitions:	Hood	Page 4-1
	Slot Velocity	Page 4-1
	Duct Velocity	Page 4-2
	Static Pressure	Page 4-3
Applicable Formulae:	Equation **#4-3**	Page 4-7
	Equation **#4-14A**	Pages 4-11 & 4-12
	Equation **#4-14C**	Pages 4-11 & 4-12
	Equation **#4-14D**	Pages 4-11 & 4-12
Solution to this Problem:	Page 4-71	

Problem Workspace

Problem Workspace Continued on the Following Page

Continuation of Workspace for Problem **#4.25**

Problem #4.26:

The exhaust duct of a compound hood was measured and found to have a total pressure of 1.3 inches of water, and a static pressure of –1.9 inches of water. The slot velocity of this hood system was measured and found to be 7,500 fpm. What is the hood static pressure, in inches of water, for this compound hood?

Applicable Definitions:	Hood	Page 4-1
	Slot Velocity	Page 4-1
	Static Pressure	Page 4-3
	Total Pressure	Page 4-3
Applicable Formulae:	Equation **#4-3**	Page 4-7
	Equation **#4-5**	Pages 4-7 & 4-8
	Equation **#4-14B**	Pages 4-11 & 4-12
	Equation **#4-14C**	Pages 4-11 & 4-12
	Equation **#4-14D**	Pages 4-11 & 4-12
Solution to this Problem:	Pages 4-72 & 4-73	

Workspace for Problem **#4.26**

Problem #4.27:

It was determined that the air handler that serviced a section of a hospital was not capable of providing an adequate flow of air. To correct this situation, it was decided to change its drive pulley in order to increase its rotational speed from 1,450 rpm to 1,750 rpm. If its adjusted discharge volume is now 10,500 cfm, what was its discharge volume before the pulley was changed?

Applicable Definitions:	Volumetric Flow Rate	Page 4-2
	Air Handler	Page 4-2
Applicable Formula:	Equation **#4-15**	Page 4-13
Solution to this Problem:	Page 4-73	

Workspace for Problem **#4.27**

Problem #4.28:

If the static discharge pressure of the fan in Problem **#4.27** was 4.6 inches of water prior to the change of pulleys, what is its static discharge pressure now?

Applicable Definitions:	Fan	Page 4-2
	Static Pressure	Page 4-3
Data from:	Problem **#4.27**	Page 4-39
Applicable Formula:	Equation **#4-16**	Page 4-13
Solution to this Problem:	Page 4-73	

Problem Workspace

Problem #4.29:

It was determined that the motor that drives the air handler of Problem **#4.27** also had to be changed in order to produce the higher fan rotational speed. If the previous motor delivered 15 bhp, what must be the brake horsepower rating of the new drive motor?

Applicable Definitions:	Air Handler	Page 4-2
Data from:	Problem **#4-27**	Page 4-39
Applicable Formula:	Equation **#4-17**	Pages 4-13 & 4-14
Solution to this Problem:	Pages 4-73 & 4-74	

Problem Workspace

Problem #4.30:

A ventilation fan that had been initially selected to deliver 25,000 cfm against 6.25 inches of water static pressure successfully did so at 1,200 rpm. Its brake horsepower rating was 50 bhp. Because new facilities are scheduled to be installed in the area serviced by this ventilation system, it has become necessary to increase the volume output of this fan by 40%. At what new rotational speed, in rpm, must the fan now be operated?

Applicable Definitions:	Fan	Page 4-2
	Brake Horsepower	Page 4-3
	Static Pressure	Page 4-3
Applicable Formula:	Equation **#4-15**	Page 4-13
Solution to this Problem:	Page 4-74	

Problem Workspace

Problem #4.31:

What will be the new static discharge pressure, in inches of water, for the fan in Problem **#4.30**?

Applicable Definitions:	Fan	Page 4-2
	Static Pressure	Page 4-3
Data from:	Problem **#4.30**	Page 4-41
Applicable Formula:	Equation **#4-16**	Page 4-13
Solution to this Problem:	Page 4-74	

Problem Workspace

Problem #4.32:

What new brake horsepower rating, in bhp, will be required for the motor that is to drive the fan in Problem **#4.30**?

Applicable Definitions:	Fan	Page 4-2
	Brake Horsepower	Page 4-3
Data from:	Problem **#4.30**	Page 4-41
Applicable Formula:	Equation **#4-17**	Pages 4-13 & 4-14
Solution to this Problem:	Page 4-74	

Problem Workspace

Problem #4.33:

The Industrial Hygienist of a large manufacturing company has recommended that one of the three exhaust ventilation systems in use at the company's West Coast plant be replaced with a new system, sized so as to be capable of handling the much larger flow volume requirements of the anticipated, expanded manufacturing capability soon to be implemented at this plant. This existing system utilizes a 6.0-inch diameter fan, and 6.0-inch diameter ducting throughout. The fan has an established discharge capacity of 28,500 cfm. The Industrial Hygienist has calculated the new discharge volume requirements of the soon to be modified ventilation system to be 67,500 cfm. What duct and fan size is this Industrial Hygienist likely to recommend for this new expanded system?

Applicable Definitions:	Duct	Page 4-2
	Volumetric Flow Rate	Page 4-2
	Fan	Page 4-2
Applicable Formula:	Equation **#4-18**	Page 4-14
Solution to this Problem:	Page 4-75	

Problem Workspace

Problem #4.34:

If the fan static discharge pressure for the existing ventilation system described in Problem **#4.33** is 7.0 inches of water, what will be the static discharge pressure of the new system?

Applicable Definitions:	Fan	Page 4-2
	Static Pressure	Page 4-3
Data from:	Problem **#4-33**	Page 4-43
Applicable Formula:	Equation **#4-19**	Page 4-14
Solution to this Problem:	Page 4-75	

Workspace for Problem **#4.34**

Problem #4.35:

The Industrial Hygienist identified in Problem **#4.33** has calculated that the new expanded system will require a fan motor of 150 bhp. What is the brake horsepower of the existing system?

Applicable Definitions:	Fan	Page 4-2
	Brake Horsepower	Page 4-3
Data from:	Problem **#4.33**	Page 4-43
Applicable Formula:	Equation **#4-20**	Page 4-15
Solution to this Problem:	Pages 4-75 & 4-76	

Problem Workspace

Problem #4.36:

A large HVAC system employing a 15.0-inch diameter fan is characterized by an air velocity of 150 fps in its 15.0-inch diameter ducts. If the fan and duct diameters are each increased by 3.0 inches, what will the new duct velocity become?

Applicable Definitions:	Duct	Page 4-2
	Duct Velocity	Page 4-2
	Fan	Page 4-2
Applicable Formulae:	Equation **#4-1**	Page 4-6
	Equation **#4-18**	Page 4-14
Solution to this Problem:	Pages 4-75 & 4-76	

Problem Workspace

Problem Workspace Continued on the Following Page

Continuation of Workspace for Problem **#4.36**

Problem #4.37:

The air handler in a ventilation system employs a fan and ducts, each of 8.0 inches diameter. This system provides a discharge air velocity of 7,525 fpm, and a discharge total pressure of 9.8 inches of water. If the fan in this system is increased in diameter by 50%, what would you forecast the fan static discharge pressure for the new system to be?

Applicable Definitions:	Duct	Page 4-2
	Duct Velocity	Page 4-2
	Fan / Air Handler	Page 4-2
	Static Pressure	Page 4-3
	Total Pressure	Page 4-3
Applicable Formulae:	Equation **#4-3**	Page 4-7
	Equation **#4-5**	Pages 4-7 & 4-8
	Equation **#4-19**	Page 4-14
Solution to this Problem:	Pages 4-77 & 4-78	

Workspace for Problem **#4.37**

Problem #4.38:

A hood exhaust system uses a 9.0-inch diameter fan and correspondingly sized ducts. Its fan motor develops 75 bhp. If the fan is decreased in size, a smaller motor could be used. If there is an available 10-bhp motor, by how much must the diameters of the fan be decreased and have it still be able to satisfy the requirements of this exhaust ventilation system?

Applicable Definitions:	Duct	Page 4-2
	Fan	Page 4-2
	Brake Horsepower	Page 4-3
Applicable Formula:	Equation **#4-20**	Page 4-15
Solution to this Problem:	Page 4-78	

Problem Workspace

Problem #4.39:

What is the brake horsepower, in bhp, of a fan that delivers 22,500 cfm at a mechanical efficiency of 65%, and at a fan total pressure of 3.4 inches of water?

Applicable Definitions:	Fan	Page 4-2
	Brake Horsepower	Page 4-3
	Total Pressure	Page 4-3
Applicable Formula:	Equation **#4-21**	Page 4-15
Solution to this Problem:	Page 4-78	

Problem Workspace

Problem #4.40:

What is the mechanical efficiency of a 20-bhp fan that delivers air at a velocity of 6,600 fpm into a 12.0-inch diameter duct at a fan total pressure of 15.0 inches of water?

Applicable Definitions:	Duct	Page 4-2
	Fan	Page 4-2
Applicable Formulae:	Equation **#4-1**	Page 4-6
	Equation **#4-21**	Page 4-15
Solution to this Problem:	Pages 4-78 & 4-79	

Problem Workspace

Problem #4.41:

The inlet duct of a fan had a combined entry and friction loss of 4.5 inches of water, and a velocity pressure of 4.2 inches of water. The fan discharge duct has friction losses of 4.8 inches of water, and a velocity pressure of 5.6 inches of water. Calculate the fan total pressure, in inches of water.

Applicable Definitions:	Duct	Page 4-2
	Fan	Page 4-2
	Velocity Pressure	Page 4-3
	Total Pressure	Page 4-3
Applicable Formulae:	Equation **#4-4**	Page 4-7
	Equation **#4-5**	Pages 4-7 & 4-8
	Equation **#4-6**	Page 4-8
	Equation **#4-22**	Page 4-16
Solution to this Problem:	Pages 4-79 & 4-80	

Problem Workspace

Problem #4.42:

The velocity pressures at both the fan inlet and fan outlet were measured at 13.4 inches of water. The static pressure at the fan inlet was 19.6 inches of water less than the static pressure at the fan outlet, and the fan total output pressure was measured at 25.7 inches of water. What is the fan total pressure?

Applicable Definitions:	Duct	Page 4-2
	Fan	Page 4-2
	Static Pressure	Page 4-3
	Velocity Pressure	Page 4-3
	Total Pressure	Page 4-3
Applicable Formulae:	Equation **#4-5**	Pages 4-7 & 4-8
	Equation **#4-22**	Page 4-16
Solution to this Problem:	Page 4-80	

Problem Workspace

Problem #4.43:

The duct leading into a ventilation fan has a velocity pressure of 6.4 inches of water, and a static pressure of –8.3 inches of water. The fan discharge duct has been measured to have a velocity pressure of 8.2 inches of water, and a total pressure of 16.3 inches of water. Calculate the fan static discharge pressure.

Applicable Definitions:	Duct	Page 4-2
	Fan	Page 4-2
	Static Pressure	Page 4-3
	Velocity Pressure	Page 4-3
	Total Pressure	Page 4-3
Applicable Formulae:	Equation **#4-5**	Pages 4-7 & 4-8
	Equation **#4-23**	Page 4-16
Solution to this Problem:	Pages 4-80 & 4-81	

Problem Workspace

Problem #4.44:

The Fan Static Discharge Pressure of an HVAC fan was determined to be 21.5 inches of water. If the Static Pressures at the fan's inlet and outlet were measured, respectively, to be –12.7 inches of water, and 14.4 inches of water, what was the Velocity Pressure at the fan's inlet?

Applicable Definitions:	Duct	Page 4-2
	Fan	Page 4-2
	Static Pressure	Page 4-3
	Velocity Pressure	Page 4-3
Applicable Formula:	Equation **#4-23**	Page 4-16
Solution to this Problem:	Page 4-81	

Workspace for Problem **#4.44**

Problem #4.45:

Consider the following tabulation of data about the three ducts, AD, BD, & CD, that meet to form duct DD. What must be done to balance this three-into-one duct junction?

Duct Designation	Q, in cfm	V, in fpm	SP, in inches of water
AD	520	3,690	1.26
BD	720	3,590	1.39
CD	950	2,580	1.44

Applicable Definitions:	Duct	Page 4-2
	Volumetric Flow Rate	Page 4-2
Applicable Formulae:	Equation **#4-24A**	Page 4-17
	Equation **#4-24B**	Pages 4-17 & 4-18
Solution to this Problem:	Pages 4-81 & 4-82	

Problem Workspace

Problem Workspace Continued on the Following Page

Continuation of Workspace for Problem **#4.45**

Problem #4.46:

A 1.50 liter glass bottle full of acetone [MW = 58.08, Vapor Pressure = 226 mm Hg @ 77°F, & density = 0.791 gms/cc] falls, during an earthquake, from its position on a shelf, breaking when it hits the floor of a room that has a volume of 1,650 ft^3. If:

(1) the conditions in the room are accurately characterized as being at NTP;

(2) the room is tightly closed when, and after, this earthquake has occurred; &

(3) all the acetone evaporates,

what will the ultimate ambient concentration of acetone be in the room? Is it reasonable to assume that all the acetone will evaporate?

Applicable Definitions:	Normal Temperature & Pressure	Page 1-14
	Duct	Page 4-2
	Fan	Page 4-2
	Vapor Pressure	Page 4-4
	Volumetric Flow Rate	Page 4-2
Applicable Formulae:	Equation **#1-1**	Page 1-16
	Equation **#1-16**	Pages 1-22 & 1-23
	Equation **#4-25**	Pages 4-18 & 4-19
Solution to this Problem:	Pages 4-82 & 4-83	

Problem Workspace

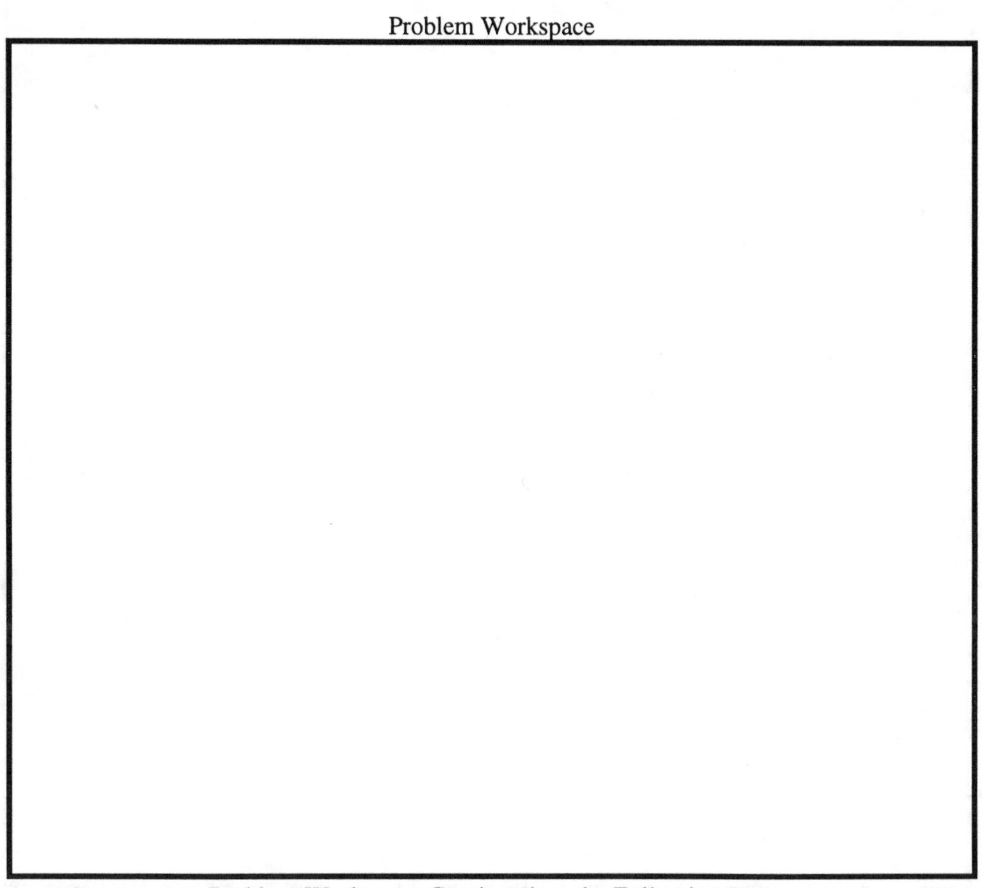

Problem Workspace Continued on the Following Page

Continuation of Workspace for Problem **#4.46**

Problem #4.47:

If the goal is to ventilate the room of Problem **#4.46** until the acetone concentration is at or below the TLV-TWA concentration, which for acetone is 750 ppm(vol), and if the room's ventilation system has a flow volume capacity of 500 cfm, how long will it take to reduce the acetone concentration to the required level?

Applicable Definitions:	Duct	Page 4-2
	Fan	Page 4-2
	Volumetric Flow Rate	Page 4-2
Data from:	Problem **#4.46**	Page 4-55
Applicable Formula:	Equation **#4-26**	Page 4-19
Solution to this Problem:	Page 4-83	

Problem Workspace

Problem #4.48:

After a long day's production run, a Tennessee Moonshiner, who is also a Certified Industrial Hygienist, has filled the bathtub in his 6-foot high, right circular cylindrical redwood tank/storeroom with one batch [one batch = 100 gal] of recently distilled "White Lightning", or pure ethanol. On the following morning, he discovers that exactly 1 gallon of his product has evaporated into the otherwise completely empty and closed tank/storeroom. Ignoring the volume of the bathtub and its remaining contents, what is the diameter, in feet, of this tank/storeroom? You may find it useful to know that the molecular weight of ethanol is 46.07 amu; its density is 0.789 gms/cc; its vapor pressure is 44.3 mm Hg at NTP, and its formula is CH_3-CH_2-OH. You should also be aware that Moonshiners always keep their storerooms at NTP.

Applicable Definitions:	Normal Temperature & Pressure	Page 1-14
	Duct	Page 4-2
	Fan	Page 4-2
	Vapor Pressure	Page 4-4
Applicable Formulae:	Equation **#1-1**	Page 1-16
	Equation **#1-16**	Pages 1-22 & 1-23
	Equation **#4-25**	Pages 4-18 & 4-19
Solution to this Problem:	Pages 4-83 through 4-85	

Workspace for Problem **#4.48**

Problem Workspace Continued on the Following Page

Continuation of Workspace for Problem **#4.48**

Problem #4.49:

The CIH of Problem **#4.48** is very careful to obey all the OSHA requirements that relate to maintaining a safe workplace environment, although he does ignore most other laws and statutes. Because of this, he wants to ventilate his vapor filled redwood cylindrical tank/storeroom until the ethanol concentration is at or below its OSHA established PEL-TWA of 1,000 ppm. If his undersized ventilation fan has a capacity of only 25 cfm, for what time period will this entrepreneurial CIH have to ventilate his tank/storeroom to achieve his goal, assuming that he first covers the bathtub so that no more evaporation occurs?

Applicable Definitions:	Normal Temperature & Pressure	Page 1-14
	Duct	Page 4-2
	Fan	Page 4-2
	Vapor Pressure	Page 4-4
Data from:	Problem **#4.48**	Page 4-57
Applicable Formula:	Equation **#4-26**	Page 4-19
Solution to this Problem:	Page 4-85	

Problem Workspace

Problem #4.50:

In general, how many "room volumes" of air must be withdrawn from — purged from — a room in order to reduce the concentration of any volatile substance in the ambient air of that room by 90%? by 99%?

Applicable Formula:	Equation **#4-27**	Pages 4-19 & 4-20
Solution to this Problem:	Pages 4-85 & 4-86	

Problem Workspace

SOLUTIONS TO THE VENTILATION PROBLEM SET

Problem #4.1:

To solve this problem, we must use what is probably the most basic relationship in the entire area of ventilation, namely, Equation **#4-1**, from Page 4-6, thus:

$$Q = AV \qquad \text{[Eqn. \#4-1]}$$

We must first transpose this relationship into a format that better suits the data provided in this problem, namely, into a format that will accommodate a duct diameter measured in "inches" rather than "feet", although the duct velocity has been provided in "fpm":

$$Q = \frac{\pi d^2 V}{4} \qquad \text{for a duct diameter expressed in feet}$$

$$Q = \frac{\pi d^2 V}{576} \qquad \text{for a duct diameter expressed in inches}$$

$$Q = \pi\frac{(10)^2(2,500)}{576} = \pi\frac{(100)(2,500)}{576} = \pi\left(\frac{250,000}{576}\right)$$

$$Q = \frac{785,398.16}{576} = 1,363.5$$

∴ The volumetric flow rate in this duct is ~ 1,364 cfm.

Problem #4.2:

This problem uses the same relationship as did Problem **#4.1**, namely, Equation **#4-1**, from Page 4-6; and in this case, too, we will use the transposed form that provides for having the duct dimensions in "inches":

$$Q = AV \qquad \& \qquad Q = \frac{\pi d^2 V}{576} \qquad \text{[Eqn. \#4-1]}$$

Now, if we transpose to solve for the diameter, d, we get:

$$d = \sqrt{\frac{576Q}{\pi V}}$$

$$d = \sqrt{\frac{(576)(25,135)}{\pi(8,000)}} = \sqrt{\frac{14,477,760}{25,132.74}} = \sqrt{576.05} = 24.00$$

∴ The diameter of this HVAC duct is ~ 24 inches.

Problem #4.3:

This extension to Problem **#4.2** requires the use of Equation **#4-2**, from Page 4-6, to develop its solution:

$$V = 1{,}096\sqrt{\frac{VP}{\rho}} \qquad \text{[Eqn. \#4-2]}$$

Transposing in order to solve directly for the velocity pressure, VP, we get:

$$VP = \frac{V^2\rho}{(1{,}096)^2}$$

$$VP = \frac{(8{,}000)^2(0.075)}{(1{,}096)^2} = \frac{4{,}800{,}000}{1{,}201{,}216} = 4.00$$

∴ The velocity pressure in this 24-inch diameter HVAC duct is ~ 4.0 inches of water.

Problem #4.4:

To solve this problem, we must use Equation **#4-3**, from Page 4-7:

$$V = 4{,}005\sqrt{VP} \qquad \text{[Eqn. \#4-3]}$$

$$V = 4{,}005\sqrt{2.32} = (4{,}005)(1.523) = 6{,}100.2$$

∴ The velocity in this duct is ~ 6,100 fpm.

Problem #4.5:

This problem, too, requires the use of Equation **#4-3**, from Page 4-7:

$$V = 4{,}005\sqrt{VP} \qquad \text{[Eqn. \#4-3]}$$

$$V = 4{,}005\sqrt{0.45} = (4{,}005)(0.671) = 2{,}686.6$$

∴ The velocity in this 8-inch diameter duct is ~ 2,687 fpm.

Problem #4.6:

This problem also requires the use of Equation **#4-3**, from Page 4-7:

$$V = 4{,}005\sqrt{VP} \qquad \text{[Eqn. \#4-3]}$$

We must first convert this relationship so that it can handle a duct velocity, which has been provided in the problem statement in units of "ft/sec", rather than the more conventional "fpm"; the modified form of Equation **#4-3**, set up to solve for the velocity pressure, VP, with the velocity in units of "ft/sec" is shown below:

$$VP = \left[\frac{V}{4{,}005}\right]^2 = \left[\frac{60V}{4{,}005}\right]^2$$

$$VP = \left[\frac{(60)(25)}{4{,}005}\right]^2 = \left[\frac{1{,}500}{4{,}005}\right]^2 = (0.375)^2 = 0.140$$

∴ The calculated velocity pressure under the stipulated conditions is 0.14 inches of water. I would guess that the recently graduated IH could stand to improve on his skill in the empirical determination of velocity pressures in ducts.

Problem #4.7:

This problem requires the application of Equation **#4-5**, from Pages 4-7 & 4-8. Since we are dealing with a duct at its exhaust point, we can safely assume that the actual static pressure must be positive — remember that static pressure can be either positive or negative. If this static pressure had been negative, air would have been flowing into, rather than exiting from, the duct at this point. We must, therefore, assume that the stated static pressure has a value of +0.01 inches of water; thus:

$$TP = SP + VP \qquad \text{[Eqn. \#4-5]}$$

Transposing so as to solve for the velocity pressure, VP, we get:

$$VP = TP - SP$$

$$VP = 0.41 - 0.01 = 0.40$$

∴ The velocity pressure at this duct exhaust point is +0.40 inches of water.

Problem #4-8:

The solution to this problem, too, will require the use of Equation **#4-5** from Pages 4-7 & 4-8. In this case, we see that the total pressure has been reported to be 0.24 inches of water; however, we must note that the point at which this measurement has been made is at the inlet to a fan. Because of this, we must assume that this total pressure is negative, i.e., TP = –0.24 inches of water. In order for air to be drawn into a fan, this total pressure must be negative. Remember that velocity pressure is always positive; thus:

$$TP = SP + VP \qquad \text{[Eqn. \#4-5]}$$

Transposing so as to solve for the static pressure, SP, we get:

$$SP = TP - VP$$

$$SP = -0.24 - 0.55 = -0.79$$

∴ The static pressure at the inlet to this fan must be –0.79 inches of water.

Problem #4.9:

This extension to the foregoing problem will require the use of Equation **#4-6**, from Page 4-8, thus:

$$SP_1 + VP_1 = SP_2 + VP_2 + \text{losses} \qquad \text{[Eqn. \#4-6]}$$

In order to apply this relationship, we will first have to determine the "losses". We are given that these losses are frictional in nature and occur at a rate of 0.02 inches of water per 20 lineal feet of duct. We have also been told that there are 220 feet of ducting between the inlet to the fan and the termination of the duct — i.e., the point from which the fan is drawing air. We can therefore determine the "losses" as follows:

$$\text{losses} = \left(\frac{220}{20}\right)(0.02) = (11)(0.02) = 0.22$$

In using Equation **#4-6** to develop the requested answer to this problem, we can assume that the velocity pressures at both points in the duct — i.e., at the point where air is being drawn from the hood and the point downstream at the fan inlet — are equal, since the gas is flowing at the same velocity at all points in this duct:

$$SP_1 = SP_2 + \text{losses}$$

$$SP_1 = -0.79 + 0.22 = -0.57$$

∴ The static pressure at the hood suction point would be –0.57 inches of water.

Problem #4-10:

The solution to this problem also requires the use of Equation **#4-6**, from Page 4-8, thus:

$$SP_1 + VP_1 = SP_2 + VP_2 + \text{losses} \qquad \text{[Eqn. \#4-6]}$$

Again, we can assume that the velocity pressures are the same at each of the two points where the engineer made measurements of static pressure; thus:

$$-0.35 = -0.41 + \text{losses}$$

$$\text{losses} = -0.35 + 0.41 = +0.06$$

Therefore, the total losses in this 75-foot section of duct are 0.06 inches of water; and the specific frictional losses per 10 lineal feet, expressed as "L", are given by:

$$\text{losses} = \left(\frac{\text{length of duct}}{\text{specific lineal measure}}\right)L$$

Transposing and solving for the specific frictional losses, L, we get:

$$L = \left(\frac{\text{specific lineal measure}}{\text{length of duct}}\right)(\text{losses})$$

$$L = \left(\frac{10}{75}\right)(-0.35 - 0.41) = (0.133)(0.06) = 0.008$$

∴ The frictional losses are at a rate of 0.008 inches of water per 10 lineal feet of duct.

Problem #4.11:

The solution to this problem depends upon Equation **#4-7**, from Pages 4-8 & 4-9 — (note, the hood opening in this situation is unflanged); thus:

$$V = \frac{Q}{10x^2 + A} \qquad \text{[Eqn. \#4-7]}$$

Transposing to solve for the volumetric flow rate, Q, we get:

$$Q = V\left(10x^2 + A\right)$$

$$Q = (100)\left[(10)(2.5)^2 + (1)(3)\right]$$

$$Q = (100)(62.5 + 3) = (100)(65.5) = 6,550$$

∴ The minimum volumetric flow rate required to be successful in capturing the irritating vapor will be ~ 6,550 cfm.

Problem #4.12:

This variation on the previous problem will employ Equation **#4-8**, which is on Page 4-9. In this case, we are dealing with a flanged hood opening:

$$V = \frac{Q}{0.75\left[10x^2 + A\right]} \qquad \text{[Eqn. \#4-8]}$$

Transposing to solve for the distance from the source of the vapor to the flanged hood opening, x, we get:

$$10x^2 + A = \frac{Q}{0.75V}$$

$$10x^2 = \frac{Q}{0.75V} - A = \frac{Q - 0.75VA}{0.75V}$$

$$x^2 = \frac{Q - 0.75VA}{(10)(0.75V)} = \frac{Q - 0.75VA}{7.5V} \quad \text{\& finally}$$

$$x = \sqrt{\frac{Q - 0.75VA}{7.5V}}$$

Now, substituting in the values provided in the problem statement, we get:

$$x = \sqrt{\frac{6,550 - (0.75)(100)(3)(1)}{(7.5)(100)}} = \sqrt{\frac{6,550 - 225}{750}}$$

$$x = \sqrt{\frac{6,325}{750}} = \sqrt{8.433} = 2.90$$

The distance that the irritating vapor source could be moved back away from the now flanged hood opening is given by the difference between the just calculated source distance, 2.90 feet, and the original distance of 2.50 feet, as listed in Problem **#4.11** [a situation wherein the hood did not have a flanged opening] and is given by:

$$d = 2.90 - 2.50 = 0.40$$

∴ The vapor source could be moved back — away from its current location in front of the now flanged hood opening — by a total of 0.40 feet, or ~ 4.9 inches.

Problem #4.13:

To solve this problem, we must again use Equation **#4-8**, from Page 4-9:

$$V = \frac{Q}{0.75\left[10x^2 + A\right]} \qquad \text{[Eqn. \#4-8]}$$

It must be remembered that the area term in the denominator of this equation, namely, A, must be in units of "square feet". Thus we must make the conversion from the units provided in the problem statement, namely, "in^2", to their equivalent in "ft^2":

$$A = (3)(12) = 36 \text{ in}^2$$

$$A = \frac{36}{144} = 0.25 \text{ ft}^2$$

$$V = \frac{8,500}{0.75\left[(10)(3)^2 + 0.25\right]} = \frac{8,500}{(0.75)(90 + 0.25)}$$

$$V = \frac{8,500}{(0.75)(90.25)} = \frac{8,500}{67.69} = 125.58$$

We have been told that a capture velocity of 2 fps is required to be able to acquire the respirable particles. Expressing this capture velocity in feet per minute, we see that it is 120 fpm. Clearly 125.58 fpm is greater than 120 fpm.

∴ The centerline velocity of this flanged hood system will be ~ 125.6 fpm, if it is operated and positioned in the manner described. It appears that it will, indeed, successfully capture the respirable dust particles.

Problem #4.14:

To solve this problem, we will have to employ Equation **#4-9**, from Page 4-9:

$$SP_h = VP_d + h_e \qquad \text{[Eqn. \#4-9]}$$

$$SP_h = 15.2 + 8.1 = 23.3$$

∴ The hood static pressure for this situation is ~ 23.3 inches of water.

Problem #4.15:

This extension of Problem **#4.14** will require the use of Equation **#4-10**, from Pages 4-9 & 4-10:

$$C_e = \sqrt{\frac{VP_d}{SP_h}} \qquad \text{[Eqn. \#4-10]}$$

Clearly, the value of 23.3 inches of water for the hood static pressure, as it was just determined for Problem **#4.14**, along with the velocity pressure of 15.2 inches of water provided in the problem statement for Problem **#4.14** are all the data that are required to obtain the desired result, thus:

$$C_e = \sqrt{\frac{15.2}{23.3}} = \sqrt{0.652} = 0.808$$

∴ The Coefficient of Entry for this hood is ~ 0.81.

Problem #4.16:

To solve this problem, we must first use Equation **#4-5**, from Pages 4-7 & 4-8, to obtain the velocity pressure in the duct, and then use this piece of information to obtain the required answer. We will accomplish this by using Equation **#4-9**, from Page 4-9, thus:

$$TP = SP + VP_d \qquad \text{[Eqn. #4-5]}$$

$$6.2 = -1.2 + VP_d$$

$$VP_d = 6.2 + 1.2 = 7.4$$

Now we can apply Equation **#4-9**, from Page 4-8 to obtain the requested result:

$$SP_h = VP_d + h_e \qquad \text{[Eqn. #4-9]}$$

$$12.5 = 7.4 + h_e$$

$$h_e = 12.5 - 7.4 = 5.1$$

∴ The hood entry losses for this hood are ~ 5.1 inches of water.

Problem #4.17:

The solution to this extension of Problem **#4.16** requires that we use Equation **#4-10**, from Pages 4-9 & 4-10:

$$C_e = \sqrt{\frac{VP_d}{SP_h}} \qquad \text{[Eqn. #4-10]}$$

$$C_e = \sqrt{\frac{7.4}{12.5}} = \sqrt{0.592} = 0.769$$

∴ The Coefficient of Entry for this hood is ~ 0.77.

Problem #4.18:

The solution to this problem will require the use of Equation **#4-11**, the *Hood Throat Suction Equation*, from Page 4-10:

$$Q = 4{,}005AC_e\sqrt{SP_h} \qquad \text{[Eqn. #4-11]}$$

Before proceeding, it is necessary to make several observations. First, the duct static pressure measurement of –1.75 inches of water was made 12 inches downstream from the end of this duct — i.e., from its unflanged opening. Since we are dealing with a 4-inch diameter duct, it is clear that this measurement was made the required three full duct diameters downstream from its unflanged opening [i.e., (3)(4-inch diameter) = 12 inches]. We can safely assume, therefore, that this measurement was downstream of any possible *vena contracta* that might possibly have existed in this duct. Thus, we can consider that the duct static pressure determined at this point will be equal to the hood static pressure, SP_h. We can proceed with substitution of the known values into Equation **#4-11**. Note, the area term, A, in this equation must be in "square feet"; therefore, we must convert the diameter that has been provided in inches into its equivalent in "feet".

$$d = 4 \text{ inches} = \frac{1}{3} \text{ foot, \&}$$

$$\text{radius} = \frac{1}{6} \text{ foot}$$

Now we can apply Equation **#4-11**, from Page 4-10:

$$Q = (4,005)(\pi)\left(\frac{1}{6}\right)^2 (0.72)\left(\sqrt{1.75}\right)$$

$$Q = \frac{(4,005)(\pi)(0.72)\left(\sqrt{1.75}\right)}{(6)^2} = \frac{(4,005)(\pi)(0.72)(1.323)}{36}$$

$$Q = \frac{11,984.06}{36} = 332.89$$

∴ The volumetric flow rate for this hood is ~ 333 cfm.

Problem #4.19:

To solve this problem, which is an extension of Problem **#4.18**, we must apply Equation **#4-1**, from Page 4-6:

$$Q = AV \qquad \text{[Eqn. \#4-1]}$$

We must now modify this equation, first, by solving for the duct velocity, V, and then by substituting for the area term, A, the formula for the area of a circle of diameter, d. Note, the diameter has been supplied in the problem statement in inches — rather than in feet — therefore, we must modify the area term, A, as it appears in Equation **#4-1** to a form that will permit us to use the duct diameter, d, in inches. Doing so, we get:

$$A_{circle} = \frac{1}{4}\pi d^2 \text{ in}^2 \qquad \& \qquad A_{circle} = \left(\frac{1}{144}\right)\left(\frac{1}{4}\right)\pi d^2 = \frac{\pi d^2}{576} \text{ ft}^2$$

therefore:

$$V = \frac{576Q}{\pi d^2}$$

$$V = \frac{(576)(332.9)}{\pi(4)^2} = \frac{(576)(332.9)}{16\pi}$$

$$V = \frac{191,744.9}{50.265} = 3,814.64$$

∴ The duct velocity for the hood system described in this problem ~ 3,815 fpm.

Problem #4.20:

To solve this problem and obtain the average duct velocity pressure, we must go back to Equation **#4-3**, from Page 4-7:

$$V = 4,005\sqrt{VP} \qquad \text{[Eqn. \#4-3]}$$

Transposing to solve for the velocity pressure, VP, we get:

$$VP = \left(\frac{V}{4,005}\right)^2$$

$$VP = \left(\frac{3,814.6}{4,005}\right)^2 = (0.952)^2 = 0.907$$

> ∴ The average duct velocity pressure for this hood will be ~ 0.91 inches of water.

Problem #4.21:

The solution to this problem will rely, first, on Equation **#4-11**, from Page 4-10, to obtain the duct velocity pressure. With this information, we will use Equation **#4-5**, from Pages 4-7 & 4-8, to develop the requested result; thus:

$$Q = 4,005A\sqrt{VP_d} \qquad \text{[Eqn. #4-11]}$$

Transposing to solve directly for the velocity pressure in the duct, VP_d, we get:

$$VP_d = \left(\frac{Q}{4,005A}\right)^2$$

Again, note that in this relationship, the area term, A, must be in units of "square feet"; therefore, we must convert the relationship listed above so that we will be able to use the dimensions of the square hood opening that were provided in the problem statement in units of "inches":

$$A = s^2, \text{ in units of in}^2 \qquad \& \qquad A = \frac{s^2}{144}, \text{ in units of ft}^2$$

$$VP_d = \left[\left(\frac{Q}{4,005}\right)\left(\frac{144}{s^2}\right)\right]^2 = \left(3.6\times10^{-2}\frac{Q}{s^2}\right)^2$$

$$VP_d = \left[\frac{(3.6\times10^{-2})(3,000)}{(6)^2}\right]^2 = \left(\frac{107.87}{36}\right)^2$$

$$VP_d = (2.996)^2 = 8.98$$

Now, with this information, we can determine the requested duct total pressure, using Equation **#4-5**, from Pages 4-7 & 4-8, as stated earlier:

$$TP = SP + VP \qquad \text{[Eqn. #4-5]}$$

$$TP = -4.5 + 8.98 = 4.48$$

> ∴ The duct total pressure for this hood is ~ 4.5 inches of water.

Problem #4.22:

To solve this problem, we must employ the *Hood Entry Loss Equation*, Equation **#4-12**, from Page 4-10; thus:

$$h_e = \left[\frac{1 - C_e^2}{C_e^2}\right] VP_h \qquad \text{[Eqn. \#4-12]}$$

$$h_e = \left[\frac{1-(0.85)^2}{(0.85)^2}\right](21.0) = \left(\frac{1-0.723}{0.723}\right)(21.0)$$

$$h_e = \left(\frac{0.278}{0.723}\right)(21.0) = (0.384)(21.0) = 8.07$$

∴ The hood entry losses for this system are ~ 8.1 inches of water.

Problem #4.23:

This problem, too, requires the use of Equation **#4-12**, from Page 4-10; however, in this case we shall have to begin by transposing this relationship so as to solve for the Coefficient of Entry:

$$h_e = \left[\frac{1 - C_e^2}{C_e^2}\right] VP_h \qquad \text{[Eqn. \#4-12]}$$

Now transposing and solving for the Coefficient of Entry, C_e, as stated above, we get:

$$C_e^2 h_e = \left(1 - C_e^2\right) VP_h = VP_h - C_e^2 VP_h$$

$$C_e^2 h_e + C_e^2 VP_h = VP_h = C_e^2\left[h_e + VP_h\right]$$

$$C_e^2 = \frac{VP_h}{h_e + VP_h}$$

$$C_e = \sqrt{\frac{VP_h}{h_e + VP_h}}$$

$$C_e = \sqrt{\frac{12.3}{6.5 + 12.3}} = \sqrt{\frac{12.3}{18.8}} = \sqrt{0.654} = 0.809$$

∴ The Coefficient of Entry for this particular hood ~ 0.81.

Problem #4.24:

The solution to this problem relies upon the *Hood Entry Loss Factor Equation*, Equation **#4-13**, from Page 4-11, thus:

$$F_h = \frac{h_e}{VP_h} \qquad \text{[Eqn. \#4-13]}$$

$$F_h = \frac{6.5}{12.3} = 0.528$$

∴ The hood entry loss factor for this particular hood ~ 0.53.

Problem #4.25:

The solution to this problem will rely on the first of the two *Compound Hood Equations*, designated as Equation **#s 4-14A**, from Pages 4-11 & 4-12. Since in this case, we have been given that the duct velocity is greater than the slot velocity, we can conclude that the relationship we will be using will be Equation **#4-14A**. To begin this overall process, we must determine the duct and the slot velocity pressures, and we can do this by employing Equation **#4-3**, from Page 4-7. Once we have made these determinations, we can then use Equation **#s 4-14C & 4-14D**, from Pages 4-11 & 4-12. We do this to determine the actual duct and slot entry losses. With these factors known, we can finally apply Equation **#4-14A**, from Pages 4-11 & 4-12:

$$V_d = 4{,}005\sqrt{VP_d} \qquad \text{[Eqn. \#4-3]}$$

Transposing and solving for the velocity pressure in the duct, VP_d, we get:

$$VP_d = \left[\frac{V_d}{4{,}005}\right]^2$$

$$VP_d = \left(\frac{5{,}600}{4{,}005}\right)^2 = (1.398)^2 = 1.955$$

We must next use the same relationship to determine the slot velocity pressure; VP_s, thus:

$$VP_s = \left[\frac{V_d}{4{,}005}\right]^2$$

$$VP_s = \left(\frac{3{,}400}{4{,}005}\right)^2 = (0.849)^2 = 0.721$$

Now using these two velocity pressure results, we can use the two entry loss approximations that apply: (1) to the slot, namely, Equation **#4-14C**, from Pages 4-11 & 4-12, and (2) to the duct, namely, Equation **#4-14D**, also from Pages 4-11 & 4-12. In doing this, we have:

(1) the slot:

$$h_{ES} = 1.78[VP_s] \qquad \text{[Eqn. \#4-14C]}$$

$$h_{ES} = (1.78)(0.721) = 1.283$$

&

(2) the duct:

$$h_{ED} = 0.25[VP_d] \qquad \text{[Eqn. \#4-14D]}$$

$$h_{ED} = (0.25)(1.955) = 0.489$$

Now, finally, we can employ Equation **#4-14A**, from Pages 4-11 & 4-12, to determine the hood static pressure; SP_h, thus:

$$SP_h = h_{ES} + h_{ED} + VP_d \qquad [\text{for } V_d > V_s] \qquad \text{[Eqn. \#4-14A]}$$

$$SP_h = 1.283 + 0.489 + 1.955 = 3.727$$

∴ For this compound hood, the hood static pressure ~ 3.7 inches of water.

Problem #4.26:

This is a slightly more complex problem that will, like its predecessor, employ one of the *Compound Hood Equations* to determine the hood static pressure. In order to determine which of these two relationships we will employ, we must first calculate some of the other parameters so as to discover whether the slot velocity is greater than the duct velocity or vice versa. To start on this track, let us use Equation **#4-5**, from Pages 4-7 & 4-8, to determine the duct velocity pressure; thus:

$$TP_d = SP_d + VP_d \qquad \text{[Eqn. \#4-5]}$$

Transposing to solve for the duct velocity pressure, VP_d, we get:

$$VP_d = TP_d - SP_d$$

$$VP_d = 1.3 - (-1.9) = 1.3 + 1.9 = 3.2$$

Since we now know the velocity pressure in this exhaust duct, we can easily determine the duct velocity by using Equation **#4-3**, from Page 4-6, thus:

$$V_d = 4{,}005\sqrt{VP_d} \qquad \text{[Eqn. \#4-3]}$$

$$V_d = 4{,}005\sqrt{3.2} = (4{,}005)(1.789) = 7{,}164.4$$

Clearly, now, we can see that the slot velocity — given as 7,500 fpm — is greater than the duct velocity — just calculated out to be 7,164 fpm — therefore, we will eventually be using Equation **#4-14B** from Pages 4-11 & 4-12 to determine the hood slot velocity. As was the case with Problem **#4.25**, we must begin by determining all of the appropriate velocity pressures. We have already calculated the velocity pressure in the duct; so, we must now calculate the velocity pressure in the slot, using the transposed form of Equation **#4-3**, from Page 4-7, which form was developed for calculating this parameter as it was required earlier in Problem **#4.25**, thus:

$$VP_s = \left(\frac{V_s}{4{,}005}\right)^2$$

$$VP_s = \left(\frac{7{,}500}{4{,}005}\right)^2 = (1.873)^2 = 3.507$$

Now using these two velocity pressure results, we can use the two entry loss approximations that apply: (1) to the slot, namely, Equation **#4-14C**, from Pages 4-11 & 4-12, and (2) to the duct, namely, Equation **#4-14D**, also from Pages 4-11 & 4-12. In doing this, we have:

(1) the slot:

$$h_{ES} = 1.78[VP_s] \qquad \text{[Eqn. \#4-14C]}$$

$$h_{ES} = (1.78)(3.507) = 6.242$$

&

(2) the duct:

$$h_{ED} = 0.25[VP_d] \qquad \text{[Eqn. \#4-14D]}$$

$$h_{ED} = (0.25)(3.2) = 0.80$$

Now, finally, we can employ Equation **#4-14B**, from Pages 4-11 & 4-12, to determine the hood static pressure, SP_h:

$$SP_h = h_{ES} + h_{ED} + VP_s \qquad [\text{for } V_s > V_d] \qquad \text{[Eqn. \#4-14B]}$$

$$SP_h = 6.242 + 0.800 + 3.507 = 10.549$$

∴ The hood static pressure for this compound hood ~ 10.5 inches of water.

Problem #4.27:

To solve this problem, we must employ the first of the relationships involving the rotational speed of a fan, namely, Equation **#4-15**, from Page 4-13:

$$\frac{CFM_1}{CFM_2} = \frac{RPM_1}{RPM_2} \qquad \text{[Eqn. \#4-15]}$$

$$\frac{CFM_1}{10,500} = \frac{1,450}{1,750}$$

$$CFM_1 = \frac{(1,450)(10,500)}{1,750} = \frac{15,225,000}{1,750} = 8,700$$

∴ The previous discharge volume of this fan was 8,700 cfm.

Problem #4.28:

The solution to this problem relies on the second of the fan rotational speed relationships, namely, Equation **#4-16**, on Page 4-13:

$$\frac{SP_{FAN_1}}{SP_{FAN_2}} = \left[\frac{RPM_1}{RPM_2}\right]^2 \qquad \text{[Eqn. \#4-16]}$$

$$\frac{SP_{FAN_2}}{4.6} = \left(\frac{1,750}{1,450}\right)^2 = (1.207)^2 = 1.457$$

$$SP_{FAN_2} = (1.457)(4.6) = 6.70$$

∴ The new fan static discharge pressure ~ 6.7 inches of water.

Problem #4.29:

The solution to this problem relies on the third of the three relationships involving the rotational speeds of fans, namely, Equation **#4-17**, from Pages 4-13 & 4-14, thus:

$$\frac{BHP_1}{BHP_2} = \left[\frac{RPM_1}{RPM_2}\right]^3 \qquad \text{[Eqn. \#4-17]}$$

$$\frac{BHP_2}{15} = \left(\frac{1,750}{1,450}\right)^3 = (1.207)^3 = 1.758$$

$$BHP_2 = (1.758)(15) = 26.37$$

∴ The new fan motor must have a minimum brake horsepower rating of ~ 26.4 bhp.

Problem #4.30:

The solution to this problem will require the use of Equation **#4-15**, from Page 4-13:

$$\frac{CFM_1}{CFM_2} = \frac{RPM_1}{RPM_2} \qquad \text{[Eqn. \#4-15]}$$

$$\frac{RPM_2}{1,200} = \frac{(25,000)(1.40)}{25,000} = 1.40$$

$$RPM_2 = (1.40)(1,200) = 1,680$$

∴ Under the new conditions, the fan rotational speed must be ~ 1,680 rpm.

Problem #4.31:

The solution to this problem will rely on Equation **#4-16**, from Page 4-13:

$$\frac{SP_{FAN_1}}{SP_{FAN_2}} = \left[\frac{RPM_1}{RPM_2}\right]^2 \qquad \text{[Eqn. \#4-16]}$$

$$\frac{SP_{FAN_2}}{6.25} = \left(\frac{1,680}{1,200}\right)^2 = (1.40)^2 = 1.96$$

$$SP_{FAN_2} = (1.96)(6.25) = 12.25$$

∴ The new fan static pressure will be ~ 12.3 inches of water.

Problem #4.32:

This problem can be solved by using Equation **#4-17**, from Pages 4-13 & 4-14, thus:

$$\frac{BHP_1}{BHP_2} = \left[\frac{RPM_1}{RPM_2}\right]^3 \qquad \text{[Eqn. \#4-17]}$$

$$\frac{BHP_1}{50} = \left(\frac{1,680}{1,200}\right)^3 = (1.40)^3 = 2.74$$

$$BHP_2 = (2.74)(50) = 137.2$$

∴ The requirements for the modified ventilation system call for a new fan motor that has a capacity of at least 137.2 bhp.

Problem #4.33:

The solution to this problem will rely on the first of the three relationships involving the diameters of fans, namely, Equation **#4-18**, from Page 4-14, thus:

$$\frac{CFM_1}{CFM_2} = \left[\frac{d_1}{d_2}\right]^3 \qquad \text{[Eqn. \#4-18]}$$

Transposing and solving for the unknown fan diameter, d_2, we get:

$$d_2 = d_1 \sqrt[3]{\frac{CFM_2}{CFM_1}}$$

$$d_2 = (6.0)\sqrt[3]{\frac{67,500}{28,500}} = [6.0](2.368)^{1/3}$$

$$d_2 = (6.00)(1.33) = 8.00$$

∴ The Industrial Hygienist will likely recommend a completely new ventilation system employing all 8.0-inch diameter components.

Problem #4.34:

The solution to this problem will rely on Equation **#4-19**, from Page 4-14, thus:

$$\frac{SP_{FAN_1}}{SP_{FAN_2}} = \left[\frac{d_1}{d_2}\right]^2 \qquad \text{[Eqn. #4-19]}$$

Transposing and solving for the fan static discharge pressure, SP_{FAN_2}, we get:

$$SP_{FAN_2} = SP_{FAN_1}\left(\frac{d_2}{d_1}\right)^2$$

$$SP_{FAN_2} = (7.00)\left(\frac{8.0}{6.0}\right)^2 = (7.00)(1.33)^2$$

$$SP_{FAN_2} = (7.00)(1.78) = 12.44$$

∴ The new fan static discharge pressure will be ~ 12.4 inches of water.

Problem #4.35:

This problem will require the application of Equation **#4-20**, from Pages 4-15:

$$\frac{BHP_1}{BHP_2} = \left[\frac{d_1}{d_2}\right]^5 \qquad \text{[Eqn. #4-20]}$$

Transposing and solving for the brake horsepower of the existing system, BHP_1, we get:

$$BHP_1 = BHP_2\left(\frac{d_1}{d_2}\right)^5$$

$$BHP_1 = (150)\left(\frac{6.0}{8.0}\right)^5 = (150)(0.75)^5$$

$$BHP_1 = (150)(0.237) = 35.60$$

∴ The brake horsepower of the fan motor that is to be replaced is ~ 35.6 bhp.

Problem #4.36:

To solve this problem we must first determine the volumetric flow rate in the 15.0-inch diameter duct, using Equation **#4-1**, from Page 4-6. Once this has been accomplished, we will use Equation **#4-18**, from Page 4-14. This second equation will be rearranged so as to accommodate the values and units provided in the problem statement. To start, let us modify Equation **#4-1** so as to use the data and units from the problem statement, namely, duct dimensions in inches, rather than feet, and the duct velocity in feet per second, rather than feet per minute. To do this we must remember that 144 in^2 = 1 ft^2, and 60 seconds = 1 minute, thus:

Area, A, in "ft^2" — with dimensions that would produce "in^2":

$$A = \left(\frac{\pi d^2}{4}\right)\left(\frac{1}{144}\right) = \frac{\pi d^2}{576}$$

Velocity, V, in "fpm" — with supplied dimensions for "ft/sec":

$$V = \left(\frac{ft}{sec}\right)(60)$$

Thus we obtain the modified form of Equation **#4-1**, from Page 4-6:

$$Q = AV = \left(\frac{\pi d^2}{576}\right)(60V) = \frac{5\pi d^2 V}{48} \qquad \text{[Eqn. #4-1]}$$

$$Q = \frac{5(\pi)(15.0)^2(150)}{48} = \frac{530,143.76}{48} = 11,044.66$$

Now, with the volumetric flow rate for the smaller system known, we can apply Equation **#4-18**, from Page 4-14, to determine the volumetric flow rate in the new larger system; thus:

$$\frac{CFM_1}{CFM_2} = \left[\frac{d_1}{d_2}\right]^3 \qquad \text{[Eqn. #4-18]}$$

Transposing and solving for CFM_2, we get:

$$CFM_2 = CFM_1\left(\frac{d_2}{d_1}\right)^3$$

$$CFM_2 = (11,044.66)\left(\frac{18.0}{15.0}\right)^3 = (11,044.66)(1.20)^3$$

$$CFM_2 = (11,044.66)(1.728) = 19,085.18$$

Now that we have the new volumetric flow rate, it will be very easy to determine the new duct velocity. We will simply reapply Equation **#4-1**, from Page 4-6, again utilizing a subset of the rewritten format as it was derived on the previous page; thus:

$$V = \frac{Q}{A} = \frac{576Q}{\pi d^2} \qquad \text{[Eqn. #4-1]}$$

$$V = \frac{(576)(19,085.18)}{\pi(18.0)^2} = \frac{10,993,061.02}{\pi(324)}$$

$$V = \frac{10,993,061.02}{1,017.88} = 10,800.00$$

$$V = \frac{10,800.00}{60} = 180.00$$

∴ The new duct velocity will become ~ 10,800 fpm = 180.00 ft/sec.

Problem #4.37:

To solve this problem, we must first determine the volumetric flow rate in the 8.0-inch diameter ducts. Again, to do this, we will simply apply Equation **#4-3**, from Page 4-7:

$$V = 4,005\sqrt{VP} \qquad \text{[Eqn. #4-3]}$$

Transposing and solving for the velocity pressure, VP, we get:

$$VP = \left[\frac{V}{4,005}\right]^2$$

$$VP = \left(\frac{7,525}{4,005}\right)^2 = (1.879)^2 = 3.530$$

Now, since we have successfully calculated the velocity pressure in the duct, and since we were given the total pressure in the duct, we can apply Equation **#4-5**, from Pages 4-7 & 4-8 to determine the static discharge pressure in the smaller sized — i.e., 8.0-inch diameter — duct and fan.

$$TP_d = SP_d + VP_d \qquad \text{[Eqn. #4-5]}$$

Transposing and solving for the static pressure in the duct, SP_d, we get:

$$SP_d = TP_d - VP_d$$

$$SP_d = 9.8 - 3.53 = 6.27$$

Now we have the information that is required to calculate the requested fan static discharge pressure for the larger fan system. To do so, we will make use of Equation **#4-19**, from Page 4-14; thus:

$$\frac{SP_{Fan_1}}{SP_{Fan_2}} = \left[\frac{d_1}{d_2}\right]^2 \qquad \text{[Eqn. #4-19]}$$

Transposing and solving for the static pressure of the larger fan, SP_{FAN_2}, we get:

$$SP_{Fan_2} = SP_{Fan_1}\left[\frac{d_2}{d_1}\right]^2$$

$$SP_{Fan_2} = (6.27)\left[\frac{(8.0)(1.5)}{8.0}\right]^2 = (6.27)(1.50)^2$$

$$SP_{Fan_2} = (6.27)(2.25) = 14.11$$

∴ The static discharge pressure for the larger fan in the new system will be ~ 14.1 inches of water.

Problem #4.38:

The solution to this problem will rely on Equation **#4-20**, on Page 4-15:

$$\frac{BHP_1}{BHP_2} = \left[\frac{d_1}{d_2}\right]^5 \qquad \text{[Eqn. #4-20]}$$

Transposing and solving for the diameter of smaller fan, d_2, we get:

$$d_2 = d_1 \sqrt[5]{\frac{BHP_2}{BHP_1}}$$

$$d_2 = 9.0\left(\frac{10}{75}\right)^{1/5} = (9.0)(0.133)^{1/5} = (9.0)(0.668) = 6.02$$

∴ The system could be downsized to an overall ~ 6.0-inch diameter configuration and still fully satisfy all stated conditions.

Problem #4.39:

To solve this problem, we will have to use Equation **#4-21**, on Page 4-15:

$$BHP = 0.0157\left[\frac{(CFM)(TP)}{FME}\right] \qquad \text{[Eqn. #4-21]}$$

$$BHP = \frac{(0.0157)(22,500)(3.4)}{65} = \frac{1,203.59}{65} = 18.52$$

∴ The brake horsepower of the subject fan is ~ 18.5 bhp.

Problem #4.40:

To solve this problem, we must first determine the volumetric flow rate for the system being evaluated. To do this, we will again rely on Equation **#4-1**, from Page 4-6. Once this has been done, we will be able to use Equation **#4-21**, on Page 4-15, to determine the required fan mechanical efficiency. As with the various earlier problems, we shall modify Equation **#4-1** to permit the direct utilization of the duct dimensions, which were given in inches in the problem statement, rather than in feet as is required by the normal format of this equation:

$$Q = AV = \frac{\pi d^2 V}{4} = \frac{d^2 V}{576}\pi \qquad \text{[Eqn. #4-1]}$$

$$Q = \frac{(12.0)^2(6,600)}{576}\pi = \frac{(144)(6,600)}{576}\pi = \frac{950,400}{576}\pi = 5,183.63$$

Now since we know the volumetric flow rate, we can apply Equation **#4-21**, and calculate the required fan mechanical efficiency; thus:

$$BHP = 0.0157\left[\frac{(CFM)(TP)}{FME}\right] \qquad \text{[Eqn. #4-21]}$$

Transposing and solving for the fan mechanical efficiency, FME, we get:

$$FME = 0.0157\left[\frac{(CFM)(TP)}{BHP}\right]$$

$$FME = (0.0157)\left[\frac{(5{,}183.6)(15.0)}{20}\right] = (0.0157)\left(\frac{77{,}754.42}{20}\right)$$

$$FME = (0.0157)(3{,}887.72) = 61.17$$

$\therefore$ The fan mechanical efficiency is ~ 61%.

Problem #4.41:

Ultimately, the solution to this problem will use Equation **#4-22**, on Page 4-16; however, we must first determine the two fan total pressure terms. To do this we will use combinations of the three *General Rule* relationships, Equation **#s 4-4**, **4-5**, & **4-6**, from Pages 4-7 & 4-8. We start by deriving the relationship for TP_{out}; thus:

$$TP_1 = TP_2 + \text{losses} \qquad \text{[Eqn. \#4-4]}$$

$$TP = SP + VP \qquad \text{[Eqn. \#4-5]}$$

$$SP_1 + VP_1 = SP_2 + VP_2 + \text{losses} \qquad \text{[Eqn. \#4-6]}$$

Let us first consider the discharge situation, calculating the discharge fan's total outlet pressure, TP_{out}, thus:

$$TP_{out} = VP_{out} + \text{losses}$$

$$TP_{out} = 5.6 + 4.8 = 10.4 \text{ inches of water}$$

Let us next consider the inlet situation, developing a solution for the inlet fan's total pressure, TP_{in}, thus:

$$TP_{in} = VP_{in} + \text{losses}$$

$$TP_{in} = 4.2 + 4.5 = 8.7 \text{ inches of water}$$

Note, here we are evaluating the fan total inlet pressure, namely the total pressure on the inlet side of the fan. Fan total inlet pressures are always negative, since they are on the suction side of the fan.

$$\therefore \quad TP_{in} = -8.7 \text{ inches of water}$$

Now, finally using Equation **#4-22**, from Page 4-16, we get:

$$TP_{Fan} = TP_{out} + TP_{in} \qquad \text{[Eqn. \#4-22]}$$

$$TP_{Fan} = 10.4 - (-8.7) = 10.4 + 8.7 = 19.1$$

$\therefore$ The fan total pressure in this case is ~ 19.1 inches of water.

Problem #4.42:

Again, we will eventually determine the fan total pressure by using Equation **#4-22**, from Page 4-16. To start with, however, we must again deal with the fan inlet side. The fan total input pressure term can be calculated by using a combination of Equation **#4-22**, from Page 4-16, and Equation **#4-5**, from Pages 4-7 & 4-8, thus:

$$TP_{Fan} = TP_{out} + TP_{in} \qquad \text{[Eqn. \#4-22]}$$

$$TP_{in} = SP_{in} + VP_{in} \qquad \text{[Eqn. \#4-5]}$$

The useful combination of these two equations is as follows:

$$TP_{Fan} = TP_{out} - SP_{in} - VP_{in}$$

Rewriting Equation **#4-5**, from Pages 4-7 & 4-8, from the perspective of the output side of the fan, rearranging it so as to provide a solution for the outlet fan static pressure, SP_{out}, and finally substituting in the values we have been given for the fan total output pressure and the velocity pressure at the fan outlet, we get:

$$SP_{out} = TP_{out} - VP_{out}$$

$$SP_{out} = 25.7 - 13.4 = 12.3 \text{ inches of water}$$

We have been given that the difference in static pressures between the fan inlet and its outlet, which we shall designate as $\Delta SP = 19.6$ inches of water, and we know that:

$$\Delta SP = SP_{out} - SP_{in}$$

$$\therefore \quad SP_{in} = SP_{out} - \Delta SP$$

$$SP_{in} = 12.3 - 19.6 = -7.3$$

Now, finally we have all the data to use the "combination" relationship developed on the previous page; thus:

$$TP_{Fan} = TP_{out} - SP_{in} - VP_{in}$$

$$TP_{Fan} = 25.7 - (-7.3) - 13.4 = 25.7 + 7.3 - 13.4 = 19.6$$

∴ The fan total pressure is ~ 19.6 inches of water.

Problem #4.43:

To solve this problem, we must eventually apply Equation **#4-23**, from Page 4-16. First, however, we must determine the static pressure at the fan outlet, and to do so, we must use Equation **#4-5**, from Pages 4-7 & 4-8:

$$TP_{out} = SP_{out} + VP_{out} \qquad \text{[Eqn. \#4-5]}$$

Rewriting to solve for the static pressure at the fan outlet, SP_{out}, we get:

$$SP_{out} = TP_{out} - VP_{out}$$

$$SP_{out} = 16.3 - 8.2 = 8.1 \text{ inches of water}$$

We can now apply Equation **#4-23**, from Page 4-16:

$$SP_{Fan} = SP_{out} - SP_{in} - VP_{in} \qquad \text{[Eqn. \#4-23]}$$

$$SP_{Fan} = 8.1 - (-8.3) - 6.4 = 8.1 + 8.3 - 6.4 = 10.0$$

∴ The fan static discharge pressure is ~ 10.0 inches of water.

Problem #4.44:

To solve this problem, we need only apply Equation **#4-23**, from Page 4-16:

$$SP_{Fan} = SP_{out} - SP_{in} - VP_{in} \qquad \text{[Eqn. \#4-23]}$$

$$21.5 = 14.4 - (-12.7) - VP_{in} \quad \&$$

$$VP_{in} = 14.4 + 12.7 - 21.5 = 5.6 \text{ inches of water}$$

∴ The velocity pressure at this fan's inlet is ~ 5.6 inches of water.

Problem #4.45:

The solution to this problem will rely on the several relationships listed on Pages 4-17 & 4-18 — the *Duct Junction Balancing System.* This problem asks us to balance a somewhat complex set of duct junctions, involving three separate ducts that join to form one single duct. Let us begin by examining two of these duct junctions, namely: [AC – DD] & [CD – DD]. Specifically, let us examine the ratio of the static pressures listed for these two junction sets — these two were selected simply because the absolute magnitude of the difference in their listed static pressures was the greatest among this group of three duct junctions — i.e., the absolute value of the difference of their listed static pressures, $|\Delta SP|$, can be expressed as follows:

$$|\Delta SP| = \left|SP_{greater} - SP_{lesser}\right| = |1.44 - 1.26| = 0.18 \text{ inches of water}$$

Thus, using Equation **#4-24A**, from Page 4-17, as stated earlier, we get:

$$R = \frac{SP_{greater}}{SP_{lesser}} \qquad \text{[Eqn. \#4-24]}$$

$$R = \frac{1.44}{1.26} = 1.143$$

For any pair of duct junctions having this ratio, $1.05 \leq R \leq 1.20$ — which is obviously true for this pair — an adequate balance can be achieved if we increase the flow volume in the branch having the lesser flow volume, according to the relationship listed under the second of the three ratio scenarios that have been tabulated under Equation **#4-24B**, on Pages 4-17 & 4-18:

$$Q_{new} = Q_{former}\sqrt{\frac{SP_{greater}}{SP_{lesser}}} = Q_{former}\sqrt{R}$$

$$Q_{new} = 520\sqrt{\frac{1.44}{1.26}} = 520\sqrt{1.143} = (520)(1.069) = 555.90$$

∴ To balance the [AC - DD] & [CD - DD] duct junction, we must increase the flow volume in duct AD from its current value of ~ 520 cfm to ~ 556 cfm.

Let us now address duct junction, [BD – DD] & [CD – DD], in the same manner; thus:

$$R = \frac{SP_{greater}}{SP_{lesser}} \quad \text{[Eqn. #4-24]}$$

$$R = \frac{1.44}{1.39} = 1.036$$

This duct junction, for which the ratio, R, is defined by R < 1.05, can be considered to be in balance.

> ∴ We can consider this duct junction to be in balance; thus, we should not attempt to adjust the flow volumes in either of the ducts making up this junction.

Problem #4.46:

To solve the final part of this problem, we will have to apply both Dalton's Law of Partial Pressures — as specified by Equation **#1-16**, from Pages 1-22 & 1-23, to provide the solution for the final part of the problem. To solve its first part, we must use Equation **#4-25**, from Pages 4-18 & 4-19, thus:

$$C = 6.24 \times 10^7 \left[\frac{V_s \rho T}{P_{atm} V_{room} (MW_s)} \right] \quad \text{[Eqn. #4-25]}$$

Before substituting values into this admittedly complex formula, we must determine that we are using the units that this relationship calls for. We note that:

(1) The solvent volume term, V, must be in milliliters [thus, we must multiply the solvent volume that was provided in the problem statement by a factor of 1,000 which is the number of milliliters in 1.0 liter].

(2) The temperature term, T, must be an absolute temperature, expressed in K — thus, we must apply Equation **#1-1**, from Page 1-16, to convert this relative metric temperature to its absolute equivalent; thus

$$t_{Metric} + 273.16 = T_{Metric} \quad \text{[Eqn. #1-1]}$$

$$25 + 273.16 = 298.16 \text{ K}$$

(3) The room volume term, V_{room}, must be in liters (thus, we must multiply the room volume that is provided in the problem statement by a factor of 28.32 liters/ft^3).

When we make these substitutions, we get:

$$C = \left(6.24 \times 10^7\right) \left[\frac{(1.5 \times 1{,}000)(0.791)(298.16)}{(760)(1{,}650 \times 28.32)(58.08)} \right]$$

$$C = \left(6.24 \times 10^7\right) \left(\frac{353{,}766.84}{2{,}062{,}611.30} \right) = \left(6.24 \times 10^7\right)\left(1.715 \times 10^{-4}\right) = 10{,}702.48$$

> ∴ The ultimate ambient concentration level of the acetone vapors in this room will be ~ 10,702 ppm(vol).

According to Dalton's Law of Partial Pressures, the equilibrium concentration of any volatile component will be the product of the vapor pressure of the pure component at the temperature in question, and the mole fraction of that component in the solution being considered. In this case we are dealing with pure acetone; thus, the mole fraction will be 1.00 =

100%; and from Dalton's Law, we see that the partial vapor pressure of acetone would simply be its pure chemical vapor pressure, or 226 mm Hg. Applying Equation **#1-16**, from Pages 1-22 & 1-23 to this fact, we get:

$$C_{acetone} = \frac{[1{,}000{,}000][PVP_{acetone}]}{P_{ambient}} \qquad \text{[Eqn. \#1-16]}$$

$$C_{acetone} = \frac{(1{,}000{,}000)(226)}{760} = 297{,}368.4$$

∴ It is reasonable to assume that <u>all</u> the acetone will evaporate, since the concentration that would be produced by the quantity of acetone in the flask that was broken during the earthquake is only a <u>small</u> <u>fraction</u> of the theoretical maximum acetone concentration that could exist in the ambient air — 10,702 ppm(vol) actual vs. 297,368 ppm(vol) theoretical maximum, with the former being only 3.6% of the latter. In fact, in a room this size, 41.7 liters of pure liquid phase acetone could reasonably be expected to evaporate completely!

Problem #4.47:

To solve this extension to the previous problem, we must apply Equation **#4-26**, from Page 4-19. For this situation, we will use the Common Log Format:

$$D_t = 2.3\left[\frac{V}{Q}\right]log\left[\frac{C_{initial}}{C_{ending}}\right] \qquad \text{[Eqn. \#4-26]}$$

$$D_t = 2.3\left(\frac{1{,}650}{500}\right)log\left(\frac{10{,}702.48}{750}\right)$$

$$D_t = (2.3)(3.30)log14.27 = (2.3)(3.30)(1.154) = 8.762$$

∴ It will take ~ 8.76 minutes to ventilate the room sufficiently well to have achieved the desired 750 ppm(vol) ambient concentration level of acetone.

Problem #4.48:

The solution to this problem also will eventually employ Equation **#4-25**, from Pages 4-18 & 4-19; however, we must first determine what the maximum ambient concentration of ethanol could be at NTP. To do this, we will again have to use Dalton's Law of Partial Pressures, listed as Equation **#1-16**, on Pages 1-22 & 1-23:

$$C_{ethanol} = \frac{[1{,}000{,}000][PVP_{ethanol}]}{P_{ambient}} \qquad \text{[Eqn. \#1-16]}$$

$$C_{ethanol} = \frac{(1{,}000{,}000)(44.3)}{760} = 58{,}289.5$$

This means that ethanol will evaporate from his batch of "White Lightning" until the concentration in this cylindrical redwood tank/storeroom has attained a value of ~ 58,290 ppm(vol). Knowing this, we can begin to work with Equation **#4-25**, from Pages 4-18 & 4-19:

$$C = 6.24 \times 10^7 \left[\frac{V_s \rho T}{P_{atm} V_{room} (MW_s)} \right] \qquad \text{[Eqn. \#4-25]}$$

Again, before substituting values into this somewhat complex formula, we must determine that we are using the units that this relationship calls for. In particular, we must note that:

(1) The solvent volume term, V, must be in milliliters [thus, we must multiply the solvent volume that is provided in the problem data by a factor of 3,785.4 ml/gallon].

(2) The temperature term, T, must be an absolute temperature expressed in K — thus, we must apply Equation **#1-1**, from Page 1-15, to convert this relative metric temperature to its absolute equivalent, thus

$$t_{Metric} + 273.16 = T_{Metric} \qquad \text{[Eqn. \#1-1]}$$

$$25 + 273.16 = 298.16 \text{ K}$$

Before making these substitutions, we should transpose the given Equation, and solve for the room volume, V_{room}, which in this case will be the volume of the cylindrical redwood tank/storeroom. When we have done this, and have made these substitutions, we get:

$$V_{room} = \left(6.24 \times 10^7\right) \left[\frac{V_s \rho T}{(MW_s) P_{atm} C} \right]$$

$$V_{room} = \left(6.24 \times 10^7\right) \left[\frac{(1 \times 3,785.4)(0.789)(298.16)}{(46.07)(760)(58,289.5)} \right]$$

$$V_{room} = \left(6.24 \times 10^7\right) \left(\frac{890,506.69}{2,040,901.92} \right) = \left(6.24 \times 10^7\right)\left(4.363 \times 10^{-4}\right) = 27,227.05$$

We now obviously have the volume of this cylindrical redwood tank/storeroom in liters (namely: ~ 27,277 liters); however, we must convert this volume into cubic feet in order to obtain the requested result in the most direct manner possible. Remember that 28.32 liters = 1.0 ft^3.

$$V_{cubic\ feet} = \frac{27,227.05}{28.32} = 961.41$$

Now we must express the desired redwood tank/storeroom diameter, d, in terms that will permit us to use this volume, $V_{storeroom}$, and the tank height, h — which has been given in the problem statement as being 6 feet — to advantage, as we proceed in obtaining the requested result:

$$V_{storeroom} = \frac{\pi d^2 h}{4}$$

Now solving this relationship for the volume of a right circular cylinder of diameter, d, and height, h, to obtain the diameter, we get:

$$d^2 = \frac{4 V_{storeroom}}{\pi h} \quad \&$$

$$d = \sqrt{\frac{4 V_{storeroom}}{\pi h}} = 2\sqrt{\frac{V_{storeroom}}{\pi h}}$$

Now, all we have to do is to substitute in the known values to obtain the requested diameter; thus:

$$d = 2\sqrt{\frac{961.41}{6\pi}} = 2\sqrt{51.004} = (2)(7.142) = 14.28$$

∴ The diameter of this Moonshiner's cylindrical redwood tank/storeroom is ~ 14.3 feet.

Problem #4.49:

To solve this problem, we must use Equation **#4-26**, from Page 4-19. Again, we will use the Common Log Format of this expression:

$$D_t = 2.3\left[\frac{V}{Q}\right] log\left[\frac{C_{initial}}{C_{ending}}\right] \quad \text{[Eqn. #4-26]}$$

$$D_t = (2.3)\left(\frac{961.41}{25}\right) log\left(\frac{59,289.5}{1,000}\right)$$

$$D_t = (2.3)(38.456)\,log(59.290) = (2.3)(38.456)(1.773) = 156.82$$

∴ He will have to run his ventilation fan for ~ 156.8 minutes — or 2 hours and 36+ minutes — in order to achieve a maximum ambient ethanol concentration in his cylindrical redwood tank/storeroom of 1,000 ppm(vol).

Problem #4.50:

To solve this problem we must apply Equation **#4-27**, from Pages 4-19 & 4-20:

$$C = C_0 e^{-\left[V_{removed}/V_{room}\right]} \quad \text{[Eqn. #4-27]}$$

For this problem, we have been asked to determine the number of Room Volumes that must be removed from some identifiable space, in order to achieve some well-defined and specific decrease in the "starting ambient concentration" of some volatile material. We must view this problem as asking for a value of "n", where "n" is the number of "room volumes" that must be purged in order to achieve the required reduction of the concentration of the undesirable vapors. In essence, the "n" we must determine is defined by the following relationship:

$$V_{removed} = (n)(V_{room})$$

$$\text{thus: } n = \frac{V_{removed}}{V_{room}}$$

Specifically, we are seeking an exponent of the "e" — or exponential — term in Equation **#4-27** that is in the following generalized format:

$$-\frac{(n)(V_{room})}{V_{room}}$$

Clearly, the V_{room} terms in the numerator and the denominator of this exponent will cancel out, and we are left with the simple exponent, n; and we can, therefore, see that the formula evolves to the following:

$$C = C_0 e^{-\left[\frac{(n)(V_{room})}{V_{room}}\right]} = C_0 e^{-n}$$

The task in this problem, then, is simply to determine the value of "n", which is the number of Room Volumes, that corresponds to:

(1) A decrease in the ambient concentration to a level that is only 10% of the starting value — i.e., the ending concentration, $C_{90\%}$, has the value $0.1C_0$.

(2) A decrease in the ambient concentration to a level that is only 1% of the starting value — i.e., the ending concentration, $C_{99\%}$, has a value $0.01C_0$.

Let us consider these two situations in order:

(1) For a 90% reduction in the concentration:

$$C_{90\%} = 0.1C_0 = C_0 e^{-n_{90\%}}$$

Notice that the C_0 coefficient terms cancel out, as this coefficient appears on both sides of the equation, thus:

$$0.1 = e^{-n_{90\%}}$$

We must next take the natural logarithm of both sides of the equation, and we get:

$$ln(0.1) = -n_{90\%} = -2.303$$

$$n_{90\%} = 2.303$$

(2) For a 99% reduction in the concentration:

$$C_{99\%} = 0.01C_0 = C_0 e^{-n_{99\%}}$$

Notice, again, that the C_0 coefficient terms cancel out, as this coefficient appears on both sides of the equation, thus:

$$0.01 = e^{-n_{99\%}}$$

Again, taking the natural logarithm of both sides of the equation:

$$ln(0.01) = -n_{99\%} = -4.605$$

$$n_{99\%} = 4.605$$

∴ To achieve specified reductions in the ambient concentrations of any volatile substance, one must purge the following number of Room Volumes to attain the target reduction in the ambient room concentration level:

Target Reduction, as a %	# of Room Volumes
90%	~ 2.3
99%	~ 4.6

Chapter 5
Thermal Stress

Thermal stress can arise from exposures to either hot or cold environments. For a wide variety of work environments it is common for there to be extremes as well as wide fluctuations in the temperature; because of this, it is important to understand thermal stress. This chapter will focus on the factors and parameters that are used to characterize thermal stress, and, in addition, will discuss the effects that these conditions can produce on individuals who, in the normal course of their work, may be exposed to them.

RELEVANT DEFINITIONS

Thermal Stress

Heat Stress

Heat Stress is a condition that arises from a variety of factors among the most important of which are:

1. the ambient temperature,
2. the relative humidity,
3. the level of effort required by the job, and
4. the clothing being worn by an exposed individual.

An individual who is experiencing **Heat Stress** will tend to exhibit an array of measurable symptoms which can include some or all of the following:

1. an increased pulse rate,
2. a greater rate of perspiration, and
3. an increase in the individual's body temperature.

Heat Stress Disorders

The five physical disorders that can arise from heat stress, listed in increasing order of severity, are as follows:

1. Heat Rash: A heat rash — also often referred to as "prickly heat" — tends to arise in an individual after a period of prolonged sweating. It is characterized by an itchy reddening of the skin and a sudden decrease in the rate of perspiration.
2. Cramps: Heat cramps arise as a result of prolonged periods of a combination of sweating and a lack of fluid and salt intake. Such a situation causes an overall body electrolyte imbalance, and the primary symptomatic manifestation is severe muscle cramps, most frequently in the abdomen.
3. Dehydration: Dehydration is the result of excessive fluid loss. Among its most common causes are: excessive sweating, vomiting, diarrhea, and/or alcohol consumption. Symptoms of dehydration are often subtle but include exhaustion, overall weakness, dry mouth, decreased work output, etc.

4. Heat Exhaustion: Heat exhaustion arises from extreme cases of dehydration. It is characterized by some or all of the following conditions or symptoms: increasing pulse rate, decreasing blood pressure, slight to moderate increases in body temperature, fatigue, increasing levels of sweating, lack of skin color, dizziness, blurred vision, headache, decreased work output, and collapse.
5. Heat Stroke: Heat stroke is usually the result of very significant overexposure to the factors of heat stress. It can also arise from drug or alcohol abuse and on occasion from genetic factors. Heat stroke is almost always accompanied by an increase in body temperature to levels greater than 104°F (40°C). Symptomatic indications include: chills, irritability, hot & dry skin, convulsions, and unconsciousness.

Cold Stress

Cold Stress differs dramatically from Heat Stress. Typically a body will adapt to conditions of Heat Stress by increasing its level of perspiration in an effort to provide increased cooling. **Cold Stress** adaptations usually involve a decrease in the blood flow to the skin and the extremities. The principal causes of **Cold Stress** are exposure to cold temperatures and vibrations, either singly or in combination.

Cold Stress Disorders

The four physical disorders that can arise from cold stress, listed in increasing order of severity, are as follows:

1. Chilblains: Chilblains usually arise as a result of inadequate clothing during periods of exposure to cold temperatures and high relative humidities. Reddening of the skin accompanied by localized itching and swelling are the principal indications of chilblains.
2. Frostnip: Frostnip, which is similar to frostbite, results from prolonged, unprotected exposures to cold temperatures above 32°F (0°C). Symptoms of frostnip are areas of pain and/or itching, and a distinct whitening of the skin.
3. Frostbite: Frostbite is produced from unprotected exposures to cold temperatures at or below freezing — i.e., ≤ 32°F or 0°C. Frostbite is characterized by the sequential change in skin color from white to gray to black [depending upon the temperature and the length of exposure], a reduction in the sensations of touch ranging from slight to total [again depending upon the temperature and the length of exposure], and numbness.
4. Hypothermia: Hypothermia results from extreme exposures to the factors of cold stress, coupled possibly with dehydration and/or exhaustion. Alcohol and/or drug abuse can also contribute to hypothermia. A person who is experiencing hypothermia will usually show some or all of the following symptoms: chills, euphoria, pain in the extremities, slow and weak pulse, body temperature of less than 95°F (35°C), fatigue, drowsiness, and unconsciousness.

Ambient Conditions Related to Thermal Stress

Dry Bulb Temperature

The **Dry Bulb Temperature** is the most direct measurement of air temperature. By definition, it is to be accomplished by the use of a capillary thermometer that is completely exposed to and/or immersed in the air mass whose temperature is to be measured. This thermometer should be shielded from sources of radiant heat.

Wet Bulb Temperature

The **Wet Bulb Temperature** of an air mass differs from the Dry Bulb Temperature measurement by the fact that the fluid reservoir bulb of the capillary thermometer that is used to make this measurement is encased in a sheath of water moistened cloth. This wet sheath provides cooling to the thermometer bulb by the evaporation of water, in most cases causing the **Wet Bulb Temperature** to be less than its Dry Bulb Temperature counterpart — the obvious exception to this is the case where the ambient relative humidity is 100%, a condition wherein evaporation, and the resulting evaporative cooling, are effectively eliminated.

There are actually two categories of **Wet Bulb Temperature**. The first is the **Natural Wet Bulb Temperature** which is obtained simply by encasing a capillary thermometer bulb in a wet cloth and then using this combination to make an air temperature measurement. The other category is described as a **Psychrometric Wet Bulb Temperature**. **Psychrometric Wet Bulb Temperatures** are obtained by the use of a sling psychrometer, a tool that is made up of a pair of identical capillary thermometers, one of which is bare while the other is sheathed in a wet cloth. To obtain a reading from a sling psychrometer, this mechanism is whirled through the air, a process that produces a maximized rate of evaporative cooling for the wet bulb. The difference in the temperatures indicated by the two thermometers of a sling psychrometer can then be used to determine the relative humidity of the air mass being measured.

Air or Wind Speed

The **Air** or **Wind Speed** is simply the rate at which a mass of air is passing an arbitrary stationary point. The direction of movement is not important since this measure is used principally in a determination of the convective heat transfer to and from the air. It is typically measured by an anemometer.

Globe Temperature

The **Globe Temperature** of an air mass arises from the combination of heat input by radiation from the surroundings coupled with the simultaneous heat loss by the convective movement of air around the **Globe Temperature** measurement device, which is a 6-inch diameter, thin-walled copper, spherical globe, painted matte black with an appropriate temperature sensor at its center.

Effective Temperature

The **Effective Temperature** is an index that is used to relate the subjective effect that the thermal environment might be expected to have on the comfort of an individual who is exposed to that environment. It is a combination of the Dry Bulb, the Wet Bulb, and the Globe Temperatures.

Wet Bulb Globe Temperature Index

The **Wet Bulb Globe Temperature Index** [usually abbreviated, **WBGT**] is the most widely used algebraic approximation of an "Effective Temperature" currently in use today. It is an Index that can be determined quickly, requiring a minimum of effort and operator skill. As an approximation to an "effective temperature", the **WBGT** takes into account virtually all the commonly accepted mechanisms of heat transfer (i.e., radiant, evaporative, etc.). It does not account for the cooling effect of wind speed. Because of its simplicity, **WBGT** has been adopted by the American Conference of Government Hygienists (ACGIH) as its principal index for use in specifying a heat stress related Threshold Limit Value (**TLV**). For outdoor use (i.e., in sunshine), the **WBGT** is computed according to the following algebraic sum:

$$\mathbf{WBGT} = 0.7\,[NWB] + 0.2\,[GT] + 0.1\,[DB].$$

For indoor use, the **WBGT** is computed according to the following slightly modified algebraic sum:

$$\mathbf{WBGT} = 0.7\,[NWB] + 0.3\,[GT].$$

where: [NWB] = Natural Wet Bulb Temperature,
[GT] = Globe Temperature, &
[DB] = Dry Bulb Temperature.

RELEVANT FORMULAE & RELATIONSHIPS

Thermal Stress (Indoor/Outdoor), With & Without Solar Load

Equation #5-1:

Equation **#5-1** is the relationship that provides the **Wet Bulb Globe Temperature Index** [**WBGT**] that is applicable only to situations for which there is no solar load (i.e., no direct solar input to the condition or circumstance of the area or space being evaluated). Obviously, most indoor situations fulfill this requirement; in addition, any outdoor circumstance wherein the sun has been shaded — i.e., where it makes no radiant contribution to the thermal or temperature environment — also fulfills this condition. This category of the **Wet Bulb Globe Temperature Index** is usually identified with an "$_{\text{Inside}}$" subscript.

$$WBGT_{Inside} = 0.7[NWB] + 0.3[GT]$$

Where:

$WBGT_{Inside}$ = the **Wet Bulb Globe Temperature Index**, applicable to any situation in which there is no Solar Load, usually measured in °C;

NWB = the **Natural Wet-Bulb Temperature**, usually also measured in °C; however, it should be noted that any temperature scale may be used for these parameters, so long as the units of every temperature parameter in the formula are consistent with the units of every other temperature parameter; &

GT = the **Globe Temperature**, also in units consistent with every other parameter in this formula.

Equation #5-2:

Equation **#5-2** is the alternative relationship that provides the **Wet Bulb Globe Temperature Index** [**WBGT**] that is applicable to situations for which there is a measurable solar load. Outdoor conditions usually require this approach; and, correspondingly, this category of the **Wet Bulb Globe Temperature Index** is usually identified with an "$_{\text{Outside}}$" subscript.

$$WBGT_{Outside} = 0.7[NWB] + 0.2[GT] + 0.1[DB]$$

Where:

$WBGT_{Outside}$ is precisely as defined, under the designation of **$WBGT_{Inside}$**, directly above on this page;

NWB is also as defined directly above on this page;

GT is as defined on the previous page, namely, Page 5-5; &

DB = the **Dry-Bulb Temperature**, usually measured in °C; however, it should be noted for this relationship, too, that any temperature scale may be used for these parameters, so long as the units of every temperature parameter in the equation are consistent with the units of every other temperature parameter.

Temperature Related Time Weighted Averages

Equation #5-3:

Equation **#5-3** provides the relationship necessary for the determination of the **Effective Time Weighted Average WBGT Index**. This Index represents an average exposure over time, and at various different **WBGT Indices**. This formula is directly analogous to every other formula that is used to determine a Time Weighted Average.

$$\mathbf{WBGT_{TWA}} = \frac{\sum_{i=1}^{n}[\mathbf{WBGT_i}][\mathbf{t_i}]}{\sum_{i=1}^{n}\mathbf{t_i}} = \frac{[\mathbf{WBGT_1}][\mathbf{t_1}]+[\mathbf{WBGT_2}][\mathbf{t_2}]+\ \ldots\ +[\mathbf{WBGT_n}][\mathbf{t_n}]}{\mathbf{t_1}+\mathbf{t_2}+\ \ldots\ +\mathbf{t_n}}$$

Where: $\mathbf{WBGT_{TWA}}$ = the **Effective Time Weighted Average WBGT Index** that corresponds to a varied thermal exposure over time, usually measured in °C or °F;

$\mathbf{WBGT_i}$ = the **ith Wet Bulb Globe Temperature Index** that was measured over the Time Interval, $\mathbf{t_i}$, usually measured in °C or °F (either "$WBGT_{Indoor}$" or "$WBGT_{Outdoor}$", but NEVER a mixture of "$WBGT_{Indoor}$" and "$WBGT_{Outdoor}$"); &

$\mathbf{t_i}$ = the **ith Time Interval**, usually measured in hours; however, it can be measured in any useful and consistent units.

THERMAL STRESS PROBLEM SET

Problem #5.1:

What would be the Wet Bulb Globe Temperature Index, in °C, for a Quarry Worker in Connecticut, who must work on a sunny summer morning when the Outdoor Dry Bulb Temperature is 88°F; the Wet Bulb Temperature, 72°F; and the Globe Temperature, 102°F?

Applicable Definition:	Wet Bulb Globe Temperature Index	Page 5-4
Applicable Formulae:	Equation **#1-3**	Page 1-16
	Equation **#5-2**	Pages 5-5 & 5-6
Solution to this Problem:	Page 5-11	

Problem Workspace

Problem #5.2:

Later in the same afternoon, at the same quarry identified in Problem **#5.1**, rain clouds have gathered, and rain has commenced to fall. The Quarry Manager has covered the work area in the quarry pit with a large tarpaulin to protect his employees. If the Wet Bulb Temperature under the tarp has increased to 78°F, while the Globe Temperature has remained unchanged, what will be the new WBGT Index for this slightly different situation?

Applicable Definition:	Wet Bulb Globe Temperature Index	Page 5-4
Data from:	Problem **#5.1**	Page 5-7
Applicable Formulae:	Equation **#1-3**	Page 1-16
	Equation **#5-1**	Page 5-5
Solution to this Problem:	Pages 5-11 & 5-12	

Workspace for Problem **#5.2**

Problem #5.3:

In a large midwest Steel Mill, the corporate Industrial Hygienist has evaluated the heat stress conditions to which each of the Mill's Open Hearth Operators is routinely exposed. These fully acclimatized workers are required, by their job description, to spend certain time periods in areas of the facility that are, quite understandably, very hot. For each time period spent in an area of substantial heat stress, these operators are required to spend a compensating time period in a cool rest area. This IH has determined site specific temperature conditions in this mill, as follows:

Location	Wet-Bulb Temperature	Globe Temperature
Open Hearth Area	117° F	175° F
Elsewhere in the Mill	102° F	93° F
Operator Rest Area	63° F	75° F

A typical Open Hearth Operator spends six 12-minute periods of each workday in the Open Hearth Area, and six corresponding 35-minute periods in the Rest Area. If these operators spend the balance of each 8-hour workday elsewhere in the Mill, what did the IH determine for the Time Weighted Average WBGT Index for a typical Open Hearth Operator?

Applicable Definition:	Wet Bulb Globe Temperature Index	Page 5-4
Applicable Formulae:	Equation **#1-3**	Page 1-16
	Equation **#5-1**	Page 5-5
	Equation **#5-3**	Page 5-6
Solution to this Problem:	Page 5-12	

Workspace for Problem **#5.3**

Problem #5-4:

The Industrial Hygienist in Problem **#5.3** from the previous two pages recommended that the Time Weighted Average WBGT Index for a typical Open Hearth Operator be reduced. His recommendation was that the Steel Mill hire an additional Open Hearth Operator for each Shift. Doing so would decrease from six to five the number of time periods that each of these operators would have to spend in the Open Hearth Area. This IH also recommended that each compensating rest period be increased from 35 minutes to 45 minutes, recognizing that the balance of the time for each of these operators would still be spent elsewhere in the Mill. If these recommendations were implemented, what would the new improved Time Weighted Average WBGT Index be for a typical Open Hearth Operator?

Applicable Definition:	Wet Bulb Globe Temperature Index	Page 5-4
Data from:	Problem **#5.3**	Page 5-8
Applicable Formula:	Equation **#1-10**	Pages 1-19 & 1-20
	Equation **#5-3**	Page 5-6
Solution to this Problem:	Page 5-13	

Problem Workspace

SOLUTIONS TO THE THERMAL STRESS PROBLEM SET

Problem #5.1:

To solve this problem, we will have to employ Equation **#5-2,** from Pages 5-5 & 5-6, since we are dealing with evaluating a WBGT Index with a solar load (i.e., the subscript, "$_{\text{Outside}}$"); thus:

$$WBGT_{Outside} = 0.7[NWB] + 0.2[GT] + 0.1[DB] \qquad \text{[Eqn. #5-2]}$$

$$WBGT_{Outside} = (0.7)(72) + (0.2)(102) + (0.1)(88)$$

$$WBGT_{Outside} = 50.40 + 20.40 + 8.8 = 79.6\,°F$$

The problem statement has asked for the Wet Bulb Globe Temperature Index in degrees Celsius; therefore, we must apply Equation **#1-3**, from Page 1-16, to convert the Fahrenheit temperature to its Celsius equivalent:

$$t_{Metric} = \frac{5}{9}\left[t_{English} - 32\right] \qquad \text{[Eqn. #1-3]}$$

$$t_{Metric} = \frac{5}{9}[79.6 - 32] = \frac{(5)(47.6)}{9} = 26.4\,°C$$

$$\therefore \text{ The } WBGT_{Outside} \text{ Index} = 79.6°\,F = 26.4°C$$

Problem #5.2:

To solve this problem we will have to employ Equation **#5-1,** from Page 5-5, since this problem involves evaluating the WBGT Index without a solar load (i.e., the subscript = "$_{\text{Inside}}$" even though the situation is outside, but under a tarp which eliminates the solar load factor).

An additional subtlety to this problem is the seeming lack of a Wet Bulb Temperature. The clue to this answer, however, is in the problem statement. When it is raining, we know that the relative humidity is 100%; thus the wet bulb temperature equals the dry bulb temperature, since there would be no evaporative cooling of the thermometer bulb by the wet sheath that covers it:

$$WBGT_{Inside} = 0.7[NWB] + 0.3[GT] \qquad \text{[Eqn. #5-1]}$$

$$WBGT_{Inside} = (0.7)(78) + (0.3)(102)$$

$$WBGT_{Inside} = 54.6 + 30.6 = 85.2°F$$

The problem statement has not specifically asked for the Wet Bulb Globe Temperature Index in degrees Celsius; however, to be fully consistent with Problem **#5.1**, we will again apply Equation **#1-3**, from Page 1-16, to convert the Fahrenheit temperature to its Celsius equivalent:

$$t_{Metric} = \frac{5}{9}\left[t_{English} - 32\right] \qquad \text{[Eqn. #1-3]}$$

$$t_{Metric} = \frac{5}{9}[85.2 - 32] = \frac{(5)(53.2)}{9} = 29.6°C$$

∴ The $WBGT_{Inside}$ Index = 85.2°F ~ 29.6°C.

Problem #5.3:

To solve this problem, we will have to employ Equation **#5-1**, from Page 5-5, first, and then Equation **#15A**, from Page 3-35; thus:

$$WBGT_{Inside} = 0.7[NWB] + 0.3[GT] \qquad \text{[Eqn. \#5-1]}$$

1. For the Open Hearth Area:

$$WBGT_{Inside_{Open\ Hearth}} = (0.7)(117) + (0.3)(175)$$

$$WBGT_{Inside_{Open\ Hearth}} = 81.9 + 52.5 = 134.4°F$$

2. For the Operator Rest Area:

$$WBGT_{Inside\ _{Rest\ Area}} = (0.7)(63) + (0.3)(75)$$

$$WBGT_{Inside\ _{Rest\ Area}} = 44.1 + 22.5 = 66.6°F$$

3. For Elsewhere in the Mill:

$$WBGT_{Inside\ _{Elsewhere}} = (0.7)(102) + (0.3)(93)$$

$$WBGT_{Inside\ _{Elsewhere}} = 71.4 + 27.9 = 99.3°F$$

With these three $WBGT_{Inside}$ Indices known, we can now employ Equation **#5-3** from Page 5-6, to obtain the final TWA value requested:

$$WBGT_{TWA} = \frac{[WBGT_1][t_1] + [WBGT_2][t_2] + \ldots + [WBGT_n][t_n]}{t_1 + t_2 + \ldots + t_n} \qquad \text{[Eqn. \#5-3]}$$

$$WBGT_{TWA} = \frac{(134.4)(72) + (66.6)(210) + (99.3)(198)}{72 + 210 + 198}$$

$$WBGT_{TWA} = \frac{9,676.8 + 13,986.0 + 19,661.4}{480} = \frac{43,324.2}{480} = 90.26°F$$

Again the problem statement has not specifically asked for the Wet Bulb Globe Temperature Index in degrees Celsius; however, we will again apply Equation **#1-3**, from Page 1-16, to convert this Fahrenheit temperature to its Celsius equivalent, providing answers in both sets of units:

$$t_{Metric} = \frac{5}{9}\left[t_{English} - 32\right] \qquad \text{[Eqn. \#1-3]}$$

$$t_{Metric} = \frac{5}{9}[90.26 - 32] = \frac{(5)(53.2)}{9} = 32.37°C$$

∴ The $WBGT_{TWA}$ for a typical Open Hearth Operator = 90.26°F ~ 32.4°C.

Problem #5.4:

The solution to this problem will utilize some of the data developed in the solution of the previous problem — specifically, the three calculated $WBGT_{Inside}$ Indices. These values will be entered into Equation **#5-3**, from Page 5-6 to develop the desired solution:

$$WBGT_{TWA} = \frac{[WBGT_1][t_1] + [WBGT_2][t_2] + \ldots + [WBGT_n][t_n]}{t_1 + t_2 + \ldots + t_n} \qquad \text{[Eqn. #5-3]}$$

Remembering, from the solution to Problem **#5.3**, on the previous page, the following data:

1. For the Open Hearth Area: $WBGT_{Inside_{Open\ Hearth}} = 134.4°F$
2. For the Operator Rest Area: $WBGT_{Inside_{Rest\ Area}} = 66.6°F$
3. For Elsewhere in the Mill: $WBGT_{Inside_{Elsewhere}} = 99.3°F$

Next, we note the modified times that each of the Open Hearth Operators must spend in each of the areas of the facility:

1. In the Open Hearth Area: 60 minutes — reduced from 72 minutes [5 periods of 12 minutes each under the new regime vs. 6 periods of 12 minutes each under the former approach].
2. In the Operator Rest Area: 225 minutes — increased from 210 minutes [5 periods of 45 minutes each under the new regime vs. 6 periods of 35 minute each under the former approach].
3. Elsewhere in the Mill: 195 minutes — reduced from 198 minutes — arrived at in each case simply by subtracting the sum of the times spent in the Open Hearth Area + the Operator Rest Area from the time of a full work day, or 480 minutes, thus: [480 – 60 – 225 = 195 minutes] under the new regime, vs. [480 – 72 – 210 = 198 minutes] under the former approach.

Now applying Equation **#5-3**, we get:

$$WBGT_{TWA} = \frac{(134.4)(60) + (66.6)(225) + (99.3)(195)}{60 + 225 + 195}$$

$$WBGT_{TWA} = \frac{8,064.0 + 14,985.0 + 19,363.5}{480} = \frac{42,412.5}{480} = 88.36$$

∴ The improved $WBGT_{TWA}$ for a typical Open Hearth Operator resulting from the modified procedures recommended by the mill's Industrial Hygienist:

88.4°F ~ 31.3°C — an improvement of only ~ 1.9°F or ~ 1.1°C

Chapter 6
Sound and Noise

The general interest in this topic stems from the steadily growing incidence in today's modern industrial workplace, of sound and/or noise-induced hearing impairments (mostly partial, but occasionally, extending even to total hearing loss). The differences between sounds and noise are subjective. Sounds are considered to be something pleasant or useful — such as music [pleasant] or speech [useful]. Noise, in contrast, is thought of as being unpleasant, consisting of such things as the sounds of a table saw cutting wood or a fingernail scratching a chalkboard. This chapter will focus on the factors, parameters, and relationships that permit an accurate assessment of the potential for physiological damage caused by the ambient noise levels that exist in the workplace.

RELEVANT DEFINITIONS

Categories of Noise

Continuous Noise

An unbroken sound, made up of one or more different frequencies of either constant or varying sound intensity/sound pressure level, is referred to as **Continuous Noise**. If such a sound is constant and unvarying in its amplitude, it would be referred to as "steady" **Continuous Noise**. The alternative to this would be "varying" **Continuous Noise**. **Continuous Noise** is a fairly common occurring phenomenon – both in the industrial and the natural environment.

In the natural environment, one might regard the sound of a waterfall as "steady" **Continuous Noise**, while the sounds of wind blowing through a forest would be in the "varying" category.

In the industrial environment, the sound of a rotating electric motor (i.e., a fan, a pump, etc.) would be "steady" **Continuous Noise**, while the operation of a floor waxer, relative to a fixed observer, would be "varying" **Continuous Noise**.

Continuous Noise is an extremely useful concept. In assessing the potential hazards of any noise filled environment, one attempts to quantify the existing noise pattern in terms of the "steady" **Continuous Noise** that could, in theory, replace it without altering any of the adverse effects that might be being experienced by a human observer. For any environment, an $L_{equivalent}$, or the "steady" equivalent **Continuous Noise** level, as described above, can usually be determined; the actual sound intensity level of this "steady" **Continuous Noise** — which equals $L_{equivalent}$ — can then be used to evaluate the overall sound hazard that is posed to individuals who must occupy that environment.

Intermittent Noise

Intermittent Noise is a broken or non-continuous sound (i.e., sound bursts or periods of time during which there are intervals of quiet [non-sound] and subsequent intervals during which there is measurable sound). **Intermittent Noise** can also be made up of one or more different frequencies of sound, of either constant or varying intensity or sound level. The sound of an operating typewriter would be considered an intermittent noise.

Although such a category of noise is more difficult to relate to in the context of an $L_{equivalent}$, such a determination can be, and frequently is, made simply by integrating, over time, the entirety of the noise regime in any setting. Most commercially available sound level meters have the capability to provide an $L_{equivalent}$ for any situation in which sound is to be measured, whether this ambient sound or noise is intermittent, continuous, or a mixture of these different categories.

Characteristics of Sounds and/or Noise

Frequency

The **Frequency** of any sound or noise is the time rate at which complete cycles of high and low pressure regions [compressions and rerefactions] are produced by the noise or sound source. The most common unit of sound **Frequency** is the Hertz (abbreviated, Hz), which is the number of complete cycles that occur in a period of one second. The frequency range over which the human ear can hear varies with age and circumstance; however, a normal hearing "young" ear will usually be able to distinguish sounds, at moderate levels, in the range 20 to 20,000 Hz = 20 to 20,000 cycles/second.

Frequency Band

A sound can be made up of a single frequency — i.e., a tuning fork set at middle "C"; however, it is far more common for a sound or a noise, coming from any source, to be made up of a combination of different frequencies. Whenever a sound or a noise consists of a set of closely related frequencies, this set can be described as a **Frequency Band**. To identify any specific **Frequency Band**, one need only identify the range of frequencies that make it up, namely, by the lowest and the highest frequency in its "inventory". These two frequencies are known as the "Upper & Lower Band-Edge Frequencies" of the particular **Frequency Band** being described.

Octave Bands & Bandwidths

Probably the most commonly used Frequency Band would be the **Octave Band**. A typical **Octave Band** is always characterized by the single frequency numerically located at its geometric center. The "Center Frequency" for any **Octave Band** is the *geometric mean* of its "Upper & Lower Band-Edge Frequencies".

A second important characteristic of an **Octave Band** is the range of frequencies in it, or its **Bandwidth**. The **Bandwidth** of any Frequency Band is the range between its "Upper & Lower Band-Edge Frequencies".

Typically, for a full **Octave Band**, this range will be up to one full octave in total **Bandwidth** — the principal characteristic of one full octave is that its highest frequency [the Upper Band-Edge Frequency] is always exactly twice its lowest frequency [the Lower Band-Edge Frequency]. A tabulation of the standard single, or full, **Octave Bands** is shown at the top of Page 6-3.

FULL OCTAVE BANDS

Center Frequency	Lower Band-Edge Frequency	Upper Band-Edge Frequency
31.3 Hz	22.1 Hz	44.2 Hz
62.5 Hz	44.2 Hz	88.4 Hz
125.0 Hz	88.4 Hz	176.8 Hz
250.0 Hz	176.8 Hz	353.6 Hz
500.0 Hz	353.6 Hz	707.1 Hz
1,000.0 Hz	707.1 Hz	1,414.2 Hz
2,000.0 Hz	1,414.2 Hz	2,828.4 Hz
4,000.0 Hz	2,828.4 Hz	5,656.9 Hz
8,000.0 Hz	5,656.9 Hz	11,313.7 Hz
16,000.0 Hz	11,313.7 Hz	22,627.4 Hz

Sound Wavelength

The **Wavelength** of a **Sound** is the precise distance required for one complete pressure cycle (i.e., one cycle of high [compressed] and low [rarefied] pressure regions) for that frequency of sound. Since sound is a periodic wave phenomenon — even though a markedly different one than the more classic example, light — it can be characterized in terms of its wavelength. This is most easily recognized by considering such things as organ pipes, the lengths of which will always relate to the wavelength of the sound that the particular pipe produces.

Pitch

The **Pitch** of a sound is the subjective auditory perception of the frequency of that sound. It, of course, depends upon the sound frequency, but also on its waveform, on the number of harmonics or overtones present, and on the overall sound pressure level.

Velocity of Sound

The **Velocity of Sound** is the speed at which the sequential regions of high and low pressure propagate away from the source of the sound. For all practical purposes this velocity can be considered to be a constant through whatever medium the sound is transiting. It varies directly as the square root of the density of the medium involved, and inversely as the compressibility of that medium. For example:

Medium	Velocity of Sound
air	~ 1,130 ft/sec
sea water	~ 4,680 ft/sec
hard wood	~ 13,040 ft/sec
steel	~ 16,550 ft/sec

Loudness

The **Loudness** of a sound is an observer's impression of its amplitude. This subjective judgment is influenced strongly by the characteristics of the ear that is doing the hearing.

Characteristic Parameters of Sound and Noise

Sound Intensity & Sound Intensity Level

The **Sound Intensity** of any sound source at any particular location is the average rate at which sound energy from that source is being transmitted through a unit area that is normal to the direction in which the sound is propagating. The most common units of measure for Sound Intensity are joules per square meter [m^2] per second, which are also equal to units of watts per square meter [m^2].

Sound Intensity is usually expressed in terms of an appropriate **Sound Intensity Level**. This parameter is determined by ratioing the Sound Intensity of some noise/sound against the accepted reference base Sound Intensity, which is 10^{-12} watts/m^2. When determined in this manner [see Equation **#6-3** on Page 6-9], the units of the Sound Intensity Level will be in decibels [dBs].

Sound Power & Sound Power Level

The **Sound Power** of any sound source is the total sound energy radiated by that source per unit time. The most common units of measure for Sound Power are watts.

Sound Power is usually expressed in terms of an appropriate **Sound Power Level**. This parameter can be calculated by ratioing the Sound Power of some noise/sound source against the accepted reference base Sound Power, which is 10^{-12} watts. When determined in this manner [see Equation **#6-4** on Page 6-9], the units of the Sound Power Level will also be in decibels [dBs].

Sound Pressure

Sound Pressure normally refers to the RMS values of the pressure changes, above and below atmospheric pressure, which are used to measure steady state or continuous noise. The most common units of measure for **Sound Pressure** are:

newtons per square meter = n/m^2 = pascals

[1 n/m^2 = 1 Pa]

dynes per square centimeter = d/cm^2

microbars

Root-Mean-Square [RMS] Sound Pressure

The **Root-Mean-Square [RMS]** value of any changing quantity, such as sound pressure, is equal to the square root of the mean of the squares of all the measured instantaneous values of that quantity.

Common Measurements of Sound Levels

Sound Pressure Level

Evaluation and/or measurement of any sound or noise, from the perspective of the characteristics of the healthy human ear, poses a difficult problem. This problem arises because of the very wide range of Sound Pressures that the human ear can hear without incurring damage. The healthy human ear can detect sounds at extremely low Sound Pressures [i.e., L_P =

20 μPa], and can survive without damage sounds having very high Sound Pressures [i.e., $L_P = 2 \times 10^8$ μPa = 200 Pa].

When evaluated at a reference frequency of 1,000 Hz, the "effective operating range" of a healthy human ear involves 7+ orders of magnitude of actual Sound Pressures. Because these "human hearing related" significant Sound Pressures vary over such an extremely wide range, the parameter that is most commonly used to describe Sound Pressures is the **Sound Pressure Level** [the **SPL**]. This parameter can be determined by ratioing the measured Sound Pressure of some noise or sound source against the reference base Sound Pressure of 2×10^{-5} n/m^2 = 2×10^{-5} Pa = 20 μPa. When determined in this manner [see Equation **#6-2** on Page 6-8], the units of the Sound Pressure Level will also be expressed in decibels [dBs]. The decibel was chosen in this situation simply because logarithmic units of measure are virtually always judged to be more useful when dealing with parameters whose values may vary over more than 4 or 5 orders of magnitude.

Threshold of Hearing

The **Threshold of Hearing** for a healthy human ear, expressed in decibels and determined at a frequency of 1,000 Hz, is **0 dB**. This is the approximate sound of a feather falling in an otherwise completely quiet room — it is doubtful that the frequency of sound produced by a falling feather would be 1,000 Hz; however, the author can find no other reference to any physical event that would produce this low level of an SPL.

Threshold of Pain

The **Threshold of Pain** for a healthy human ear, also expressed in decibels and also determined at a frequency of 1,000 Hz, is approximately **140** to **145 dB**. At this SPL, an exposed individual would likely experience both permanent damage to his or her hearing and in addition experience actual pain, thus the name. The sound of a commercial jet plane taking off, 25 feet from the unprotected observer, would produce this approximate SPL.

Sound and/or Noise Measurement Time Weightings

In the quantification of various Sound Pressure Levels [in decibels], there are four different commonly used averaging periods, or **Time Weightings**, that are part of the standard RMS detection method. These four are: Peak, Impulse, Fast, and Slow Noise Weightings.

Peak Noise

A burst of sound having a duration of less than 100 milliseconds is considered to be in the **Peak Noise** category. Such sounds will also fall under the category of Impulsive or Impact Noise.

Impulsive or Impact Noise

The types of noise produced by such things as a gun being fired, the operation of an industrial punch press, or the use of a hammer to drive a nail are all highly transient sound phenomena, and are usually treated as **Impulsive or Impact Noises**. This type of noise is defined to be any sound having an amplitude rise time of 35 milliseconds or less, and a fall time of 1,500 milliseconds or less.

Fast Time Weighted Noise

Sound pressure level measurements using a 125-millisecond moving average time weighting period are said to have been determined using **Fast Time Weighting.**

Slow Time Weighted Noise

Sound pressure level measurements using a 1.0-second moving average time weighting period are said to have been determined using **Slow Time Weighting.**

Sound and/or Noise Measurement Frequency Weightings

Linear Frequency Weighting

Any measurement of a sound pressure level can be thought of as the unique "sum" of the ten discrete sound pressure levels of the standard Octave Bands that have made up the sound being monitored. If these measurements are developed without the application of any internal "adjustments" by the sound level meter that is being used — i.e., the meter neither increases nor decreases its measured decibel level of any of the Octave Bands before developing its overall "sum" measurement — then the result is said to have been produced using **Linear Frequency Weighting**. Whenever one attempts to characterize a noise, for the purpose of designing or implementing some sort of sound mitigation, the measurements will probably employ a Linear Weighting approach. If it is desired to measure any one single Octave Band, **Linear Frequency Weighting** will always be employed.

A-Frequency Weighting Scale

The **A-Frequency Weighting Scale** [covered in complete quantitative detail in the next sub-section of this Chapter, namely, **RELEVANT FORMULAE & RELATIONSHIPS**, as the first part of Equation **#6-10**] is a set of measurement weightings that must be applied to the decibel reading for each of the standard Octave Bands that make up the sound being measured. The application of these weighted adjustments — by the internal A-Weighting network in the sound level meter that is making the measurement — ensures that the resultant indicated overall Sound Pressure Level measurement will be of a magnitude that constitutes the very best approximation to what a normal human ear would have perceived. The A-Weighting Scale is usually thought to apply to noises having relatively low level intensities. Sound Pressure Level measurements made using this weighting are always identified by the inclusion of the letter "**A**" after the "dB" unit; thus, **dBA**. The **A-Frequency Weighting Scale** is the most commonly used and widely accepted frequency weighting scale employed in sound pressure level measurements today.

B-Frequency Weighting Scale

The **B-Frequency Weighting Scale** [also covered in complete quantitative detail in the next sub-section of this Chapter, namely, **RELEVANT FORMULAE & RELATIONSHIPS**, as the second part of Equation **#6-10**] is one of the two alternative sets of measurement weightings that must also be applied to each of the standard Octave Bands that make up any sound. The application of these particular weightings — again by a sound level meter's internal B-Weighting network — is designed to produce a result that approximates what a normal human ear would have perceived to noises having relatively moderate, in contrast to low, intensities. Sound Pressure Level measurements made using this category of weighting are always identified by the inclusion of the letter "**B**" after the dB unit, thus, **dBB**. The **B-Frequency Weighting Scale** is not in very wide use today.

C-Frequency Weighting Scale

The **C-Frequency Weighting Scale** [also covered in complete quantitative detail in the next sub-section of this Chapter, namely, **RELEVANT FORMULAE & RELATIONSHIPS**, as the third and final part of Equation **#6-12**] is the second of the two alternative sets of measurement weightings that must also be applied to each of the standard Octave Bands that make up any sound. The application of these particular weightings — again by a sound level meter's internal C-Weighting network — is designed to produce a measurement that approximates what a normal human ear would have perceived as high intensity noises. Sound Pressure Level measurements made using this category of weighting are always identified by the inclusion of the letter "**C**" after the dB unit, thus, **dBC**. The **C-Frequency Weighting Scale** is only rarely used today.

RELEVANT FORMULAE & RELATIONSHIPS

Approximate Velocity of Sound in Air

The velocity of sound in the earth's atmosphere varies directly as the square root of the density of the air. The most easily measured parameter that affects the velocity of sound in the air is its prevailing ambient temperature.

Equation #6-1:

The following relationship, Equation **#6-1**, was empirically derived; however, it has proven to be very accurate for calculating the **Velocity of Sound in Air** over a very wide range of ambient temperatures.

$$\mathbf{V = 49\sqrt{t + 459}}$$

Where:

V = the Velocity of Sound in Air, measured in feet per second; &

t = the Ambient Air Temperature, measured in relative English Units, namely, °F.

Basic Sound Measurements — Definitions

Equation #6-2:

The following relationship, Equation **#6-2**, constitutes the definition of a **Sound Pressure Level**. The expression relates the measured Analog Sound Pressure Level to a "Base Reference Analog Sound Pressure Level", defining the common logarithm of this ratio to be the **Sound Pressure Level**. Because of the potential for extremely wide variations in the measurable analog Sound Pressures, the unit of measure for the **Sound Pressure Level** is the decibel, which, as stated above, is logarithmic and, as such, is better suited as a measure of numeric values, the magnitude of which can vary over several orders of magnitude.

$$\mathbf{L_P = 20\,log\left[\frac{P}{P_0}\right] = 20\,log\left[\frac{P}{2\times 10^{-5}}\right] = 20\,log\,P + 93.98}$$

Where:

$\mathbf{L_P}$ = the **Sound Pressure Level**, measured in decibels (dBs);

P = the measured Analog Sound Pressure Level, in units of newtons/square meter (nt/m^2); &

$\mathbf{P_0}$ = the "Base Reference Analog Sound Pressure Level", which has been defined to have a value of 2×10^{-5} nt/m^2.

Equation #6-3:

This relationship, Equation **#6-3**, is the definition of a **Sound Intensity Level**. In this case, the equation relates the measured Analog Sound Intensity Level to a "Base Reference Analog Sound Intensity Level". As was true in the preceding case, this parameter also is measured in units of decibels, and again for the very same reasons.

$$\mathbf{L_I = 10\,log\left[\frac{I}{I_0}\right] = 10\,log\left[\frac{I}{10^{-12}}\right] = 10\,log\,I + 120}$$

Where:

$\mathbf{L_I}$ = the **Sound Intensity Level**, measured in decibels (dBs);

$\mathbf{I}$ = the Analog Sound Intensity Level, measured in watts/square meter (wts/m^2); &

$\mathbf{I_0}$ = the "Base Reference Analog Sound Intensity Level", which has been defined to have a value of 10^{-12} wts/m^2.

Equation #6-4:

This final analogous relationship, Equation **#6-4**, constitutes the definition of a **Sound Power Level**. As is the case for its two previous close relatives, this expression relates the measured Analog Sound Power Level to a "Base Reference Analog Sound Power Level". Like the two preceding equations, this one also provides **Sound Power Levels** in units of decibels, since the magnitudes of these values can also vary over a number of orders of magnitude.

$$\mathbf{L_P = 10\,log\left[\frac{P}{P_0}\right] = 10\,log\left[\frac{P}{10^{-12}}\right] = 10\,log\,P + 120}$$

Where:

$\mathbf{L_P}$ = the **Sound Power Level**, measured in decibels (dBs);

$\boldsymbol{P}$ = the measured Analog Sound Power Level, measured in units of watts (wts); &

$\boldsymbol{P_0}$ = the "Base Reference Analog Sound Power Level", which has been set to have a value of 10^{-12} wts.

Sound Pressure Levels of Noise Sources in a Free Field

Equation #6-5:

The following expression, Equation **#6-5**, identifies and relates the specific factors that must be accounted for when one determines the **Effective Sound Pressure Level** of any noise source in a "Free Field". For reference, a "Free Field" is any region (within which the noise source is located) that can be characterized as being free or void of any and all objects other than the noise source itself. Such a region permits the unhindered propagation of sound from the source in ALL directions. Because noise sources in the real world are seldom, if ever, in a true "Free Field", this expression has, as its final additive factor, a logarithmic term that effectively adjusts the resultant **Effective Sound Pressure Level** for any asymmetry that may exist in a real world "Non-Free Field" situation. Such factors effectively modify the "Free Fieldness" of the region where the noise source is located. As an example, a bell mounted on a wall would not be able to radiate sound in the direction of the wall; rather, it would effectively radiate sound only into a single spatial hemisphere. The factor that is used to achieve this result modification is called the Directionality Factor, and is defined below.

$$\mathbf{L_{P\text{-}Effective} = L_{P\text{-}Source} - 20\log r - 0.5 + 10\log Q}$$

Where:

$L_{P\text{-}Effective}$ = the **Effective Sound Pressure Level**, evaluated at a point that is "r" feet distant from the noise source itself, measured in decibels (dBs);

$L_{P\text{-}Source}$ = the source Sound Pressure Level, also measured in decibels (dBs);

r = the distance from the point where the **Effective Sound Pressure Level** is to be measured, to the noise source, measured in feet (ft); &

Q = the Directionality Factor, a dimensionless parameter — as defined and valued below:

$Q = 1$	for "spherical omnidirectional" radiating sources;
$Q = 2$	for "single hemisphere" radiating sources;
$Q = 4$	for "single quadrant" radiating sources;
$Q = 8$	for "single octant" radiating sources

Addition of Sound Pressure Levels from Several Independent Sources

Equation #6-6:

The following expression, Equation **#6-6**, is one of the most frequently employed relationships in all of acoustical engineering. It provides the basic methodology for determining the cumulative effect of several noise sources (each producing noise at an identifiable Sound Pressure Level) on an observer. This relationship provides for the determination of the **Ef-**

fective Sound Pressure Level that would be experienced by an observer from the several noise sources.

For this determination, we assume the perspective of an observer whose relative location — among two or more noise sources — causes him or her to experience an overall noise exposure — i.e., an **Effective Sound Pressure Level** — that will be: (1) obviously greater than would have been for a situation involving only a single noise source, but (2) certainly not simply the sum of the several sound pressure levels. As a descriptive example, consider the starter at a drag race. Assume this person is standing midway between two mufflerless dragsters, each producing sound at 130 dBA. Clearly this person would be exposed to a sound pressure level greater than 130 dBA, but not simply the sum value of 260 dBA. **Equation #6-6** provides the solution to the addition of sound pressure levels from several different separate and independent sources.

$$L_{total} = 10\log\left[\sum_{i=1}^{n} 10^{\left[L_i/10\right]}\right]$$

OR

$$L_{total} = 10\log\left[10^{\left[L_1/10\right]} + 10^{\left[L_2/10\right]} + \ldots + 10^{\left[L_n/10\right]}\right]$$

Where:

L_{total} = the total **Effective Sound Pressure Level** resulting from the "n" different noise sources, measured in decibels (dBs); &

L_i = the Sound Pressure Level of the **i**th of "n" different noise sources, also measured in decibels (dBs).

Calculations Involving Sound Pressure Level "Doses"

Equation #6-7:

The following expression, Equation **#6-7**, provides for the determination of the **Maximum Time Period** any worker may be safely exposed to some specifically quantified and/or **Equivalent Sound Pressure Level** from any number of noise sources.

$$T_{max} = \frac{8}{2^{\left[(L-90)/5\right]}}$$

Where:

T_{max} = the **Maximum Time Period** — at any Equivalent Sound Pressure Level, L — to which a worker may be exposed during a normal 8-hour workday, measured in some convenient time unit, usually hours; &

L = the Sound Pressure Level (or Equivalent Sound Pressure Level) being evaluated for this situation, measured in decibels (dBs).

Equation #6-8:

The second relationship involving "doses" is Equation **#6-8**, which provides the basis for calculating the **Effective Daily Dose** that an individual would have experienced as a result of his or her having been exposed to several different well-quantified sound pressure levels, each of which occurred for some specific Time Period or Time Interval.

$$D = \sum_{i=1}^{n} \frac{C_i}{T_{max_i}} = \frac{C_1}{T_{max_1}} + \frac{C_2}{T_{max_2}} + \ldots + \frac{C_n}{T_{max_n}}$$

Where:

D = the **Effective Daily Dose** (of noise) to which an individual who has been exposed to a series of "n" different sound pressure levels, with each of these exposures lasting for a known Time Period or Time Interval, C_i; the **Effective Daily Dose**, **D**, is a dimensionless decimal number;

C_i = the overall ith Time Interval or Time Period during which the individual being considered was exposed to the ith sound pressure level; these time intervals will always have to be measured in some consistent unit of time, usually in hours; &

T_{max_i} = the **Maximum Time Period** that would be permitted for the ith specific sound pressure level to which an individual could be exposed; as defined by Equation **#6-7**, on Page 6-10.

> Note: For any value of **Effective Daily Dose**, **D** $\leq$ 1.00, the individual who has experienced this dose will have accumulated neither an excessive nor a harmful amount of noise. On the other hand, if this parameter assumes a value greater than 1.00, then the exposure would have to be classified as potentially harmful.

Equation #6-9:

The following expression, Equation **#6-9**, provides the relationship for determining the **Equivalent Sound Pressure Level** that corresponds to any identified **Daily Dose**.

$$L_{equivalent} = 90 + 16.61 \log D$$

Where:

$L_{equivalent}$ = the **Equivalent Sound Pressure Level** that corresponds to any Daily Dose, with this parameter measured in decibels (dBs); &

D = the Daily Dose, as defined on the previous page, namely Page 6-11, by **Equation #6-8**.

Definitions of the Three Common Frequency Weighting Scales

Equation #6-10:

The following tabular listing serves as the defining descriptor for each of the three commonly used Frequency Weightings, as these weightings are applied to the measurement of any sound or noise. These weightings, which are applied to the specific Full or Unitary Octave Bands that make up the sound that is being monitored, are designed to provide a measured result that approximates the response of the human ear to sounds or noises of various intensities. Specifically, the **A-Weighting Scale** is thought to provide a result that approximates the response of the human ear to low intensity sounds or noises. The **B-Weighting Scale** provides a human ear based response to sounds or noises having a moderate or medium intensity. Finally, the **C-Weighting Scale** is thought to provide a similar result when applied to high intensity sounds or noises. The **A-Weighting Scale** is, by far, the most widely used of these three; the other two are now only rarely used.

All three Frequency Weighting Sales are in the form of "additions to" OR "deductions from" the Full or Unitary Octave Bands that make up the sound that is being monitored. Whenever the sound level meter being used to monitor some sound or noise has been set up to provide a result to which one of these Frequency Weightings has been applied, the resultant units of the measurement must have — as applicable — an "A", a "B", or a "C" appended to it — i.e., dBA for an **A-Weighting Scale** measurement, dBB for the **B-Weighting Scale**, and/or dBC for the **C-Weighting Scale**.

The following tabulation shows the additions OR deductions that must be applied to the various Octave Bands in order to make the required Frequency Weighting adjustments.

Full Octave Band, in Hertz	⟨Deductions⟩ OR Increments, in decibels		
	A-Scale	B-Scale	C-Scale
31 Hz	⟨39⟩	⟨17⟩	⟨3⟩
63 Hz	⟨26⟩	⟨9⟩	⟨1⟩
125 Hz	⟨16⟩	⟨4⟩	0
250 Hz	⟨9⟩	⟨1⟩	0
500 Hz	⟨3⟩	0	0
1,000 Hz	0	0	0
2,000 Hz	1	0	0
4,000 Hz	1	⟨1⟩	⟨1⟩
8,000 Hz	⟨1⟩	⟨3⟩	⟨3⟩

Various Octave Band Relationships

Equation #s 6-11 & 6-12:

The following two relationships, namely, Equation **#s 6-11 & 6-12**, identify the specific relationships that apply to any Full or Unitary Octave Band, as specified by ANSI S1.11-1966 (R1975). Note, the nine Full or Unitary Octave Bands listed in the tabulation on the previous page, namely, Page 6-13 — as part of the Definitions of the three Frequency Weighting Scales — are the commonly accepted Full or Unitary Octave Bands. For the overall set of Full or Unitary Octave Bands, the Center Frequency of the "Middle" Octave Band is 1,000 Hz. Each member of this set of nine Full or Unitary Octave Bands will have the following characteristics:

(1) Each band will be one full octave in total Bandwidth — i.e., the Band's Lower Band-Edge Frequency will always be half of its Upper Band-Edge Frequency.

(2) The Center Frequency of each band will be the "Geometric Mean" of its Lower and its Upper Band-Edge Frequencies.

(3) Each band will have a Center Frequency that will be one half of the Center Frequency of the next higher Octave Band, and twice the Center Frequency of the next lower one.

The first two of these three overall relationships can be expressed quantitatively and are shown below.

Equation #6-11:

$$f_{upper-1/1} = 2\left(f_{lower-1/1}\right)$$

Equation #6-12:

$$f_{center-1/1} = \sqrt{\left(f_{upper-1/1}\right)\left(f_{lower-1/1}\right)} = \text{the "Geometric Mean"}$$

Where: $\mathbf{f_{upper-1/1}}$ = the Upper Band-Edge Frequency for the specific Full or Unitary Octave Band being considered;

$\mathbf{f_{lower-1/1}}$ = the Lower Band-Edge Frequency for the specific Full or Unitary Octave Band being considered; &

$\mathbf{f_{center-1/1}}$ = the Center Frequency for the specific Full or Unitary Octave Band being considered.

Equation #s 6-13 & 6-14:

The following two relationships, Equation **#s 6-13 & 6-14**, as shown on Page 6-15, identify the specifics of a Standard Half Octave Band, also as specified by ANSI S1.11-1966 (R1975). For the overall set of Half Octave Bands, the Center Frequency of the "Middle" Octave Band is, like its Full or Unitary Octave Band counterpart, at 1,000 Hz. For the overall set of Half Octave Bands, the following set of characteristics always applies:

(1) Each Half Octave Band will be $1/\sqrt{2}$ Octaves in total Bandwidth, i.e., the Lower Band-Edge Frequency of each Half Octave Band will always be $1/\sqrt{2}$ of its Upper Band-Edge Frequency.
(2) The Center Frequency of each Half Octave Band will be the "Geometric Mean" of its Lower and its Upper Band-Edge Frequencies.
(3) Each Half Octave Band will have a Center Frequency that will be $1/\sqrt{2}$ of the Center Frequency of the next higher Half Octave Band, and $\sqrt{2}$ times the Center Frequency of the next lower one.

Again the first two of these three overall relationships can be expressed quantitatively and are shown below.

Equation #6-13:

$$f_{upper-1/2} = \left(\sqrt{2}\right)\left(f_{lower-1/2}\right)$$

Equation #6-14:

$$f_{center-1/2} = \sqrt{\left(f_{upper-1/2}\right)\left(f_{lower-1/2}\right)} = \text{the "Geometric Mean"}$$

Where:

$f_{upper-1/2}$ = the Upper Band-Edge Frequency for the specific Half Octave Band being considered;

$f_{lower-1/2}$ = the Lower Band-Edge Frequency for the specific Half Octave Band being considered; &

$f_{center-1/2}$ = the Center Frequency for the specific Half Octave Band being considered.

Equation #s 6-15 & 6-16:

The following two relationships, **Equation #s 6-15 & 6-16**, as shown on the following page, identify the specifics of a "$1/n^{th}$" Octave Band, also as specified by ANSI S1.11-1966 (R1975).. For the overall set of $1/n^{th}$ Octave Bands, the Center Frequency of the "Middle" Band is, like its other counterparts, at 1,000 Hz. For the overall set of $1/n^{th}$ Octave Bands, the following set of characteristics always applies:

(1) Each $1/n^{th}$ Octave Band will be $1/\sqrt[n]{2}$ Octaves in total Bandwidth, i.e., the Lower Band-Edge Frequency of each $1/n^{th}$ Octave Band will always be $1/\sqrt[n]{2}$ of its Upper Band-Edge Frequency.
(2) The Center Frequency of each $1/n^{th}$ Octave Band will be the "Geometric Mean" of its Lower and its Upper Band-Edge Frequencies.
(3) Each $1/n^{th}$ Octave Band will have a Center Frequency that will be $1/\sqrt[n]{2}$ of the Center Frequency of the next higher Half Octave Band, and $\sqrt[n]{2}$ times the Center Frequency of the next lower one.

Again the first two of these three overall relationships can be expressed quantitatively and are shown on the following page.

Equation #6-15:

$$\mathbf{f_{upper-1/n} = \left(\sqrt[n]{2}\right)\left(f_{lower-1/n}\right)}$$

Equation #6-16:

$$f_{center-1/n} = \sqrt[n]{\left(f_{upper-1/n}\right)\left(f_{lower-1/n}\right)} = \text{the "Geometric Mean"}$$

Where: $\mathbf{f_{upper-1/n}}$ = the Upper Band-Edge Frequency for the specific 1/n^th Octave Band being considered;

$\mathbf{f_{lower-1/n}}$ = the Lower Band-Edge Frequency for the specific 1/n^th Octave Band being considered; &

$\mathbf{f_{center-1/n}}$ = the Center Frequency for the specific 1/n^th Octave Band being considered.

SOUND & NOISE PROBLEM SET

Problem #6.1:

The average noontime, unshaded summer temperature in the Mojave Desert is 129°F. What will be the approximate speed of sound in the Mojave Desert under these conditions?

Applicable Definitions:	Velocity of Sound	Page 6-3
Applicable Formula:	Equation **#6-1**	Page 6-8
Solution to this Problem:	Page 6-31	

Problem Workspace

Problem #6.2:

On a calm day in January of any year, in Fairbanks, AK, the noontime temperature will typically be –35°C. What will be the speed of sound in air under such conditions?

Applicable Definitions:	Velocity of Sound	Page 6-3
Applicable Formulae:	Equation **#1-3**	Page 1-16
	Equation **#6-1**	Page 6-8
Solution to this Problem:	Page 6-31	

Problem Workspace

Problem #6.3:

The maximum continuous noise level that is permitted by OSHA is an SPL of 115 dBA (at this level, the maximum permitted duration of this sort of noise is limited to 7.5 minutes). What is the analog Sound Pressure, in Pascals, of noise at this SPL?

Applicable Definitions:	Sound Pressure	Page 6-4
Applicable Formula:	Equation **#6-2**	Page 6-8
Solution to this Problem:	Pages 6-31 & 6-32	

Problem Workspace

Problem #6.4:

The average analog Sound Intensity, measured at a distance of 2.0 meters, of a hummingbird hovering has been measured to be 2.45×10^{-7} watts/cm^2. What is the corresponding Sound Intensity Level, in dB, at this distance?

Applicable Definitions:	Sound Intensity & Intensity Level	Page 6-4
Applicable Formula:	Equation **#6-3**	Page 6-9
Solution to this Problem:	Page 6-32	

Problem Workspace

Problem #6.5:

The Sound Power Level of a top fuel dragster (at maximum engine and supercharger RPM) is 134 dB. To what analog Sound Power, expressed in watts, does this measured Sound Power Level correspond?

Applicable Definitions:	Sound Power & Power Level	Page 6-4
Applicable Formula:	Equation **#6-4**	Page 6-9
Solution to this Problem:	Page 6-32	

Problem Workspace

Problem #6.6:

At a distance of 300 feet, what sound pressure level, in dBA, would a ground observer, without hearing protection, experience if he were to witness and listen to the takeoff of a U.S. Navy F8U single jet fighter-interceptor? At takeoff, the sound pressure level of this aircraft's afterburner-assisted jet engine, which can be regarded as being directly on the ground, is 165 dBA.

Applicable Definitions:	Sound Pressure Level	Pages 6-4 & 6-5
Applicable Formula:	Equation **#6-5**	Page 6-10
Solution to this Problem:	Pages 6-32 & 6-33	

Problem Workspace

Workspace Continued on the Next Page

Continuation of Workspace for Problem **#6.6**

Problem #6.7:

At what altitude, measured in feet, would the fighter listed in Problem **#6.6** have to pass (measured to a point directly above the observer), in order for that observer to experience the identical sound pressure level that was calculated for the previous problem — in that case from the F8U's afterburner-assisted jet engine while on the ground taking off?

Applicable Definitions:	Sound Pressure Level	Pages 6-4 & 6-5
Applicable Formula:	Equation **#6-5**	Page 6-10
Solution to this Problem:	Page 6-33	

Problem Workspace

Problem #6.8:

The Foreman of a Machine Shop has his work station located an equal distance from six separate grinders, each of which produces noise at 106 dBA when in operation. What Sound Pressure Level, in dBA, would the Foreman experience if all six grinders were operated simultaneously?

Applicable Definitions:	Sound Pressure Level	Pages 6-4 & 6-5
Applicable Formula:	Equation **#6-6**	Pages 6-10 & 6-11
Solution to this Problem:	Page 6-33	

Problem Workspace

Problem #6.9:

The Director of a 20-member bagpipe band experiences a total Sound Pressure Level of 109 dBA when he directs his ensemble. Assuming that every bagpipe produces music (??) at the same sound pressure level as every other one, what must be the Sound Pressure Level of each instrument?

Applicable Definitions:	Sound Pressure Level	Pages 6-4 & 6-5
Applicable Formula:	Equation **#6-6**	Pages 6-10 & 6-11
Solution to this Problem:	Page 6-34	

Problem Workspace

Workspace Continued on the Next Page

Continuation of Workspace for Problem **#6.9**

Problem #6.10:

How much longer is an individual, without hearing protection, permitted to work at a location where the noise level has just been reduced from 104 dBA to 92 dBA?

Applicable Definitions:	Sound Pressure Level	Pages 6-4 & 6-5
Applicable Formula:	Equation **#6-7**	Page 6-11
Solution to this Problem:	Pages 6-34 & 6-35	

Problem Workspace

Problem #6.11:

Standard ear plugs can reduce the sound of a band saw by 24 dBA. Ear muffs can reduce the sound of this saw by 31 dBA. A Band Saw Operator wearing ear plugs can safely operate her band saw for 4.6 hours per day. If she changes to using ear muffs, for how long a period will she be able to operate her band saw?

Applicable Definitions:	Sound Pressure Level	Pages 6-4 & 6-5
Applicable Formula:	Equation **#6-7**	Page 6-11
Solution to this Problem:	Pages 6-35 & 6-36	

Problem Workspace

Problem #6.12:

What is the Daily Dose, expressed as a percentage, for a worker who operates a lathe for 1.5 hours per day, sets his lathe up for 4.5 hours per day, performs administrative tasks for 1 hour per day, and spends the balance of his 8-hour workday either at breaks or eating his lunch? The average noise levels given below were determined by a competent Industrial Hygienist:

Task	Average Sound Pressure Level
Operating the Lathe	95 dBA
Setting up the Lathe	90 dBA
Breaks, Lunch, etc.	84 dBA
Administrative Tasks	82 dBA

Applicable Definitions:	Sound Pressure Level	Pages 6-4 & 6-5
Applicable Formula:	Equation **#6-7**	Page 6-11
	Equation **#6-8**	Page 6-12
Solution to this Problem:	Pages 6-36 & 6-37	

Problem Workspace

Problem #6.13:

What is the Equivalent 8-hour Sound Pressure Level experienced by the Lathe Operator in Problem **#6.12** on the previous page?

Applicable Definitions:	Sound Pressure Level	Pages 6-4 & 6-5
Applicable Formula:	Equation **#6-9**	Pages 6-12 & 6-13
Solution to this Problem:	Page 6-37	

Problem Workspace

Problem #6.14:

Four Printers work on a printing production floor where there are three offset presses. The A-Weighted Sound Pressure Levels, as a function of the number of these presses that are in operation, were determined to be as follows:

Number of Presses Operating	Average Sound Pressure Level	Average Daily Time in Operation
0	81 dBA	4.5 hrs
1	93 dBA	2.1 hrs
2	96 dBA	1.0 hrs
3	98 dBA	0.4 hrs

What is the Daily Dose that these Printers are experiencing? Is their Printing Company employer in violation of any OSHA Sound Pressure Level PEL?

Applicable Definitions:	Sound Pressure Level	Pages 6-4 & 6-5
Applicable Formula:	Equation **#6-7**	Page 6-11
	Equation **#6-8**	Page 6-12
Solution to this Problem:	Pages 6-37 & 6-38	

Workspace for Problem **#6.14**

Problem #6.15:

What is the Equivalent 8-hour Sound Pressure Level experienced by the three Printers listed above in Problem **#6.14**?

Applicable Definitions:	Sound Pressure Level	Pages 6-4 & 6-5
Applicable Formula:	Equation **#6-9**	Pages 6-12 & 6-13
Solution to this Problem:	Page 6-38	

Problem Workspace

Problem #6.16:

A monochromatic tuning fork, operating at "C-below-Middle-C" [for which the frequency is 261 Hz], is observed to produce this tone at an analog Sound Pressure Level of 71 dB, measured on the linear scale. What would a well-calibrated Sound Level Meter, operating on the A-Weighting Scale, indicate as the Sound Pressure Level of this tuning fork?

Applicable Definitions:	Frequency	Page 6-2
	Octave Bands & Bandwidths	Pages 6-2 & 6-3
	Sound Pressure Level	Pages 6-4 & 6-5
	A-Frequency Weighting Scale	Page 6-6
Applicable Formula:	Equation **#6-10**	Page 6-13
Solution to this Problem:	Page 6-38	

Workspace for Problem **#6.16**

Problem #6.17:

What are the Upper and Lower Band-Edge Frequencies of the only Octave band on the A-Weighting Scale that does not have a Sound Pressure Level adjustment?

Applicable Definitions:	Frequency	Page 6-2
	Octave Bands & Bandwidths	Pages 6-2 & 6-3
	Sound Pressure Level	Pages 6-4 & 6-5
	A-Frequency Weighting Scale	Page 6-6
Applicable Formula:	Equation **#6-10**	Page 6-13
	Equation **#6-11**	Page 6-14
	Equation **#6-12**	Page 6-14
Solution to this Problem:	Pages 6-38 & 6-39	

Problem Workspace

Problem #6.18:

What is the Center Frequency of the Standard Unitary Octave Band, for which the Lower Band-Edge Frequency is 2,828 kHz? Justify your choice quantitatively.

Applicable Definitions:	Frequency	Page 6-2
	Octave Bands & Bandwidths	Pages 6-2 & 6-3
Applicable Formula:	Equation **#6-11**	Page 6-14
	Equation **#6-12**	Page 6-14
Solution to this Problem:	Page 6-39	

Problem Workspace

Problem #6.19:

What are the Upper and Lower Band-Edge Frequencies of the Standard One Half Octave Band that has a Center Frequency of 354 Hz?

Applicable Definitions:	Frequency	Page 6-2
	Octave Bands & Bandwidths	Pages 6-2 & 6-3
Applicable Formula:	Equation **#6-13**	Pages 6-14 & 6-15
	Equation **#6-14**	Pages 6-14 & 6-15
Solution to this Problem:	Page 6-40	

Workspace for Problem **#6.19**

Problem #6.20:

What is the Center Frequency of the Standard One Third Octave Band for which the Lower Band-Edge Frequency is 1,122 Hz?

Applicable Definitions:	Frequency	Page 6-2
	Octave Bands & Bandwidths	Pages 6-2 & 6-3
Applicable Formula:	Equation **#6-15**	Pages 6-15 & 6-16
	Equation **#6-16**	Pages 6-15 & 6-16
Solution to this Problem:	Pages 6-40 & 6-41	

Problem Workspace

SOLUTIONS TO THE SOUND & NOISE PROBLEM SET

Problem #6.1:

To solve this problem, we must use Equation **#6-1**, from Page 6-8:

$$V = 49\sqrt{t + 459} \qquad \text{[Eqn. \#6-1]}$$

$$V_{129°F} = 49\sqrt{129 + 459} = 49\sqrt{588}$$

$$V_{129°F} = (49)(24.25) = 1{,}188.19 \text{ ft/sec}$$

∴ The Speed of Sound at 129°F in the Mojave Desert is 1,188.2 ft/sec.

Problem #6.2:

For the solution to this problem, we must again use Equation **#6-1**, from Page 6-8:

$$V = 49\sqrt{t + 459} \qquad \text{[Eqn. \#6-1]}$$

We must, of course, first convert this temperature, given in relative Metric units, namely, °C, to its corresponding temperature, in relative English units, namely, °F. To do this, we must use **Equation #1-3**, from Page 1-16:

$$t_{Metric} = \frac{5}{9}\left[t_{English} - 32°\right] \qquad \text{[Eqn. \#1-16]}$$

We must now rewrite this expression to solve for the relative English temperature, thus:

$$t_{English} = \frac{9}{5}\left[t_{Metric}\right] + 32°$$

$$t_{English} = \frac{9}{5}(-35°) + 32° = -63° + 32° = -31°F$$

Now that we have the relative English System temperature, we can apply Equation **#6-1**, from Page 6-8:

$$V_{-35°C} = 49\sqrt{(-31° + 459°)} = 49\sqrt{428°}$$

$$V_{-35°C} = (49)(20.69) = 1{,}013.72 \text{ ft/sec}$$

∴ The Speed of Sound at –35°C in Anchorage, AK, is 1,013.7 ft/sec.

Problem #6.3:

This problem relies upon the Definition of an Analog Sound Pressure Level, as listed in Equation **#6-2**, from Page 6-8:

$$L_P = 20\,log\,P + 93.98 \qquad \text{[Eqn. \#6-2]}$$

$$115 = 20\,log\,P + 93.98$$

$$20\log P = 115 - 93.98 = 21.02$$

$$\log P = \frac{21.02}{20} = 1.05$$

We next must take the antilogarithm of both sides of this equation, thus:

$$P = 11.25 \text{ nt/m}^2 = 11.25 \text{ Pa}$$

∴ The Analog Sound Pressure that results in a 115 dBA sound = 11.3 Pa.

Problem #6.4:

This solution to this problem relies on the Definition of an Analog Sound Intensity Level, as listed in Equation **#6-3**, from Page 6-9:

$$L_I = 10\log I + 120 \qquad \text{[Eqn. \#6-3]}$$

$$L_I = 10\log\left(2.45\times10^{-7}\right) + 120$$

$$L_I = (10)(-6.611) + 120 = -66.11 + 120 = 53.89 \text{ dB}$$

∴ The Analog Sound Intensity Level of a hovering hummingbird at a distance of 2 meters is 53.9 dB.

Problem #6.5:

This solution to this problem relies upon the Definition of an Analog Sound Power Level, as listed in **Equation #6-4**, from Page 6-9:

$$L_P = 10\log[P] + 120 \qquad \text{[Eqn. \#6-4]}$$

$$134 = 10\log[P] + 120$$

$$10\log[P] = 134 - 120 = 14$$

$$\log[P] = \frac{14}{10} = 1.4$$

We next must take the antilogarithm of both sides of this equation:

$$P = 25.12 \text{ watts}$$

∴ The Sound Power of a Top Fuel Dragster (at maximum engine and supercharger RPM) is 25.1 watts.

Problem #6.6:

The solution to this problem will require the use of Equation **#6-5**, from Page 6-10. We must first observe that this aircraft's jet engine is producing sound as a "single hemisphere" radiating source — i.e., since its jet engine can be considered to be "directly on the ground", it radiates sound only into the air (it is, therefore, only into a "single hemisphere" radiating source); consequently, we must use a directionality factor, **Q = 2**, thus:

$$L_{P\text{-}Effective} = L_{P\text{-}Source} - 20\,log[r] - 0.5 + 10\,log[Q] \qquad \text{[Eqn. \#6-5]}$$

$$L_{P\text{-}Effective} = 165 - 20\,log(300) - 0.5 + 10\,log(2)$$

$$L_{P\text{-}Effective} = 165 - (20)(2.477) - 0.5 + (10)(0.301)$$

$$L_{P\text{-}Effective} = 165 - 49.54 - 0.5 + 3.01 = 117.97 \text{ dBA}$$

∴ The effective Sound Pressure Level experienced by the Ground Observer listed in this problem would be 118 dBA — a very uncomfortably loud sound.

Problem #6-7:

The solution to this problem will also require the use of Equation **#6-5**, from Page 6-10. In this case, we observe that this aircraft's jet engine is producing sound as a "spherical omnidirectional" radiating source — i.e., since the F8U is now airborne, its jet engine is now radiating sound in all directions; consequently, we must use a directionality factor, **Q = 1**, thus:

$$L_{P\text{-}Effective} = L_{P\text{-}Source} - 20\,log[r] - 0.5 + 10\,log[Q] \qquad \text{[Eqn. \#6-5]}$$

$$117.97 = 165 - 20\,log[r] - 0.5 + 10\,log[1]$$

$$20\,log[r] = 165 - 117.97 - 0.5 + (10)(0) = 46.53$$

$$log[r] = \frac{46.53}{20} = 2.33$$

Next taking the antilogarithm of both sides of this equation, we get:

$$r = 212.14 \text{ feet}$$

∴ The F8U will deliver a ~ 118 dBA sound when it is at an altitude of approximately 212 feet directly above the Ground Observer.

Problem #6-8:

This is the classic problem involving the addition of several different quantified (in dBA) noise sources with the goal of obtaining a single overall equivalent noise source. It will require the use of Equation **#6-6**, from Pages 6-10 & 6-11:

$$L_{Total} = 10\,log\left[\sum_{i=1}^{n} 10^{L_i/10}\right] = 10\,log\left[10^{L_1/10} + 10^{L_2/10} + \ldots + 10^{L_n/10}\right] \qquad \text{[Eqn. \#6-6]}$$

$$L_{Total} = 10\,log\left[\left(10^{106/10}\right)(6)\right] = 10\,log\left[\left(3.981\times10^{10}\right)(6)\right]$$

$$L_{Total} = 10\,log\left[2.39\times10^{11}\right] = (10)(11.38) = 113.78 \text{ dBA}$$

∴ The Foreman of this Machine Shop will experience noise, at a combined level of 113.8 dBA, from the simultaneous operation of all six grinders.

Problem #6.9:

This problem is completely analogous to Problem **#6.8**; it deals with the combined effect of several different sound sources; it, too, will require the use of Equation **#6-6**, from Pages 6-10 & 6-11:

$$L_{Total} = 10\,log\left[\sum_{i=1}^{n} 10^{L_i/10}\right] = 10\,log\left[10^{L_1/10} + 10^{L_2/10} + \ldots + 10^{L_n/10}\right] \quad \text{[Eqn. #6-6]}$$

$$109 = 10\,log\left[(20)\left(10^{L_{Bagpipe}/10}\right)\right]$$

$$log\left[(20)\left(10^{L_{Bagpipe}/10}\right)\right] = \frac{109}{10} = 10.9$$

We next must take the antilogarithm of both sides of this equation, thus:

$$(20)\left(10^{L_{Bagpipe}/10}\right) = 7.943 \times 10^{10}$$

$$10^{L_{Bagpipe}/10} = \frac{7.943 \times 10^{10}}{20} = 3.972 \times 10^{9}$$

Now we must take the common logarithm of both sides of this equation:

$$\frac{L_{Bagpipe}}{10} = 9.599$$

$$L_{Bagpipe} = (10)(9.599) = 95.99 \text{ dBA}$$

∴ Each Bagpipe produces music/noise at approximately 96 dBA.

Problem #6.10:

To answer this question, we must first determine the OSHA permitted duration, in hours, for each of the two identified noise levels. Once this has been accomplished, we simply employ Equation **#6-7**, from Page 6-11, to obtain the requested result:

$$T_{max} = \frac{8}{2^{[L-90]/5}} \quad \text{[Eqn. #6-7]}$$

1. For an SPL of 104 dBA:

$$T_{max\ @\ 104\ dBA} = \frac{8}{2^{(104-90)/5}} = \frac{8}{2^{14/5}}$$

$$T_{max\ @\ 104\ dBA} = \frac{8}{2^{2.8}} = \frac{8}{6.964} = 1.149 \text{ hours}$$

2. For an SPL of 92 dBA:

$$T_{max\ @\ 92\ dBA} = \frac{8}{2^{(92-90)/5}} = \frac{8}{2^{2/5}}$$

$$T_{max\ @\ 92\ dBA} = \frac{8}{2^{0.4}} = \frac{8}{1.32} = 6.063 \text{ hours}$$

The additional time permitted at the lesser noise level of 92 dBA, ΔT_{max}, is simply the difference between these two OSHA permitted time intervals; thus:

$$\Delta T_{max} = 6.063 - 1.149 = 4.914 \text{ hours}$$

∴ This individual can spend an additional 4.9 hours (or ~ 4 hours, 54 minutes) at a 92 dBA noise level than would have been permitted at a 104 dBA level.

Problem #6.11:

This problem can be solved by using Equation **#6-7**, from Page 6-11, first to identify the effective equivalent Sound Pressure Level that is being experienced by the Band Saw Operator when she is wearing her ear plugs; then reapply the same relationship to determine the additional time permitted when she uses ear muffs:

$$T_{max} = \frac{8}{2^{[L-90]/5}} \qquad \text{[Eqn. \#6-7]}$$

We must begin by first rearranging this equation so as to solve for the required SPL:

$$2^{[L-90]/5} = \frac{8}{T_{max}}$$

Next, we must take the common logarithm of both sides of this expression:

$$\text{og}\left[2^{[L-90]/5}\right] = \log\left[\frac{8}{T_{max}}\right]$$

$$\log(2)\left(\frac{L-90}{5}\right) = \log(8) - \log(T_{max})$$

$$(0.301)\left(\frac{L-90}{5}\right) = 0.903 - \log(T_{max})$$

$$\frac{L-90}{5} = \frac{0.903 - \log(T_{max})}{0.301}$$

$$L - 90 = \frac{5}{0.301}\left[0.903 - \log(T_{max})\right]$$

$$L = \frac{5}{0.301}\left[0.903 - \log(T_{max})\right] + 90$$

Finally, now, we have an expression that will permit the direct determination of the equivalent "attenuated" band saw SPL (we are dealing here with an attenuated SPL, not the actual operating SPL of the Band Saw). We know the permitted time interval the Operator can work using ear plugs; thus we can determine the effective equivalent SPL she must be exposed to while using her ear plugs, thus:

$$L = \left(\frac{5}{0.301}\right)\left[0.903 - \log(4.6)\right] + 90 = (16.61)(0.903 - 0.663) + 90$$

$$L = (16.61)(0.240) + 90 = 3.992 + 90 = 93.992 \sim 94 \text{ dBA}$$

Therefore, this Band Saw Operator experiences a noise level of 94 dBA while she operates the band saw wearing ear plugs. If she wears ear muffs, she will experience a noise level 7 dBA lower than this level (i.e., there is a 31 dBA reduction with ear muffs vs. 24 dBA reduction with ear plugs). The new reduced noise level will, therefore, become approximately 87 dBA, and the maximum time permitted at this SPL will be given by Equation **#6-7**, from Page 6-11:

$$T_{max} = \frac{8}{2^{[L-90]/5}} \qquad \text{[Eqn. \#6-7]}$$

$$T_{max\ @\ 87\ dBA} = \frac{8}{2^{[87-90]/5}} = \frac{8}{2^{-3/5}} = \frac{8}{0.660} = 12.126 \text{ hours}$$

> ∴ Using ear muffs, this Band Saw Operator will be able to operate her band saw, without the danger of suffering any hearing loss, for up to 12.1 hours (or ~ 12 hours, 6 minutes) per day.

Problem #6.12:

The solution to this problem will require the use of both Equation **#6-7**, from Page 6-11, and Equation **#6-8**, from Page 6-12, and in that order; thus:

$$T_{max} = \frac{8}{2^{[L-90]/5}} \qquad \text{[Eqn. \#6-7]}$$

1. For an average SPL of 95 dBA:

$$T_{max\ @\ 95\ dBA} = \frac{8}{2^{[95-90]/5}} = \frac{8}{2^{5/5}} = \frac{8}{2} = 4 \text{ hours}$$

2. For an average SPL of 90 dBA:

$$T_{max\ @\ 90\ dBA} = \frac{8}{2^{[90-90]/5}} = \frac{8}{2^{0/5}} = \frac{8}{1} = 8 \text{ hours}$$

3. For an average SPL of 84 dBA:

$$T_{max\ @\ 84\ dBA} = \frac{8}{2^{[84-90]/5}} = \frac{8}{2^{-6/5}} = \frac{8}{0.435} = 18.379 \text{ hours}$$

4. For an average SPL of 82 dBA:

$$T_{max\ @\ 82\ dBA} = \frac{8}{2^{[82-90]/5}} = \frac{8}{2^{-8/5}} = \frac{8}{0.330} = 24.251 \text{ hours}$$

Now with each of these T_{max}s at the various average SPLs, we can apply Equation **#6-8**, from Page 6-12, to obtain the requested result:

$$D = \sum_{i=1}^{n} \frac{C_i}{T_{max_i}} = \frac{C_1}{T_{max_1}} + \frac{C_2}{T_{max_2}} + \ldots + \frac{C_n}{T_{max_n}} \qquad \text{[Eqn. \#6-8]}$$

$$D_{Lathe\ Operator} = \frac{1.5}{4.00} + \frac{4.5}{8.00} + \frac{1}{18.379} + \frac{1}{24.251}$$

$$D_{Lathe\ Operator} = 0.375 + 0.563 + 0.054 + 0.041 = 1.033$$

Now, expressing this result as a percentage, as required by the problem statement, we have:

$$D_{Lathe\ Operator} = 103.3\%$$

∴ This Lathe Operator's daily noise dose, expressed as a percentage, is 103.3%.

Problem #6.13:

The solution to this problem, which is an extension of Problem **#6.12**, requires the use of Equation **#6-9**, from Pages 6-12 & 6-13:

$$L_{equivalent} = 90 + 16.61 log[D] \qquad \text{[Eqn. #6-9]}$$

$$L_{equivalent} = 90 + 16.61 log(1.033)$$

$$L_{equivalent} = 90 + (16.61)(0.014) = 90 + 0.234 = 90.234$$

$$L_{equivalent} = 90.234 \text{ dBA}$$

∴ This Lathe Operator experienced an equivalent SPL of ~ 90.24 dBA.

Problem #6.14:

Like Problem **#6.12**, earlier, the solution to this problem will require the use of both Equation **#6-7**, from Page 6-11, and Equation **#6-8**, from Page 6-12, and in that order:

$$T_{max} = \frac{8}{2^{[L-90]/5}} \qquad \text{[Eqn. #6-7]}$$

1. For an average SPL of 81 dBA, over a duration of 4.5 hours:

$$T_{max\ @\ 81\ dBA} = \frac{8}{2^{[81-90]/5}} = \frac{8}{2^{-9/5}} = \frac{8}{0.287} = 27.858 \text{ hours}$$

2. For an average SPL of 93 dBA, over a duration of 2.1 hours:

$$T_{max\ @\ 93\ dBA} = \frac{8}{2^{[93-90]/5}} = \frac{8}{2^{3/5}} = \frac{8}{1.516} = 5.278 \text{ hours}$$

3. For an average SPL of 96 dBA, over a duration of 1.0 hours:

$$T_{max\ @\ 96\ dBA} = \frac{8}{2^{[96-90]/5}} = \frac{8}{2^{6/5}} = \frac{8}{2.297} = 3.482 \text{ hours}$$

4. For an average SPL of 98 dBA, over a duration of 0.4 hours:

$$T_{max\ @\ 98\ dBA} = \frac{8}{2^{[98-90]/5}} = \frac{8}{2^{8/5}} = \frac{8}{3.031} = 2.639 \text{ hours}$$

Now with these four T_{max}s at each of the various average SPLs, we can apply Equation **#6-8**, from Page 6-12, to obtain the requested result; thus:

$$D = \sum_{i=1}^{n} \frac{C_i}{T_{max_i}} = \frac{C_1}{T_{max_1}} + \frac{C_2}{T_{max_2}} + \ldots + \frac{C_n}{T_{max_n}} \qquad \text{[Eqn. #6-8]}$$

$$D_{Printer} = \frac{4.5}{27.858} + \frac{2.1}{5.278} + \frac{1.0}{3.482} + \frac{0.4}{2.639}$$

$$D_{Printer} = 0.162 + 0.398 + 0.287 + 0.152 = 0.998$$

Now, expressing this result as a percentage as required by the problem statement, we have:

$$D_{Printer} = 98.82\%$$

∴ These Printers' daily noise dose expressed as a percentage = 99.8%.

& The Printing Company that employs these four Printers is not in violation of any established OSHA SPL dosage standards.

Problem #6.15:

The solution to this problem, which is an extension of Problem **#6.14**, will require the use of Equation **#6-9**, from Pages 6-12 & 6-13:

$$L_{equivalent} = 90 + 16.61\,log[D] \qquad \text{[Eqn. \#6-9]}$$

$$L_{equivalent} = 90 + 16.61\,log(0.998)$$

$$L_{equivalent} = 90 + (16.61)(-0.001) = 90 - 0.013 = 89.987 \sim 90 \text{ dBA}$$

∴ These Printers experience an equivalent SPL of ~ 90 dBA.

Problem #6.16:

The solution to this problem requires only the Descriptive Definition of the A-Frequency Weighting Scale Factors, as shown in the Tabulation associated with Equation **#6-10**, from Page 6-13. Clearly this note ("C-below-Middle-C" — reference the piano scale) falls into the 250-Hz Octave Band. For this Octave Band, we must deduct 9 dB from every identified Linear Sound Pressure level; therefore:

∴ The SPL of this "C-Below-Middle-C" Tuning Fork = 71 dB_{Linear} = 62 dBA.

Problem #6-17:

Again from the Descriptive Definition of the A-Frequency Weighting Scale Factors, as shown in the Tabulation associated with Equation **#6-10**, from Page 6-13, we can see that the only Octave Band that does not have a SPL adjustment is the 1,000-Hz Octave Band. To determine the Upper and Lower Band-Edge Frequencies, we must employ the following two Equations, namely, Equation **#s 6-11 & 6-12**, both from Page 6-14:

$$f_{upper-1/1} = 2f_{lower-1/1} \qquad \text{[Eqn. \#6-11]}$$

$$f_{center-1/1} = \sqrt{[f_{upper-1/1}][f_{lower-1/1}]} \qquad \text{[Eqn. \#6-12]}$$

Combining these two expressions, we obtain the useful relationship shown below:

$$f_{center-1/1} = \sqrt{[2f_{lower-1/1}][f_{lower-1/1}]} = \sqrt{2}(f_{lower-1/1})$$

Now, solving for the Lower Band-Edge Frequency, given the Center Frequency, we get:

$$f_{lower-1/1} = \frac{f_{center-1/1}}{\sqrt{2}} = \frac{\sqrt{2}}{2}\left(f_{center-1/1}\right)$$

Now, substituting in the known values, we get:

$$f_{lower-1/1} = \frac{\sqrt{2}}{2}(1,000) = (0.707)(1,000) = 707.1 \text{ Hz}$$

And since we know from Equation **#6-11** that the Upper Band-Edge Frequency is twice the Lower Band-Edge Frequency, we can obtain the final result asked for in the problem statement:

$$f_{upper-1/1} = 2f_{lower-1/1} \qquad \text{[Eqn. \#6-11]}$$

$$f_{upper-1/1} = 2\left(f_{lower-1/1}\right) = 1,414.2 \text{ Hz}$$

∴ The Upper and Lower Band-Edge Frequencies of the 1,000-Hz Standard Unitary Octave Band are as follows:

Upper Band-Edge Frequency = 1,414 Hz
Lower Band-Edge Frequency = 707 Hz

Problem #6.18:

To develop the solution for this problem, we must employ the following two Equations, namely, Equation **#s 6-11 & 6-12**, both from Page 6-14. In this case, we must determine the Center Frequency of a Full Octave Band for which we know only the Lower Band-Edge Frequency.

$$f_{upper-1/1} = 2f_{lower-1/1} \qquad \text{[Eqn. \#6-11]}$$

$$f_{center-1/1} = \sqrt{\left[f_{upper-1/1}\right]\left[f_{lower-1/1}\right]} \qquad \text{[Eqn. \#6-12]}$$

Again combining these two relationships, as was the case for Problem **#6.17**, we can develop the following equation:

$$f_{center-1/1} = \sqrt{2}\left(f_{lower-1/1}\right)$$

Now, substituting in the known value for the Lower Band-Edge Frequency, we can directly determine the Center Frequency, thus:

$$f_{center-1/1} = \sqrt{2}(2,828) = 3,999.4 \sim 4,000 \text{ Hz}$$

∴ The Center Frequency of this Full Octave Band is 4,000 Hz (rounding up from the result of 3,999.4 Hz). This is the Standard 4,000 Hz = 4 kHz Octave Band.

Problem #6.19:

To solve this problem, we will have to apply the two Equations that make up the relationships for Half Octave Bands, namely, Equation **#s 6-13 & 6-14**, from Pages 6-14 & 6-15:

$$f_{upper-1/2} = \left(\sqrt{2}\right)\left(f_{lower-1/2}\right) \qquad \text{[Eqn. \#6-13]}$$

$$f_{center-1/2} = \sqrt{\left(f_{upper-1/2}\right)\left(f_{lower-1/2}\right)} \qquad \text{[Eqn. \#6-14]}$$

Again combining these two relationships, as was the case for Problem **#s 6.17 & 6.19**, we get the following useful equation:

$$f_{center-1/2} = \sqrt{\left(f_{lower-1/2}\right)\left(\left[\sqrt{2}\right]\left[f_{lower-1/2}\right]\right)} = \left(f_{lower-1/2}\right)\sqrt{\sqrt{2}} = \left(f_{lower-1/2}\right)\sqrt[4]{2}$$

We can rearrange this relationship to solve for the Lower Band-Edge Frequency:

$$f_{lower-1/2} = \frac{f_{center-1/2}}{\sqrt[4]{2}}$$

Using this relationship, we can solve directly for the Lower Band-Edge Frequency of this Half Octave Band:

$$f_{lower-1/2} = \frac{354}{\sqrt[4]{2}} = \frac{354}{1.189} = 297.7 \text{ Hz}$$

With the Lower Band-Edge Frequency known, we can use Equation **#6-13**, from Pages 6-14 & 6-15, to calculate the required Upper Band-Edge Frequency directly:

$$f_{upper-1/2} = \left(\sqrt{2}\right)\left(f_{lower-1/2}\right)$$

$$f_{upper-1/2} = \left(\sqrt{2}\right)(297.7) = 421.0 \text{ Hz}$$

> ∴ The Upper and Lower Band-Edge Frequencies of the 354-Hz Standard Half Octave Band are as follows:
>
> Upper Band-Edge Frequency ~ 421 Hz
>
> Lower Band-Edge Frequency ~ 298 Hz

Problem #6.20:

To solve this problem, we will have to apply the two Equations that make up the relationships for any 1/nth Octave Bands, as shown in Equation **#s 6-15 & 6-16**, from Pages 6-15 & 6-16, respectively. We must start out by determining the Upper Band-Edge Frequency for this 1/3rd Octave Band; using Equation **#6-15**:

$$f_{upper-1/n} = \sqrt[n]{2}\left(f_{lower-1/n}\right) \qquad \text{[Eqn. \#6-15]}$$

Using this relationship we can directly determine the Upper Band-Edge Frequency for this 1/3rd Octave Band:

$$f_{upper-1/3} = \sqrt[3]{2}\left(f_{lower-1/3}\right) = \sqrt[3]{2}(1,122) = (1.26)(1,122) = 1,413.6 \text{ Hz}$$

Finally, now we can apply Equation **#6-16** to determine the requested Center Frequency of this 1/3rd Octave Band:

$$f_{center-1/n} = \sqrt{\left(f_{upper-1/n}\right)\left(f_{lower-1/n}\right)} \qquad \text{[Eqn. \#6-16]}$$

$$f_{center-1/3} = \sqrt{(1,122)(1,413.6)} = \sqrt{1,586,094.5} = 1.259.4 \text{ Hz}$$

∴ The Standard 1/3rd Octave Band, for which the Lower Band-Edge Frequency is 1,122 Hz, has a Center Frequency of 1,259 Hz.

Chapter 7
Ionizing & Non-ionizing Radiation

Interest in this area of potential human hazard stems, in part, from the magnitude of harm or damage that an individual who is exposed can experience. It is widely known that the risks associated with exposures to ionizing radiation are significantly greater than comparable exposures to non-ionizing radiation. This fact notwithstanding, it is steadily becoming more widely accepted that non-ionizing radiation exposures also involve risks to which one must pay close attention. This chapter will focus on the fundamental characteristics of the various types of ionizing and non-ionizing radiation, as well as on the factors, parameters, and relationships whose application will permit accurate assessments of the hazard that might result from exposures to any of these physical agents.

RELEVANT DEFINITIONS

Electromagnetic Radiation

Electromagnetic Radiation refers to the entire spectrum of photonic radiation, from wavelengths of less than 10^{-5} Å (10^{-15} meters) to those greater than 10^{8} meters — a dynamic wavelength range of more than 22+ decimal orders of magnitude! It includes all of the segments that make up the two principal sub-categories of this overall spectrum, which are the "Ionizing" and the "Non-Ionizing" radiation sectors. Photons having wavelengths shorter than 0.4 μ (400 nm or 4,000 Å) fall under the category of Ionizing Radiation; those with longer wavelengths will all be in the Non-Ionizing group. In addition, the overall Non-Ionizing Radiation sector is further divided into the following three sub-sectors:

Optical Radiation Band *	0.1 μ to 2,000 μ, or 0.0001 to 2.0 mm
Radio Frequency/Microwave Band	2.0 mm to 10,000,000 mm, or 0.002 to 10,000 m
Sub-Radio Frequency Band	10,000 m to 10,000,000+ m, or 10 km to 10,000+ km

* *It must be noted that the entirety of the ultraviolet sector [0.1 μ to 0.4 μ wavelengths] is listed as a member of the Optical Radiation Band, and appears, therefore, to be a Non-Ionizing type of radiation. This is not true. UV radiation is indeed ionizing; it is just categorized incorrectly insofar as its group membership among all the sectors of **Electromagnetic Radiation**.*

Although the discussion thus far has focused on the wavelengths of these various bands, this subject also has been approached from the perspective of the frequencies involved. Not surprisingly, the dynamic range of the frequencies that characterize the entire **Electromagnetic Radiation** spectrum also covers 22+ decimal orders of magnitude — ranging from 30,000 exahertz or 3×10^{22} hertz [for the most energetic cosmic rays] to approximately 1 or 2 hertz [for the longest wavelength ELF photons]. The energy of any photon in this overall spectrum will be directly proportional to its wavelength — i.e., photons with the highest frequency will be the most energetic.

The most common **Electromagnetic Radiation** bands are shown in a tabular listing on the following page. This tabulation utilizes increasing wavelengths, or λs, as the basis for identifying each spectral band.

Electromagnetic Radiation Bands

Spectral Band	Photon Wavelength, λ, for each Band Band Min. λ	Band Max. λ
IONIZING RADIATION		
Cosmic Rays	<0.00005 Å	0.005 Å
γ-Rays	0.005 Å	0.8 Å
X-Rays — hard	0.8 Å	5.0 Å
X-Rays — soft	5.0 Å 0.5 nm	80 Å 8.0 nm
NON-IONIZING RADIATION Optical Radiation Bands		
Ultraviolet — UV-C	8.0 nm 0.008 μ	250 nm 0.25 μ
Ultraviolet — UV-B	250 nm 0.25 μ	320 nm 0.32 μ
Ultraviolet — UV-A	320 nm 0.32 μ	400 nm 0.4 μ
Visible Light	0.4 μ	0.75 μ
Infrared — Near or IR-A	0.75 μ	2.0 μ
Infrared — Mid or IR-B	2.0 μ	20 μ
Infrared — Far or IR-C	20 μ 0.02 mm	2,000 μ 2 mm
Radio Frequency/Microwave Bands		
Extremely High Frequency [EHF] *Microwave* Band	1 mm	10 mm
Super High Frequency [SHF] *Microwave* Band	10 mm	100 mm
Ultra High Frequency [UHF] *Microwave* Band	100 mm 0.1 m	1,000 mm 1 m
Very High Frequency [VHF] *Radio Frequency* Band	1 m	10 m
High Frequency [HF] *Radio Frequency* Band	10 m	100 m
Medium Frequency [MF] *Radio Frequency* Band	100 m 0.1 km	1,000 m 1 km
Low Frequency [LF] Band	1 km	10 km
Sub-Radio Frequency Bands		
Very Low Frequency [VLF] Band	10 km	100 km
Ultra Low Frequency [ULF] Band	100 km 0.1 Mm	1,000 km 1 Mm
Super Low Frequency [SLF] Band	1 Mm	10 Mm
Extremely Low Frequency [ELF] *Power Freq.* Band	10 Mm	>100 Mm

Ionizing Radiation

Ionizing Radiation is any photonic (or particulate) radiation — either produced naturally or by some man-made process — that is capable of producing or generating ions. Only the shortest wavelength [highest energy] segments of the overall electromagnetic spectrum are capable of interacting with other forms of matter to produce ions. Included in this grouping are most of the ultraviolet band [even though this band is catalogued in the Non-Ionizing sub-category of Optical Radiation], as well as every other band of photonic radiation having wavelengths shorter than those in the UV band.

Ionizations produced by this class of electromagnetic radiation can occur either "directly" or "indirectly". "Directly" ionizing radiation includes:

(1) electrically charged particles [i.e., electrons, positrons, protons, α-particles, etc.], &

(2) photons/particles of sufficiently great kinetic energy that they produce ionizations by colliding with atoms and/or molecules present in the matter.

In contrast, "indirectly" ionizing particles are always uncharged [i.e., neutrons, photons, etc.]. They produce ionizations indirectly, either by:

(1) liberating one or more "directly" ionizing particles from matter with which these particles have interacted or are penetrating, or

(2) initiating some sort of nuclear transition or transformation [i.e., radioactive decay, fission, etc.] as a result of their interaction with the matter through which these particles are passing.

Protection from the adverse effects of exposure to various types of **Ionizing Radiation** is an issue of considerable concern to the occupational safety and health professional. Certain types of this class of radiation can be very penetrating [i.e., γ-Rays, X-Rays, & neutrons]; that is to say these particles will typically require very substantial shielding in order to ensure the safety of workers who might otherwise become exposed. In contrast to these very penetrating forms of **Ionizing Radiation**, α- and β-particles are far less penetrating, and therefore require much less shielding.

Categories of Ionizing Radiation

Cosmic Radiation

Cosmic Radiation [cosmic rays] makes up the most energetic — therefore, potentially the most hazardous — form of Ionizing Radiation. **Cosmic Radiation** consists primarily of high speed, very high energy protons [protons with velocities approaching the speed of light] — many or even most with energies in the billions or even trillions of electron volts. These particles originate at various locations throughout space, eventually arriving on the earth after traveling great distances from their "birthplaces". Cataclysmic events, or in fact any event in the universe that liberates large amounts of energy [i.e., supernovae, quasars, etc.], will be sources of **Cosmic Radiation**. It is fortunate that the rate of arrival of cosmic rays on Earth is very low; thus the overall, generalized risk to humans of damage from cosmic rays is also relatively low.

Nuclear Radiation

Nuclear Radiation is, by definition, terrestrial radiation that originates in, and emanates from, the nuclei of atoms. From one perspective then, this category of radiation probably should not be classified as a subset of electromagnetic radiation, since the latter is made up of photons of pure energy, whereas **Nuclear Radiation** can be either energetic photons or particles possessing mass [i.e., electrons, neutrons, helium nuclei, etc.]. It is clear, however, that this class of "radiation" does belong in the overall category of Ionizing Radiation; thus it will be discussed here. In addition, according to Albert Einstein's Relativity Theory, energy and mass are equivalent — simplistically expressed as $\mathbf{E = mc^2}$ — this fact further solidifies the inclusion of **Nuclear Radiation** in this area.

Nuclear events such as radioactive decay, fission, etc. all serve as sources for **Nuclear Radiation**. Gamma rays, X-Rays, alpha particles, beta particles, protons, neutrons, etc., as stated on the previous page, can all be forms of **Nuclear Radiation**. Cosmic rays should also be included as a subset in this overall category, since they clearly originate from a wide variety of nuclear sources, reactions, and/or disintegrations; however, since they are extraterrestrial in origin, they are not thought of as **Nuclear Radiation**. Although of interest to the average occupational safety and health professional, control and monitoring of this class of ionizing radiation usually falls into the domain of the Health Physicist.

Gamma Radiation

Gamma Radiation — Gamma Rays [γ-Rays] — consists of very high energy photons that have originated, most probably, from one of the following four sources:

(1) nuclear fission [i.e., the explosion of a simple "atomic bomb", or the reactions that occur in a power generating nuclear reactor],
(2) nuclear fusion [i.e., the reactions that occur during the explosion of a fusion based "hydrogen bomb", or the energy producing mechanisms of a star, or the operation of one of the various experimental fusion reaction pilot plants, the goal of which is the production of a self-sustaining nuclear fusion-based source of power],
(3) the operation of various fundamental particle accelerators [i.e., electron linear accelerators, heavy ion linear accelerators, proton synchrotrons, etc.], or
(4) the decay of a radionuclide.

While there are clearly four well-defined source categories for **Gamma Radiation**, the one upon which we will focus will be the decay of a radioactive nucleus. Most of the radioactive decays that produce γ-Rays also produce other forms of ionizing radiation [β^--particles, principally]; however, the practical uses of these radionuclides rest mainly on their γ-Ray emissions. The most common application of this class of isotope is in the medical area. Included among the radionuclides that have applications in this area are: ${}^{125}_{53}I$ & ${}^{131}_{53}I$ [both used in thyroid therapy], and ${}^{60}_{27}Co$ [often used as a source of high energy γ-Rays in radiation treatments for certain cancers].

Gamma rays are uncharged, highly energetic photons possessing usually 100+ times the energy, and less than 1% of the wavelength, of a typical X-Ray. They are very penetrating, typically requiring a substantial thickness of some shielding material [i.e., lead, steel reinforced concrete, etc.].

Alpha Radiation

Alpha Radiation — Alpha Rays [α-Rays, α-particles] — consists solely of the completely ionized nuclei of helium atoms, generally in a high energy condition. As such, α-Rays are particulate and not simply pure energy; thus they should not be considered to be electromagnetic radiation — see the discussion under the topic of Nuclear Radiation, beginning on the previous page.

These nuclei consist of two protons and two neutrons each, and as such, they are among the heaviest particles that one ever encounters in the nuclear radiation field. The mass of an α-particle is 4.00 atomic mass units, and its charge is +2 [twice the charge of the electron, but positive — the basic charge of an electron is -1.6×10^{-19} coulombs]. The radioactive decay of many of the heaviest isotopes in the periodic table frequently involves the emission of α-particles. Among the nuclides included in this grouping are: ${}^{238}_{92}U$, ${}^{226}_{88}Ra$, and ${}^{222}_{86}Rn$.

Considered as a member of the nuclear radiation family, the α-particle is the least penetrating. Typically, **Alpha Radiation** can be stopped by a sheet of paper; thus, shielding individuals from exposures to α-particles is relatively easy. The principal danger to humans arising from exposures to α-particles occurs when some alpha active radionuclide is ingested and becomes situated in some vital organ in the body where its lack of penetrating power is no longer a factor.

Beta Radiation

Beta Radiation constitutes a second major class of directly ionizing charged particles; and again because of this fact, this class or radiation should not be considered to be a subset of electromagnetic radiation.

There are two different β-particles — the more common negatively charged one, the β^- [the electron], and its positive cousin, the β^+ [the positron]. **Beta Radiation** most commonly arises from the radioactive decay of an unstable isotope. A radioisotope that decays by emitting β-particles is classified as being beta active. Among the most common beta active [all β^- active] radionuclides are: ${}^{3}_{1}H$ (tritium), ${}^{14}_{6}C$, and ${}^{90}_{38}Sr$.

Most **Beta Radiation** is of the β^- category; however, there are radionuclides whose decay involves the emission of β^+ particles. β^+ emissions inevitably end up falling into the Electron Capture [EC] type of radioactive decay simply because the emitted positron — as the antimatter counterpart of the normal electron, or β^- particle — annihilates immediately upon encountering its antiparticle, a normal electron. Radionuclides that are β^+ active include: ${}^{22}_{11}Na$ and ${}^{18}_{9}F$.

Although more penetrating than an α-particle, the β-particle is still not a very penetrating form of nuclear radiation. β-particles can generally be stopped by very thin layers of any material of high mass density [i.e., 0.2 mm of lead], or by relatively thicker layers of more common, but less dense materials [i.e., a 1-inch thickness of wood]. As is the case with α-particles, β-particles are most dangerous when an ingested beta active source becomes situated in some susceptible organ or other location within the body.

Neutron Radiation

Although there are no naturally occurring neutron sources, this particle still constitutes an important form of nuclear radiation; and again since the neutron is a massive particle, it should not simply be considered to be a form of electromagnetic radiation. As was the case with both α- and β-particles, neutrons can generate ions as they interact with matter; thus they definitely are a subset of the overall class of ionizing radiation. The most important source of **Neutron Radiation** is the nuclear reactor [commercial, research, and/or military]. The characteristic, self-sustaining chain reaction of an operating nuclear reactor, by definition, generates a steady supply of neutrons. Particle accelerators also can be a source of **Neutron Radiation**.

Protecting personnel from exposures arising from **Neutron Radiation** is one of the most difficult problems in the overall area of radiation protection. Neutrons can produce considerable damage in exposed individuals. Unlike their electrically charged counterparts [α- and β-particles], uncharged neutrons are not capable, either directly or indirectly, of producing ionizations. Additionally, neutrons do not behave like high energy photons [γ-Rays and/or X-Rays] as they interact with matter. These relatively massive uncharged particles simply pass through matter without producing anything until they collide with one of the nuclei that are resident there. These collisions accomplish two things simultaneously:

(1) they reduce the energy of the neutron, and

(2) they "blast" the target nucleus, usually damaging it in some very significant manner — i.e., they mutate this target nucleus into an isotope of the same element that has a higher atomic weight, one that will likely be radioactive. Alternatively, if neutrons are passing through some fissile material, they can initiate and/or maintain a fission chain reaction, etc.

Shielding against **Neutron Radiation** always involves processes that reduce the energy or the momentum of the penetrating neutron to a point where its collisions are no longer capable of producing damage. High energy neutrons are most effectively attenuated [i.e., reduced in energy or momentum] when they collide with an object having approximately their same mass. Such collisions reduce the neutron's energy in a very efficient manner. Because of this fact, one of the most effective shielding media for neutrons is water, which obviously contains large numbers of hydrogen nuclei, or protons which have virtually the same mass as the neutron.

X-Radiation

X-Radiation — X-Rays — consists of high energy photons that, by definition, are man-made. The most obvious source of **X-Radiation** is the X-Ray Machine, which produces these energetic photons as a result of the bombardment of certain heavy metals — i.e., tungsten, iron, etc. — with high energy electrons. X-Rays are produced in one or the other of the two separate and distinct processes described below:

(1) the acceleration (actually, negative acceleration or "deceleration") of a fast moving, high energy, negatively charged electron as it passes closely by the positively charged nucleus of one of the atoms of the metal matrix that is being bombarded [energetic X-Ray photons produced by this mechanism are known as "Bremsstrahlung X-Rays", and their energy ranges will vary according to the magnitude of the deceleration experienced by the bombarding electron]; and

(2) the de-excitation of an ionized atom — an atom that was ionized by a bombarding, high energy electron, which produced the ionization by "blasting" out one of the target atom's own inner shell electrons — the de-excitation occurs when one of the target atom's remaining outer shell electrons "falls" into (transitions into) the vacant inner shell position, thereby producing an X-Ray with an energy precisely equal to the energy difference between the beginning and ending states of the target atom [energetic X-Ray photons produced in this manner are known as "Characteristic X-Rays" because their energies are always precisely known].

The principal uses of **X-Radiation** are in the areas of medical and industrial radiological diagnostics. The majority of the overall public's exposure to ionizing radiation occurs as a result of exposure to X-Rays.

Like their γ-Ray counterparts, X-Rays are uncharged, energetic photons with substantial penetrating power, typically requiring a substantial thickness of some shielding material [i.e., lead, iron, steel reinforced concrete, etc.] to protect individuals who might otherwise be exposed.

Ultraviolet Radiation

Photons in the **Ultraviolet Radiation**, or UV, spectral band have the least energy that is still capable of producing ionizations. As stated earlier, all of the UV band has been classified as being a member of the *Optical Radiation Band*, which — by definition — is Non-Ionizing. This is erroneous, since UV is indeed capable of producing ionizations in exposed matter. Photoionization detection, as a basic analytical tool, relies on the ability of certain wavelengths of UV radiation to generate ions in certain gaseous components.

"Black Light" is a form of **Ultraviolet Radiation**. In the industrial area, UV radiation is produced by plasma torches, arc welding equipment, and mercury discharge lamps. The most prominent source of UV is the Sun.

Ultraviolet Radiation has been further classified into three sub-categories by the *Commission Internationale d'Eclairage* (CIE). These CIE names are: UV-A, UV-B, and UV-C. The wavelengths associated with each of these "CIE Bands" are shown in the tabulation on Page 7-2.

The UV-A band is the least dangerous of these three, but it has been shown to produce cataracts in exposed eyes. UV-B and UV-C are the bands responsible for producing injuries such as photokeratitis [i.e., welder's flash, etc.], and erythema [i.e., sunburn, etc.]. A variety of protective measures are available to individuals who may become exposed to potentially harmful UV radiation. Included among these methods are glasses or skin ointments designed to block harmful UV-B and/or UV-C photons.

Categories of Non-Ionizing Radiation

Visible Light

Visible Light is that portion of the overall electromagnetic spectrum to which our eyes are sensitive. This narrow spectral segment is the central member of the *Optical Radiation Band*. The hazards associated with **Visible Light** depend upon a combination of the energy of the source and the duration of the exposure. Certain combinations of these factors can pose very significant hazards [i.e., night and color vision impairments]. In cases of extreme exposure, blindness can result. As an example, it would be very harmful to an individual's vision for that individual to stare, even for a very brief time period, at the sun without using some sort of eye protection. In the same vein, individuals who must work with visible light lasers must always wear protective glasses — i.e., glasses with appropriate optical density characteristics.

For reference, the retina, which is that part of the eye that is responsible for our visual capabilities, can receive the entire spectrum of visible light as well as the near infrared — which will be discussed under the next definition. It is the exposure to these bands that can result in vision problems for unprotected individuals.

Infrared Radiation

Infrared Radiation, or IR, is the longest wavelength sector of the overall *Optical Radiation Band*. The IR spectral band, like its UV relative, is usually thought of as being divided into three sub-segments, the near, the mid, and the far. These three sub-bands have also been designated by the *Commission Internationale d'Eclairage* (CIE), respectively, as IR-A, IR-B, and IR-C. The referenced non-CIE names, "near", "mid", and "far", refer to the relative position of the specific IR band with respect to visible light — i.e., the near IR band has wavelengths that are immediately adjacent to the longest visible light wavelengths, while the far IR photons, which have the greatest infrared wavelengths, are most distant from the visible band. In general, we experience **Infrared Radiation** as radiant heat.

As stated earlier in the discussion for visible light, the anterior portions of the eye [i.e., the lens, the vitreous humor, the cornea, etc.] are all largely opaque to the mid and the far IR;

only the photons of the near IR can penetrate all the way to the retina. Near IR photons are, therefore, responsible for producing retinal burns. Mid and far IR band photons, for which the anterior portions of the eye are relatively opaque, will typically be absorbed in these tissues and are, therefore, responsible for injuries such as corneal burns.

Microwave Radiation

General agreement holds that **Microwave Radiation** involves the EHF, SHF, & UHF Bands, plus the shortest wavelength portions of the VHF Band — basically, the shortest wavelength half of the *Radio Frequency/Microwave Band* sub-group. All the members of this group have relatively short wavelengths — the maximum λ is in the range of 3 meters.

Virtually all the adverse physiological effects or injuries that accrue to individuals who have been exposed to harmful levels of **Microwave Radiation** can be understood from the perspective of the "radiation" rather than the "electric and/or magnetic field" characteristics of these physical agents [see the discussion of the differences between these two characteristic categories, as well as the associated concepts of the "Near Field" and the "Far Field", later on Pages 7-10 & 7-11, under the heading, Radiation Characteristics vs. Field Characteristics]. Physiological injuries to exposed individuals, to the extent that they occur at all, are simply the result of the absorption — within the body of the individual who has been exposed to the **Microwave Radiation** — of a sufficiently large amount of energy to produce significant heating in the exposed organs or body parts. The long-term health effects of exposures that do not produce any measurable heating [i.e., increases in the temperature of some organ or body part] are unknown at this time.

Some of the uses/applications that make up each of the previously identified **Microwave Radiation** bands are listed in the following tabulation:

Band	Wavelength	Frequency	Use or Application
EHF	1 to 10 mm	300 to 30 GHz	Satellite Navigational Aids & Communications, Police 35 GHz *K Band* Radar, Microwave Relay Stations, Radar: *K (partial), L & M Bands* (military fire control), High Frequency Radio, etc.
SHF	10 to 100 mm	30 to 3 GHz	Police 10 & 24 GHz *J & K Band* Radars, Satellite Communications, Radar: *F, G, H, I, J, & K (partial) Bands* (surveillance, & marine applications), etc.
UHF	0.1 to 1.0 m	3,000 to 300 MHz	UHF Television [Channels 14 to 84], certain CB Radios, Cellular Phones, Microwave Ovens, Radar: *B (partial), D, & E Bands* (acquisition & tracking, + air traffic control), Taxicab Communications, Spectroscopic Instruments, some Short-wave Radios, etc.
VHF	1.0 to 3.0 m	300 to 100 MHz	Higher Broadcast Frequency Standard Television [174 to 216 MHz: Channels 7 to 13], Radar *B Band*, Higher Frequency FM Radio [100+ MHz], walkie-talkies, certain CB Radios, Cellular Telephones, etc.

Radio Frequency Radiation

Radio Frequency Radiation makes up the balance of the *Radio Frequency/Microwave Band* sub-group. The specific segments involved are the longest wavelength half of the VHF Band, plus all of the HF, MF, & LF Bands. In general, all of the wavelengths involved in this sub-group are considered to be long to very long, with the shortest λ being 3+ meters and the longest, approximately 10 km, or just less than 6.25 miles.

The adverse physiological effects or injuries, if any, that result from exposures to **Radio Frequency Radiation** can be understood from the perspective of the "electric and/or magnetic field", rather than the "radiation" characteristics of these particular physical agents [again, see the discussion of the differences between these two characteristic categories, as well as the associated concepts of the "Near Field" and the "Far Field", later on Pages 7-10 & 7-11, under the heading, Radiation Characteristics vs. Field Characteristics]. Injuries to exposed individuals, to the extent that they have been documented at all, are also the result of the absorption by some specific organ or body part of a sufficiently large amount of energy to produce highly localized heating. As was the case with Microwave Radiation exposures, the long-term health effects of exposure events that do not produce any measurable heating are unknown at this time.

Some of the uses/applications that make up each of the previously identified **Radio Frequency Radiation** bands are listed in the following tabulation:

Band	Wavelength	Frequency	Use or Application
VHF	3.0 to 10.0 m	100 to 30 MHz	Lower Frequency Broadcast Standard Television [54 to 72, & 76 to 88 MHz: Channels 2 to 6], Lower Frequency FM Radio [88 to 100 MHz], Dielectric Heaters, Diathermy Machines, certain CB Radios, certain Cellular Telephones, etc.
HF	10 to 100 m	30 to 3 MHz	Plasma Processors, Dielectric Heaters, various types of Welding, some Short-wave Radios, Heat Sealers, etc.
MF	0.1 to 1.0 km	3,000 to 300 kHz	Plasma Processors, AM Radio, various types of Welding, some Short-wave Radios, etc.
LF	1 to 10 km	300 to 30 kHz	Cathode Ray Tubes or Video Display Terminals

Sub-Radio Frequency Radiation

This final portion of the overall electromagnetic spectrum is comprised of its longest wavelength members. **Sub-Radio Frequency Radiation** makes up its own "named" category, namely, the *Sub-Radio Frequency Band*, as the final sub-group of the overall category of Non-Ionizing Radiation.

At the time that this paragraph is being written, there is little agreement as to the adverse physiological effects that might result from exposures to **Sub-Radio Frequency Radiation**. Again, and to the extent that human hazards do exist for this class of physical agent, these hazards can be best understood from the perspective of the "electric and/or magnetic field", rather than the "radiation" characteristics of **Sub-Radio Frequency Radiation** [again, see the discussion of the differences between these two characteristic categories, as well as the associated concepts of the "Near Field" and the "Far Field", on this page and the next, under the heading, Radiation Characteristics vs. Field Characteristics].

Primary concern in this area seems generally to be related to the strength of either or both the electric and the magnetic fields that are produced by sources of this class of radiation. The American Conference of Government Industrial Hygienists [ACGIH] has published the following expressions that can be used to calculate the appropriate 8-hour TLV-TWA — each as a function of the frequency, **f**, of the **Sub-Radio Frequency Radiation** source being considered. The relationship for electric fields provides a field strength TLV expressed in volts/meter [V/m]; while the relationship for magnetic fields produces a magnetic flux density TLV in milliteslas [mT].

Electric Fields

$$E_{TLV} = \frac{2.5 \times 10^6}{f}$$

Magnetic Fields

$$B_{TLV} = \frac{60}{f}$$

Finally, one area where there does appear to be very considerable, well-founded concern about the hazards produced by **Sub-Radio Frequency Radiation** is in the area of the adverse impacts of the electric and magnetic fields produced by this class of source on the normal operation of cardiac pacemakers. An electric field of 2,500 volts/meter [2.5 kV/m] and/or a magnetic flux density of 1.0 gauss [1.0 G, which is equivalent to 0.1 milliteslas or 0.1 mT] each clearly has the potential for interrupting the normal operation of an exposed cardiac pacemaker, virtually all of which operate at roughly these same frequencies.

Some of the uses/applications that make up each of the previously identified **Sub-Radio Frequency Radiation** bands are listed in the following tabulation:

Band	Wavelength	Frequency	Use or Application
VLF	10 to 100 km	30 to 3 kHz	Cathode Ray Tubes or Video Display Terminals [video flyback frequencies], certain Cellular Telephones, Long-Range Navigational Aids [LORAN], etc.
ULF	0.1 to 1 Mm	3,000 to 300 Hz	Induction Heaters, etc.
SLF	1 to 10 Mm	300 to 30 Hz	Standard Electrical Power [60 Hz], Home Appliances, Underwater Submarine Communications, etc.
ELF	10 to 100 Mm	30 to 3 Hz	Underwater Submarine Communications, etc.

Radiation Characteristics vs. Field Characteristics

All of the previous discussions have been focused on the various categories and subcategories of the electromagnetic spectrum [excluding, in general, the category of particulate nuclear radiation]. It must be noted that every band of electromagnetic radiation — from the extremely high frequencies of Cosmic Rays [frequencies often greater than 3×10^{21} Hz or 3,000 EHz] to the very low end frequencies characteristic of normal electrical power in the United States [i.e., 60 Hz] — will consist of photons of **radiation** possessing both electric and magnetic **field** characteristics.

That is to say, we are dealing with **radiation** phenomena that possess **field** [electric and magnetic] characteristics. The reason for considering these two different aspects or factors is that measuring the "strength" or the "intensity" of any radiating source is a process in which only rarely will both the **radiation** and the **field** characteristics be easily quantifiable. The vast majority of measurements in this field will, of necessity, have to be made on only one or the other of these two characteristics. It is the frequency and/or the wavelength being considered that determines whether the measurements will be made on the **radiation** or the **field** characteristics of the source involved.

When the source frequencies are relatively high — i.e., **f** > 100 MHz [with λ < 3 meters] — it will almost always be easier to treat and measure such sources as simple **radiation** sources. For these monitoring applications [with the exception of situations that involve lasers], it will be safe to assume that the required "strength" and/or "intensity" characteristics will behave like and can be treated as if they were **radiation** phenomena — i.e., they vary according to the inverse square law.

In contrast, when the source frequencies fall into the lower ranges — i.e., **f** $\leq$ 100 MHz [with $\lambda \geq$ 3 meters] — then it will be the **field** characteristics that these sources produce [electric and/or magnetic] that will be relatively easy to measure. While it is certainly true that these longer wavelength "photons" do behave according to the inverse square law — since they are, in fact, radiation — their relatively long wavelengths make it very difficult to measure them as radiation phenomena.

These measurement problems relate directly to the concepts of the Near and the Far Field. The Near Field is that region that is close to the source — i.e., no more than a very few wavelengths distant from it. The Far Field is the entire region that exists beyond the Near Field.

Field measurements [i.e., separate electric and/or magnetic field measurements] are usually relatively easy, so long as the measurements are completed in the Near Field. It is in this region where specific, separate, and distinct measurements of either of these two **fields** can be made. The electric **fields** that exist in the Near Field are produced by the voltage characteristics of the source, while the magnetic **fields** in this region result from the source's electrical current. Electric field strengths will typically be expressed in one of the following three sets of units: (1) volts/meter — v/m; (2) volts^2/meter^2 — v^2/m^2; or (3) milliwatts/cm^2 — mW/cm^2. Magnetic field intensities will typically be expressed in one of the following four sets of units: (1) amperes/meter — A/m; (2) milliamperes/meter — mA/m; (3) Amperes^2/meter^2 — A^2/m^2; or (4) milliwatts/cm — mW/cm.

Radiation measurements, in contrast, are typically always made in the Far Field. As an example, let us consider a 75,000 volt X-Ray Machine — i.e., one that is producing X-Rays with an energy of 75 keV. For such a machine, the emitted X-Rays will have a frequency of 1.81×10^{19} Hz and a wavelength of 1.66×10^{-11} meters, or 0.166 Å [from Planck's Law]. Clearly for such a source, it would be virtually impossible to make any measurements in the Near Field — i.e., within a very few wavelengths distant from the source — since even a six wavelength distance would be only 1 Å away [a 1 Å distance is less than the diameter of a methane molecule!!]. Measurements made in the Far Field of the strength or intensity of a radiating source then will always be **radiation** measurements, usually in units such as millirem/hour — mRem/hr. As stated earlier, **radiation** behaves according to the inverse square law, a relationship that states that radiation intensity decreases as the square of the distance between the point of measurement and the source.

Sources of Ionizing Radiation

Radioactivity

Radioactivity is the process by which certain unstable atomic nuclei undergo a nuclear disintegration. In this disintegration, the unstable nucleus will typically emit one or more of: (1) the common sub-atomic particles [i.e., the α-Particle, the β-Particles, etc.], and/or (2) photons of electromagnetic energy, [i.e., γ-Rays, etc.].

Radioactive Decay

Radioactive Decay refers to the actual process — involving one or more separate and distinct steps — by which some specific radioactive element, or radionuclide, undergoes the transition from its initial condition, as an "unstable" nucleus, ultimately to a later generation "unstable" radioactive nucleus, or — eventually — a "stable" non-radioactive nucleus. In the process of this **Radioactive Decay**, the originally unstable nucleus will very frequently experience a change in its basic atomic number. Whenever this happens, its chemical identity will change — i.e., it will become an isotope of a different element. As an example, if an unstable nucleus were to emit an electron [i.e., a β^--particle], its atomic number would increase by one — i.e., an unstable isotope of calcium decays by emitting an electron, and in so doing becomes an isotope of scandium, thus:

$${}^{45}_{20}Ca \rightarrow {}^{45}_{21}Sc + {}^{0}_{-1}e, \text{ which could also be written as follows: } {}^{45}_{20}Ca \rightarrow {}^{45}_{21}Sc + \beta^-$$

A second example would be the **Radioactive Decay** of the only naturally occurring isotope of thorium, which involves the emission of an α-particle:

$${}^{232}_{90}Th \rightarrow {}^{228}_{88}Ra + {}^{4}_{2}He, \text{ which could also be written as follows: } {}^{232}_{90}Th \rightarrow {}^{228}_{88}Ra + {}^{4}_{2}\alpha$$

In this situation, the unstable thorium isotope was converted into an isotope of radium.

Radioactive Decay can occur in any of nine different modes. These nine are listed below, in each case with an example of a radioactive isotope that undergoes radioactive decomposition — in whole or in part — following the indicated decay mode:

Decay Mode	Example
Alpha Decay [α-decay]	${}^{235}_{92}U \rightarrow {}^{231}_{90}Th + {}^{4}_{2}He$
Beta Decay [β^--decay]	${}^{90}_{38}Sr \rightarrow {}^{90}_{39}Y + {}^{0}_{-1}e$
Positron Decay [β^+-decay]	${}^{22}_{11}Na \rightarrow {}^{22}_{10}Ne + {}^{0}_{+1}e + \gamma$ [simultaneous β^+ & γ-decay]
Gamma Decay [γ-decay]	${}^{60}_{27}Co \rightarrow {}^{60}_{28}Ni + {}^{0}_{-1}e + \gamma$ [simultaneous β^- & γ-decay]
Neutron Decay [*n*-decay]	${}^{252}_{98}Cf \rightarrow {}^{107}_{42}Mo + {}^{141}_{56}Ba + 4{}^{1}_{0}n$ [simultaneous *n*-decay & SF]
Electron Capture [EC]	${}^{125}_{53}I + {}^{0}_{-1}e \rightarrow {}^{125}_{52}Te + \gamma$ [simultaneous EC & γ-decay]
Internal Conversion [IC]	${}^{125}_{52}Te \rightarrow {}^{0}_{-1}e$ [following the simultaneous EC & γ-decay reaction shown above; the electron is ejected — i.e., IC — from one of the technetium atom's innermost electron subshells]
Isomeric Transition [IT]	${}^{121m}_{50}Sn \rightarrow {}^{121}_{50}Sn + \gamma$ [simultaneous IT & γ-decay]
Spontaneous Fission [SF]	${}^{252}_{98}Cf \rightarrow {}^{107}_{42}Mo + {}^{141}_{56}Ba + 4{}^{1}_{0}n$ [simultaneous SF & *n*-decay]

Radioactive Decay Constant

The **Radioactive Decay Constant** is the isotope specific "time" coefficient that appears in the exponent term of Equation **#7-4** on Page 7-18. Equation **#7-4** is the widely used relationship that always serves as the basis for determining the quantity [atom count or mass] of any as yet undecayed radioactive isotope. This exponential relationship is used to evaluate remaining quantities at any time interval after a starting determination of an "initial" quantity. By definition, all radioactive isotopes decay over time, and the **Radioactive Decay Constant** is an empirically determined factor that effectively reflects the speed at which the decay process has occurred or is occurring.

Mean Life

The **Mean Life** of any radioactive isotope is simply the average "lifetime" of a single atom of that isotope. Quantitatively, it is the reciprocal of that nuclide's Radioactive Decay Constant — see Equation **#7-6**, on Page 7-19. **Mean Lives** can vary over extremely wide ranges of time; as an example of this wide variability, the following are the **Mean Lives** of two fairly common radioisotopes, namely, the most common naturally occurring isotope of uranium and a fairly common radioactive isotope of beryllium:

For an atom of ${}^{238}_{92}U$, the **Mean Life** [α-decay] is 6.44×10^9 years

For an atom of ${}^{7}_{4}Be$, the **Mean Life** [EC decay] is 76.88 days

Half-Life

The **Half-Life** of any radioactive species is the time interval required for the population of that material to be reduced, by radioactive decay, to one half of its initial level. The **Half-Lives** of different isotopes, like their Mean Lives, can vary over very wide ranges. As an example, for the two radioactive decay schemes described under the definition of Radioactive Decay on the previous page, namely, Page 7-12, the **Half-Lives** are as follows

For ${}^{45}_{20}Ca$, the **Half-Life** is 162.7 days

For ${}^{232}_{90}Th$, the **Half-Life** is 1.4×10^{10} years

As can be seen from these two **Half-Lives**, this parameter can assume values over a very wide range of times. Although the thorium isotope listed above certainly has a very long **Half-Life**, it is by no means the longest. On the short end of the scale, consider another thorium isotope, ${}^{218}_{90}Th$, which has a **Half-Life** of 0.11 microseconds.

Nuclear Fission

Nuclear Fission, as the process that will be described here, differs from the Spontaneous Fission mode that was listed on Page 7-12 under the description of Radioactive Decay as one of the nine radioactive decay modes. This class of **Nuclear Fission** is a nuclear reaction in which a fissile isotope — i.e., an isotope such as ${}^{235}_{92}U$ or ${}^{239}_{94}Pu$ — upon absorbing a free neutron undergoes a fracture which results in the conversion of the initial isotope into:

1. two daughter isotopes,
2. two or more additional neutrons,
3. several very energetic γ-rays, and
4. considerable additional energy, usually appearing in the form of heat.

Nuclear Fission reactions are the basic energy producing mechanisms used in every nuclear reactor, whether it is used to generate electric power, or to provide the motive force for a nuclear submarine. One of the most important characteristics of this type of reaction is that by regenerating one or more of the particles [i.e., neutrons] that initiated the process, the reaction can become self-sustaining. Considerable value can be derived from this process if the chain reactions involved can be controlled. In theory, control of these chain reactions occurs in such things as nuclear power stations. An example of an uncontrolled **Nuclear Fission** reaction would be the detonation of an atomic bomb.

An example of a hypothetically possible **Nuclear Fission** reaction might be:

$$^{235}_{92}U + ^{1}_{0}n \rightarrow ^{109}_{44}Ru + ^{123}_{48}Cd + 3\,^{1}_{0}n + 3\,\gamma s + \text{considerable energy}$$

In this hypothetical fission reaction, the sum of the atomic masses of the two reactants to the left of the arrow is 236.052589 amu, whereas the sum of atomic masses of all the products to the right of this arrow is 234.856015 amu. Clearly there is a mass discrepancy of 1.196574 amu or 1.987×10^{-24} grams. It is this mass that was converted into the several γ-rays that were created and emitted, as well as the very considerable amount of energy that was liberated. It appears that Albert Einstein was correct: mass and energy are simply different forms of the same thing.

Since **Nuclear Fission** reactions are clearly sources for a considerable amount of ionizing radiation, they are of interest to occupational safety and health professionals.

Radiation Measurements

The Strength or Activity of a Radioactive Source

The most common measure of **Radiation Source Strength** or **Activity** is the number of radioactive disintegrations that occur in the mass of radioactive material per unit time. There are several basic units that are employed in this area; they are listed below, along with the number of disintegrations per minute that each represents:

Unit of Source Activity	Abbreviation	Disintegrations/min
1 Curie	Ci	2.22×10^{12}
1 Millicurie	mCi	2.22×10^{9}
1 Microcurie	μCi	2.22×10^{6}
1 Picocurie	pCi	2.22
1 Becquerel	Bq	60

Exposure

Exposure is a unit of measure of radiation that is currently falling into disuse. The basic definition of **Exposure** — usually designated as X — is that it is the sum number of all the ions, of either positive or negative charge — usually designated as ΣQ — that are produced in a mass of air — which has a total mass, Σm — by some form of ionizing radiation that, in the course of producing these ions, has been totally dissipated. Quantitatively, it is designated by the following formula:

$$X = \frac{\sum Q}{\sum m}$$

The unit of **Exposure** is the roentgen, or R. There is no *SI* unit for **Exposure**; thus as stated above this measure is now only rarely encountered. References to **Exposure** are now only likely to be found in older literature.

Dose

Dose, or more precisely **Absorbed Dose**, is the total energy imparted by some form of ionizing radiation to a known mass of matter that has been exposed to that radiation. Until the mid 1970s the most widely used unit of **Dose** was the rad, which has been defined to be equal to 100 ergs of energy absorbed into one gram of matter. Expressed as a mathematical relationship:

$$1.0 \text{ rad} = 100 \ \frac{\text{ergs}}{\text{gram}} = 100 \ \text{ergs} \cdot \text{grams}^{-1}$$

At present, under the *SI System*, a new unit of **Dose** has come into use. This unit is the gray, which has been defined to be the deposition of 1.0 joule of energy into 1.0 kilogram of matter. Expressed as a mathematical relationship:

$$1.0 \text{ gray} = 1.0 \ \frac{\text{joule}}{\text{kilogram}} = 1.0 \ \text{joule} \cdot \text{kilogram}^{-1}$$

The gray is steadily replacing the rad although the latter is still in fairly wide use. For reference, 1 gray = 100 rad [1 Gy = 100 rad], or 1 centigray = 1 rad [1 cGy = 1 rad]. For most applications, Doses will be measured in one of the following "sub-units": (1) millirad — mrads; (2) microrads — μrads; (3) milligrays — mGys; or (4) micrograys — μGys. These units are — as their prefixes indicate — either 10^{-3} or 10^{-6} multiples of the respective basic Dose unit.

Dose, as a measurable quantity, is always represented by the letter "D".

Dose Equivalent

The **Dose Equivalent** is the most important measured parameter insofar as the overall subject of radiation protection is concerned. It is basically the product of the Absorbed Dose and an appropriate Quality Factor, a coefficient that is dependent upon the type of ionizing particle involved — see Equation **#7-12** on Pages 7-22 & 7-23. This parameter is usually represented by the letter "H". There are two cases to consider, and they are as follows:

1. If the Dose or Absorbed Dose, D, has been given in units of rads [or mrads, or μrads], then the units of the Dose Equivalent, H, will be rem [or mrem, or μrem] as applicable.
2. If the Dose or Absorbed Dose, D, has been given in units of grays [or mGy, or μGy], then the units of the Dose Equivalent, H, will be sieverts [or mSv, or μSv] as applicable.

It is very important to note that since **1 Gray = 100 rads**, it follows that **1 sievert = 100 rem**.

Finally, if it is determined that a Dose Equivalent > 100 mSv, there is almost certainly a very serious situation with a great potential for human harm; thus, in practice, for Dose Equivalents above this level, the unit of the sievert is rarely, if ever, employed.

RELEVANT FORMULAE & RELATIONSHIPS

Basic Relationships for Electromagnetic Radiation

Equation #7-1:

For any photon that is a part of the overall electromagnetic spectrum, the relationship between that photon's wavelength, its frequency, and/or its wavenumber is given by the following expression, Equation **#7-1**, which is shown below in two equivalent forms:

$$c = \lambda \nu$$

$$c = \frac{\nu}{k}$$

Where:

- **c** = the speed of light in a vacuum, which is 2.99792458×10^8 meters/second [frequently approximated as 3.0×10^8 meters/second];
- λ = the wavelength of the photon in question, in units of meters [actually meters/cycle];
- ν = the frequency associated with the photon in question, in units of reciprocal seconds — sec^{-1} — [actually cycles/second or Hertz]; &
- **k** = the wavenumber of the photon in question, in units of reciprocal meters — $meters^{-1}$ — [actually cycles/meter].

Equation #7-2:

The relationship between the wavelength and the wavenumber of any electromagnetic photon is given by the following expression, Equation **#7-2**:

$$\lambda = \frac{1}{k}$$

Where:

- λ = the wavelength of the photon in question, in units of meters [actually meters/cycle], as defined above for Equation **#7-1**; &
- **k** = the wavenumber of the photon in question, in units of reciprocal meters — $meters^{-1}$ — [actually cycles/meter], also as defined above for Equation **#7-1**. Note: wavenumbers are very frequently expressed in units of reciprocal centimeters — cm^{-1} — and when expressed in these units, the photon is said to be at "xxx" wavenumbers [i.e., a 3,514 cm^{-1} photon is said to be at 3,514 wavenumbers].

Equation #7-3:

Equation **#7-3** expresses the relationship between the energy of any photon in the electromagnetic spectrum, and the wavelength of that photon. This relationship is Planck's Law, which was the first specific, successful, quantitative relationship ever to be applied in the area of quantum mechanics. This Law, as the first significant result of Planck's basic research in this area, formed one of the main foundation blocks upon which modern physics and/or quantum mechanics was built.

$$\mathbf{E = h\nu}$$

Where:

E = the energy of the electromagnetic photon in question, in some suitable energy unit — i.e., joules, electron volts, etc.;

h = Planck's Constant, which has a value of 6.626×10^{-34} joule · seconds, and/or 4.136×10^{-15} electron volt · seconds; &

ν = the frequency associated with the photon in question, in units of reciprocal seconds [actually cycles/second or Hertz] — as defined on the previous page for Equation **#7-1**.

Calculations Involving Radioactive Decay

Equation #7-4:

For any radioactive isotope, the following Equation, **#7-4**, identifies the current **Quantity** or amount of the isotope that would be present at any incremental time period after the initial or starting mass or number of atoms had been determined [i.e., the mass or number of atoms that has not yet undergone radioactive decay]. With any radioactive decay, the number of disintegrations or decays per unit time will be exponentially proportional to both the Radioactive Decay Constant for that nuclide, and the actual numeric count of the nuclei that are present [i.e., the **Quantity**].

$$N_t = N_0 e^{-kt}$$

Where:

N_t = the **Quantity** of any radioactive isotope present at any time, **t**; this **Quantity** is usually measured either in mass units [mg, μg, etc.] OR as a specific numeric count of the as yet undecayed nuclei remaining in the sample [i.e., 3.55×10^{19} atoms];

N_0 = the **Initial Quantity** of that same radioactive isotope — i.e., the **Quantity** that was present at the time, $t = t_0$ [i.e., 0 seconds, 0 minutes, 0 hours, 0 days, or whatever unit of time is appropriate to the units in which the Radioactive Decay Constant has been expressed]. This is the "Starting" or **Initial Quantity** of this isotope, and it is always expressed in the same units as N_t, which is described above;

k = the **Radioactive Decay Constant**, which measures number of nuclear decays per unit time; in reality, the "number of nuclear decays" is a simple integer, and as such, is effectively dimensionless; thus this parameter should be thought of as being measured in reciprocal units of time [i.e., $seconds^{-1}$, $minutes^{-1}$, $hours^{-1}$, $days^{-1}$, or even $years^{-1}$, etc.]; &

t = the **Time Interval** that has passed since the Initial Quantity of material was determined. This **Time Interval** must be expressed in an appropriate unit of time — i.e., the units of "**k**" and "**t**" must be mutually consistent; thus the units of "**k**" must be: seconds, minutes, hours, days, years, etc.

Equation #7-5:

The following Equation, **#7-5**, provides the relationship between the **Half-Life** of a radioactive isotope and its **Radioactive Decay Constant**. The **Half-Life** of any radioactive nuclide is the statistically determined time interval required for exactly half of the isotope to decay, effectively leaving the other half of the isotope in its original form.

$$T_{1/2} = \frac{0.693}{k}, \text{ or}$$

$$k = \frac{0.693}{T_{1/2}}$$

Where:

$T_{1/2}$ = the **Half-Life** of the radioactive isotope under consideration; this parameter must be expressed in the same units of time that are used as reciprocal time units for the Radioactive Decay Constant; &

k = the **Radioactive Decay Constant**, measured in reciprocal units of time [i.e., $seconds^{-1}$, $minutes^{-1}$, $hours^{-1}$, $days^{-1}$, or even $years^{-1}$, etc.], as defined on the previous page, namely Page 7-18, for Equation **#7-4**.

Equation #7-6:

The **Mean Life** of any radioactive isotope is the measure of the average 'lifetime" of a single atom of that isotope. It is simply the reciprocal of that nuclide's Radioactive Decay Constant. Equation **#7-6** provides the quantitative relationship that is involved in calculating this parameter.

$$\tau = \frac{1}{k} = \frac{T_{1/2}}{0.693} = 1.443T_{1/2}$$

Where:

τ = the **Mean Life** of some specific radionuclide, expressed in units of time [i.e., seconds, minutes, hours, days, or years, etc.]

k = the **Radioactive Decay Constant**, measured in consistent reciprocal units of time [i.e., $seconds^{-1}$, $minutes^{-1}$, $hours^{-1}$, $days^{-1}$, or even $years^{-1}$, etc.]; &

$T_{1/2}$ = the **Half-Life** of the radioactive isotope under consideration; this parameter must be expressed in the same units of time as the Mean Life, and as the reciprocal of the time units in which the Radioactive Decay Constant is expressed.

Equation #s 7-7 & 7-8:

The **Activity** of any radioisotope is defined to be the number of radioactive disintegrations that occur per unit time. Equation **#s 7-7 & 7-8** are two simplified forms of the relationship that can be used to calculate the **Activity** of any radioactive nuclide.

Equation #7-7:

$$\mathbf{A_b = kN}$$

Equation #7-8:

$$A_c = \frac{kN}{3.70 \times 10^{10}} = \left[2.703 \times 10^{-11}\right] kN$$

Where:

$\mathbf{A_b}$ = the **Activity** of the radionuclide, expressed in becquerels,

OR

$\mathbf{A_c}$ = the **Activity** of the radionuclide, expressed in curies;

k = the **Radioactive Decay Constant**, measured in reciprocal units of time [i.e., seconds^{-1}, minutes^{-1}, hours^{-1}, days^{-1}, or even years^{-1}, etc.]; &

N = the **Quantity** of the radioactive isotope that is present in the sample at the time when the evaluation of the **Activity** is to be made, measured as a specific numeric count of the as yet undecayed nuclei remaining in the sample [i.e., 3.55×10^{19} atoms];

Equation #s 7-9 & 7-10:

The following two Equations, **#s 7-9 & 7-10**, provide the two more general forms of the relationship for determining the **Activity** of any radioactive nuclide.

Equation #7-9:

$$\mathbf{A_t = kN_0 e^{-kt}}$$

Equation #7-10:

$$\mathbf{A_t = \left[\frac{0.693}{T_{1/2}}\right] N_0 e^{-(0.693)t/T_{1/2}}}$$

Where:

A_t = the **Activity** of any radioactive nuclide at any time, **t**. The units of this calculated parameter will be becquerels;

k = the **Radioactive Decay Constant**, measured in reciprocal units of time [i.e., $seconds^{-1}$, $minutes^{-1}$, $hours^{-1}$, $days^{-1}$, or even $years^{-1}$, etc.];

N_0 = the **Initial Quantity** of that same radioactive isotope — i.e., the **Quantity** that was present at the time, $t = t_0$ [i.e., 0 seconds, 0 minutes, 0 hours, 0 days, or zero of whatever unit of time is appropriate to the dimensionality in which the Radioactive Decay Constant has been expressed] — this is the "Starting" or **Initial Quantity** of this isotope, measured as a specific numeric count of the as yet undecayed nuclei remaining in the sample [i.e., 3.55×10^{19} atoms];

$T_{1/2}$ = the **Half-Life** of the radioactive isotope under consideration; this parameter must be expressed in the same units of time that appear as reciprocal time units for the Radioactive Decay Constant; &

t = the **Time Interval** that has passed since the Initial Quantity of material was determined; this **Time Interval** must be expressed in an appropriate unit of time — i.e., the units of "**k**" and "**t**" must be consistent with each other.

Dose and/or Exposure Calculations

Equation #7-11:

The following Equation, **#7-11**, is applicable only to **Dose Exposure Rates** caused by high energy X-Rays and/or γ-Rays [as well as — hypothetically, at least, but certainly not practically — any other photons such as a Cosmic Ray, which have a still shorter wavelength]. Determinations of these **Dose Exposure Rates** are largely limited to medical applications. In order to be able to make these determinations, some very specific and unique source-based radiological data [i.e., the Radiation Constant of the source] must be known. In addition, the Radiation Source Activity, and the distance from the source to the point at which **Dose Exposure Rate** is to be measured, must also be known.

$$\mathbf{E} = \frac{\mathbf{\Gamma A}}{\mathbf{d^2}}$$

Where:

E = the **Dose Exposure Rate** that has resulted from an individual's exposure to some specific X- or γ-radiation source, for which the specific Radiation Constant, Γ, is known; this dose rate is commonly expressed in units such as Rads/hour;

Γ = the **Radiation Constant** for the X- or γ-Ray active nuclide being considered, expressed in units of [Rads · centimeters]^2 per millicurie · hour, or

$$\left[\frac{\text{Rad} \cdot \text{cm}^2}{\text{mCi} \cdot \text{hr}}\right];$$

A = the **Radiation Source Activity**, measured usually in millicuries [mCi's]; &

d = the **Distance** between the "Target" and the radiation source, measured in centimeters [cm].

Equation #7-12:

This Equation, **#7-12**, provides for the conversion of an **Absorbed Radiation Dose**, expressed either in Rads or in Grays, to a more useful form — useful from the perspective of measuring the magnitude of the overall impact of the dose on the individual who has been exposed. This alternative, and more useful, form of Radiation Dose is called the **Dose Equivalent** and is expressed either in **rems** or in **sieverts**, both of which measure the "Relative Hazard" caused by the energy transfer that results from an individual's exposure to various different types or categories of radiation. The **rem** and/or the **sievert**, therefore, is dependent upon two specific factors: (1) the specific type of radiation that produced the exposure, and (2) the amount or physical dose of the radiation that was involved in the exposure.

To make these determinations, a "Quality Factor" is used to adjust the measurement that was made in units of **rads** or **grays** — both of which are independent of the radiation source — into an equivalent in **rems** and/or **sieverts**.

This Quality Factor [QF] is a simple multiplier that adjusts for the effective ***Linear Energy Transfer*** (***LET***) that is produced on a target by each type or category of radiation. The higher the ***LET***, the greater will be the damage that can be caused by the type of radiation being considered; thus, this alternative **Dose Equivalent** measures the overall biological effect, or impact, of an otherwise "simple" measured Radiation Dose.

The "range" of β- and/or α-rays is, as stated earlier, very limited — i.e., the "range" is the distance that any form of radiation is capable of traveling through solid material, such as metal, wood, human tissue, etc. before it is stopped. Because of this, Quality Factors as they apply to alpha and beta particles are only considered from the perspective of internal **Dose Equivalent** problems. Quality factors for neutrons, X-, and γ-rays apply both to internal and external **Dose Equivalent** situations.

$$\mathbf{H_{Rem} = D_{Rad}[QF]} \quad \&$$

$$\mathbf{H_{Sieverts} = D_{Grays}[QF]}$$

Where: $\mathbf{H_{Rem}}$ or $\mathbf{H_{Sievert}}$ = the adjusted **Dose Equivalent** in the more useful "effect related" form, measured in either rems or *sieverts [SI Units]*;

$\mathbf{D_{Rad}}$ or $\mathbf{D_{Gray}}$ = the **Absorbed Radiation Dose**, which is independent of the type of radiation, and is measured in either rads or *grays [SI Units]*; &

QF = the **Quality Factor**, which is a properly dimensioned coefficient — either in units of rems/rad or sieverts/gray, as applicable — that is, itself, a function of the type of radiation being considered [see the following Tabulation].

Tabulation of Quality Factors [QFs] by Radiation Type

Types of Radiation	Quality Factors — QFs	Internal/External
X-Rays or γ-Rays	1.0	Both
β-Rays [positrons or electrons]	1.0	Internal Only
Thermal Neutrons	5.0	Both
Slow Neutrons	4.0 - 22.0	Both
Fast Neutrons	3.0 - 5.0	Both
Heavy, Charged Particles [Alphas, etc.]	20.0	Internal Only

Calculations Involving the Reduction of Radiation Intensity Levels

Equation #7-13:

This Equation, **#7-13**, identifies the effect that shielding materials have in reducing the intensity level of a beam of ionizing radiation. The **Radiation Emission Rate** produced by such a beam can be reduced either by interposing shielding materials between the radiation source and the receptor, or by increasing the source-to-receptor distance. Obviously, the **Radiation Emission Rate** could be decreased still further by using both approaches simultaneously.

The approach represented by Equation **#7-13** deals solely with the use of shielding materials [i.e., it does not consider the effect of increasing source-to-receptor distances]. This approach involves the use of the Half-Value Layer [HVL] concept. A **Half-Value Layer** represents the thickness of any shielding material that would reduce, by one half, the intensity level of incident X- or γ-radiation. This expression is provided in two forms:

$$\mathbf{ER_{goal}} = \frac{\mathbf{ER_{source}}}{2^{x/HVL}} \quad \text{or}$$

$$x = \frac{\log\left[\frac{\mathbf{ER_{source}}}{\mathbf{ER_{goal}}}\right][HVL]}{\log 2} = 3.32\log\left[\frac{\mathbf{ER_{source}}}{\mathbf{ER_{goal}}}\right][HVL]$$

Where:

$\mathbf{ER_{goal}}$ = the target **Radiation Emission Rate**, measured in units of radiation dose per unit time [i.e., Rads/hour];

$\mathbf{ER_{source}}$ = the observed **Radiation Emission Rate** to be reduced by interposing Shielding Materials, in the same units as $\mathbf{ER_{goal}}$;

x = the **Thickness** of shielding material required to reduce the measured **Radiation Emission Rate** to the level desired, usually measured in units of centimeters or inches [cm or in]; &

HVL = the **Half-Value Thickness** of the Shielding Material being evaluated (i.e., the **Thickness** of this material that will halve the Intensity Level of incident X- or γ-radiation), measured in the same units as "**x**", above.

Equation #7-14:

The following Equation, **#7-14**, is the relationship that describes the effect of increasing the distance between a point source of X- or γ-radiation and a receptor, as an alternative method for decreasing the incident radiation intensity on the receptor. The relationship involved is basically geometric, and is most commonly identified or referred to as **The Inverse Squares Law**.

$$\frac{ER_a}{ER_b} = \frac{S_b^2}{S_a^2} \quad \text{or}$$

$$ER_a S_a^2 = ER_b S_b^2$$

Where:

ER_a = the **Radiation Emission Rate**, or **Radiation Intensity**, in units of radiation dose per unit time [i.e., Sieverts/hour], measured at a distance, "**a**" units from the radiation source;

ER_b = the **Radiation Emission Rate**, or **Radiation Intensity**, in the same units as, ER_a, above, measured at a different distance, "**b**" units from the radiation source;

S_a = the "**a**" **Distance**, or the distance between the radiation source and the first position of the Receptor; this distance is measured in some appropriate unit of length [i.e., meters, feet, etc.]; &

S_b = the "**b**" **Distance**, or the distance between the radiation source and the second — usually more distant — position of the Receptor; this distance is also measured in some appropriate unit of length, and most importantly in the same units of length as S_a, above [i.e., meters, feet, etc.].

Calculations Involving Optical Densities

Equation #7-15:

The following Equation, **#7-15**, describes the relationship between the absorption of monochromatic visible light [i.e., laser light], and the length of the path this beam of light must follow through some absorbing medium. This formula relies on the fact that each incremental thickness of this absorbing medium will absorb the same fraction of the incident radiation as will each other identical incremental thickness of this same medium.

The logarithm of the ratio of the **Incident Beam Intensity** to the **Transmitted Beam Intensity** is used to calculate the **Optical Density** of the medium. This relationship, then, is routinely used to determine the intensity diminishing capabilities [i.e., the **Optical Density**] of the protective goggles that must be worn by individuals who must operate equipment that makes use of high intensity monochromatic light sources, such as lasers.

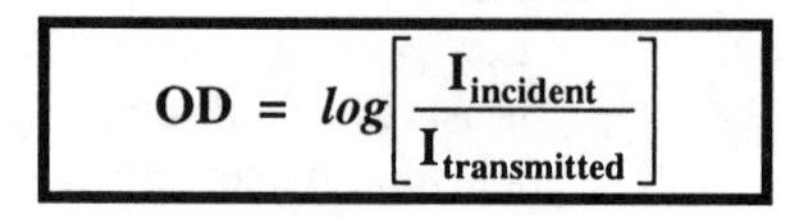

$$OD = \log\left[\frac{I_{incident}}{I_{transmitted}}\right]$$

Where:

OD = the measured **Optical Density** of the material being evaluated, this parameter is dimensionless;

$I_{incident}$ = the **Incident Laser Beam Intensity**, measured in units of power/unit area [i.e., W/cm^2); &

$I_{transmitted}$ = the **Transmitted Laser Beam Intensity**, measured in the same units as $I_{incident}$, above.

Relationships Involving Microwaves

Equation #7-16:

The following Equation, **#7-16**, provides the necessary relationship for determining the **Distance to the *Far Field*** for any radiating circular microwave antenna. The ***Far Field*** is that region that is sufficiently distant [i.e., more than 2 or 3 wavelengths away] from the radiating antenna, that there is no longer any interaction between the electrical and the magnetic fields being produced by this source. In the ***Near Field*** the interactions between the two electromagnetic fields being produced by any source require a different approach to the measurement of the effects, etc. The ***Near Field*** is every portion of the radiation field that is not included in the ***Far Field*** — i.e., it is that area that is closer to the source antenna than is the ***Far Field***.

$$r_{FF} = \frac{A}{2\lambda} = \frac{\pi D^2}{8\lambda}$$

Where:

r_{FF} = the **Distance to the *Far Field*** from the microwave radiating antenna [all distances equal to or greater than r_{FF} are considered to be in the ***Far Field***; all distances less than this value will be in the ***Near Field***], these distances are usually measured in centimeters [cm];

A = the **Area** of the radiating circular antenna, measured in square centimeters [cm^2] — for reference, this area can be calculated according to the following relationship,

$$\text{Circular Area} = \frac{\pi D^2}{4};$$

D = the circular microwave antenna **Diameter**, measured in centimeters [cm]; &

λ = the **Wavelength** of microwave energy being radiated by the circular antenna, also measured in centimeters [cm].

Equation #7-17:

The following Equation, **#7-17**, provides the relationship for determining the ***Near Field* Microwave Power Density** levels that are produced by a circular microwave antenna, radiating at a known **Average Power Output**.

$$W_{NF} = \frac{4P}{A} = \frac{16P}{\pi D^2}$$

Where:

W_{NF} = the ***Near Field* Microwave Power Density**, measured in milliwatts/cm^2 [mW/cm^2];

P = the **Average Power Output** of the microwave radiating antenna, measured in milliwatts [mW];

A = the **Area** of the radiating circular antenna, measured in square centimeters [cm^2] — for reference, this area can be calculated according to the following relationship,

$$\text{Circular Area} = \frac{\pi D^2}{4}; \; \&$$

D = the circular microwave antenna **Diameter**, measured in centimeters [cm].

Equations #s 7-18 & 7-19:

The following two Equations, **#s 7-18 & 7-19**, provide the basic approximate relationships that are used for calculating either microwave **Power Density Levels** in the ***Far Field*** [Equation **#7-18**], OR, alternatively, for determining the actual ***Far Field* Distance** from a radiating circular microwave antenna at which one would expect to find some specific **Power Density Level** [Equation **#7-19**].

Unlike the Equation at the top of this page [i.e., Equation **#7-17**], these two formulae have been empirically derived; however, they may both be regarded as sources of reasonably accurate values for the **Power Density Levels** at points in the ***Far Field*** [Equation **#7-18**], or for various ***Far Field* Distances** [Equation **#7-19**].

Equation #7-18:

$$W_{FF} = \frac{AP}{\lambda^2 r^2} = \frac{\pi D^2 P}{4\lambda^2 r^2}$$

Equation #7-19:

$$r = \frac{1}{\lambda}\sqrt{\frac{AP}{W_{FF}}} = \frac{D}{2\lambda}\sqrt{\frac{\pi P}{W_{FF}}}$$

Where:

W_{FF} = the **Power Density Level** at a point in the ***Far Field*** that is "**r**" centimeters distant from the circular microwave antenna, with this Power Density Level measured in milliwatts/cm^2 [mW/cm^2];

r = the ***Far Field*** **Distance** [from the point where the **Power Density Level** is being evaluated] to the radiating circular microwave antenna, also measured in centimeters [cm];

D = the circular microwave antenna's **Diameter**, measured in centimeters [cm].

A = the **Area** of the radiating circular antenna, measured in square centimeters [cm^2] — for reference, this area can be calculated according to the following relationship,

$$\text{Circular Area} = \frac{\pi \mathbf{D}^2}{4};$$

λ = the **Wavelength** of microwave energy being radiated by the circular antenna, also measured in centimeters [cm]; &

P = the **Average Power Output** of the microwave radiating antenna, measured in milliwatts [mW].

IONIZING AND NON-IONIZING RADIATION PROBLEM SET

Problem #7.1:

The mid-infrared wavelength at which the carbon-hydrogen bond absorbs energy [i.e., the "carbon-hydrogen stretch"] is at approximately 3.35 μ [i.e., 35 microns]. What is the frequency of a photon having this wavelength?

Applicable Definitions:	Electromagnetic Radiation	Page 7-1
	Infrared Radiation	Pages 7-7 & 7-8
Applicable Formula:	Equation **#7-1**	Page 7-16
Solution to this Problem:	Page 7-56	

Problem Workspace

Problem #7.2:

What is the energy, in electron volts, of one of these "carbon-hydrogen stretch" photons? Remember, the wavelength of these photons is 3.35 μ.

Applicable Definitions:	Electromagnetic Radiation	Page 7-1
	Infrared Radiation	Pages 7-7 & 7-8
Applicable Formula:	Equation **#7-3**	Page 7-17
Solution to this Problem:	Page 7-56	

Problem Workspace

Problem #7.3:

What is the wavenumber of the mid-infrared photon that is readily absorbed by a carbon-hydrogen bond [i.e., a photon with a wavelength of 3.35 μ — see Problem **#7.1**, on Page 7-30]?

Applicable Definitions:	Electromagnetic Radiation	Page 7-1
	Infrared Radiation	Pages 7-7 & 7-8
Applicable Formulae:	Equation **#7-2**	Page 7-16
OR	Equation **#7-1**	Page 7-16
Solution to this Problem:	Page 7-56	

Problem Workspace

Problem #7.4:

One of the two γ-ray photons that are emitted during the decay of ${}^{60}_{27}Co$ has frequency, ν, of 2.84×10^{14} MHz. What is the wavelength, in microns, of this photon?

Applicable Definitions:	Electromagnetic Radiation	Page 7-1
	Gamma Radiation	Page 7-4
	Radioactive Decay	Page 7-12
Applicable Formula:	Equation **#7-1**	Page 7-16
Solution to this Problem:	Page 7-57	

Problem Workspace

Problem #7.5:

What is the wavenumber, in cm^{-1}, of the γ-ray photon identified in Problem **#7.4**, on Page 7-31? Remember, this photon has a frequency of 2.84×10^{14} MHz.

Applicable Definitions:	Electromagnetic Radiation	Page 7-1
	Gamma Radiation	Page 7-4
Applicable Formulae:	Equation **#7-2**	Page 7-16
OR	Equation **#7-1**	Page 7-16
Solution to this Problem:	Page 7-57	

Problem Workspace

Problem #7.6:

What is the energy, in electron volts, of the γ-ray photon emitted during the decay of ${}^{60}_{27}Co$, as described in Problem **#7.5** above on this page? Remember, the frequency, ν, of this photon is 2.84×10^{14} MHz.

Applicable Definitions:	Electromagnetic Radiation	Page 7-1
	Gamma Radiation	Page 7-4
	Radioactive Decay	Page 7-12
Applicable Formula:	Equation **#7-3**	Page 7-17
Solution to this Problem:	Page 7-57	

Problem Workspace

Problem #7.7:

An atom is observed, in order:

(1) to absorb an ultraviolet [UV-B] photon having a wavelength, $\lambda_{UV\text{-}B}$, of 274 nm, and then subsequently

(2) to emit a visible light photon having a wavelength, λ_{Vis}, of 0.46 μ.

What was the net energy absorbed by this atom during this process? If the ionization energy of this atom is known to be 1.2 ev, did this process ionize this atom?

Applicable Definitions:	Electromagnetic Radiation	Page 7-1
	Ultraviolet Radiation	Pages 7-6 & 7-7
	Visible Light	Page 7-7
Applicable Formulae:	Equation **#7-1**	Page 7-16
	Equation **#7-3**	Page 7-17
Solution to this Problem:	Page 7-58	

Problem Workspace

Problem #7.8:

The radioactive isotope, $^{131}_{53}I$, is frequently used in the treatment of thyroid cancer. It has a Radioactive Decay Constant of 0.0862 days^{-1}. A local hospital received its order of 2.0 μg of this isotope on January 1st. How much of this isotope will remain on January 20th of the same year? How much will remain on the one year anniversary [not a leap anniversary] of the receipt of the 2.0 μg of the $^{131}_{53}I$ isotope?

Applicable Definitions:	Radioactivity	Page 7-12
	Radioactive Decay	Page 7-12
	Radioactive Decay Constant	Page 7-13
Applicable Formula:	Equation **#7-4**	Page 7-18
Solution to this Problem:	Page 7-59	

Problem Workspace

Problem #7.9:

What is the Half-Life of $^{131}_{53}I$? What is the Mean Life of an $^{131}_{53}I$ atom? Remember, $^{131}_{53}I$ has a Radioactive Decay Constant of 0.0862 days^{-1} — see Problem **#7.8**, on the previous page.

Applicable Definitions:	Radioactivity	Page 7-12
	Radioactive Decay	Page 7-12
	Half-Life	Page 7-13
	Mean Life	Page 7-13
	Radioactive Decay Constant	Page 7-13
Applicable Formulae:	Equation **#7-5**	Page 7-19
	Equation **#7-6**	Page 7-19
Solution to this Problem:	Pages 7-59 & 60	

Problem Workspace

Problem #7.10:

What would be the measured Activity of the ${}^{131}_{53}I$ isotope mentioned in Problem **#7.8**, on Page 7-34, if measurement were made: (1) on January 1st — i.e., the day when it was received at the Hospital; (2) on January 20th of that same year; and (3) on January 1st of the following year [not a leap year]. Remember, the Radioactive Decay Constant of ${}^{131}_{53}I$ is 0.0862 days^{-1}, and its mass on January 1st, the day it was received at the Hospital, was 2.0 μg. If it is of any use to you, the atomic weight of the ${}^{131}_{53}I$ isotope is 130.9061 amu.

Applicable Definitions:	Amount of Any Substance	Page 1-3
	Radioactivity	Page 7-12
	Radioactive Decay	Page 7-12
	Radioactive Decay Constant	Page 7-13
	Activity of a Radioactive Source	Page 7-14
Applicable Formulae:	Equation **#1-10**	Pages 1-19 & 1-20
	Equation **#1-11**	Page 1-20
	Equation **#7-7**	Page 7-20
	Equation **#7-9**	Pages 7-20 & 7-21
Solution to this Problem:	Pages 7-60 & 7-61	

Problem Workspace

Workspace Continued on the Next Page

Continuation of Workspace for Problem **#7.10**

Problem #7.11:

The smallest amount of $^{245}_{99}Es$ that can be detected or utilized in any type of experimentation, is 1.5×10^{-10} ng. The Radioactive Decay Constant for $^{245}_{99}Es$ is 0.502 minutes^{-1}. If 8.8×10^{-6} ng of this material was successfully accumulated by a research scientist, how much time will this scientist have available to her as she performs experiments with her supply of this isotope of einsteinium — i.e., how much time will pass before this quantity has decayed to the "barely detectable" level?

Applicable Definitions:	Radioactivity	Page 7-12
	Radioactive Decay	Page 7-12
	Radioactive Decay Constant	Page 7-13
Applicable Formula:	Equation **#7-4**	Page 7-18
Solution to this Problem:	Page 7-61	

Problem Workspace

Problem #7.12:

The heaviest hydrogen isotope, tritium, $^{3}_{1}H$, is radioactive. Its Half-Life is 12.26 years. What is its Radioactive Decay Constant?

Applicable Definitions:	Radioactivity	Page 7-12
	Radioactive Decay	Page 7-12
	Radioactive Decay Constant	Page 7-13
	Half-Life	Page 7-13
Applicable Formula:	Equation **#7-5**	Page 7-19
Solution to this Problem:	Page 7-62	

Workspace for Problem **#7.12**

Problem #7.13:

What are the Half-Lives of the two isotopes, $^{131}_{53}I$ and $^{245}_{99}Es$, that were identified, respectively, in Problem **#s 7.8 & 7.11**, on Pages 7-34 & 7-38? The Radioactive Decay Constants for these two isotopes are as follows: for $^{131}_{53}I$, k = 0.0862 days^{-1}, and for $^{245}_{99}Es$, k = 0.502 minutes^{-1}.

Applicable Definitions:	Radioactivity	Page 7-12
	Radioactive Decay	Page 7-12
	Radioactive Decay Constant	Page 7-12
	Half-Life	Page 7-13
Applicable Formula:	Equation **#7-5**	Page 7-19
Solution to this Problem:	Page 7-62	

Problem Workspace

Problem #7.14:

$^{241}_{95}Am$ is one of the most commonly used and readily available radioactive isotopes. As an example, it is widely used as the ionization source in most commercial smoke detectors. Its Half-Life is 432.2 years. A functional smoke detector must have at least 1.75 μg of this isotope in order to operate properly. If there are 4.0×10^{15} Americium atoms in each microgram of this isotope, what is the minimum number of disintegrations per second that are required to operate a smoke detector?

Applicable Definitions:	Radioactivity	Page 7-12
	Radioactive Decay	Page 7-12
	Radioactive Decay Constant	Page 7-13
	Half-Life	Page 7-13
Applicable Formula:	Equation **#7-10**	Pages 7-20 & 7-21
Solution to this Problem:	Pages 7-62 & 7-63	

Problem Workspace

Problem #7.15:

If each commercial smoke detector is manufactured with 1.80 μg of $^{241}_{95}Am$, how long will it be before the minimum required level of radioactive disintegrations per second has been reached — i.e., how long will it take for this isotope's radioactive decay to reduce the mass of $^{241}_{95}Am$ to 1.75 μg? Is it your opinion that the manufacturer has successfully produced a product with guaranteed obsolescence?

Applicable Definitions:	Radioactivity	Page 7-12
	Radioactive Decay	Page 7-12
	Radioactive Decay Constant	Page 7-13
	Half-Life	Page 7-13
Applicable Formulae:	Equation **#7-4**	Page 7-18
	Equation **#7-5**	Page 7-19
Solution to this Problem:	Pages 7-63 & 7-64	

Problem Workspace

Problem #7.16:

What is the Dose Exposure Rate, in mrads/day, that would be produced on a target that is at a distance of 1.0 meter from a 550 mCi $^{24}_{11}Na$ source? The Radiation Constant for the $^{24}_{11}Na$ isotope is 18.81 $\frac{rad \cdot cm^2}{mCi \cdot hr}$.

Applicable Definitions:	Radioactivity	Page 7-12
	Radioactive Decay	Page 7-12
	Dose	Page 7-15
Applicable Formula:	Equation **#7-11**	Page 7-22
Solution to this Problem:	Page 7-64	

Problem Workspace

Problem #7.17:

A Dose Exposure Rate of 48.0 mrads/hour was determined for a 440 μCi source of $^{226}_{88}Ra$ at a distance of 300 mm. What is the Radiation Constant for this isotope, expressed in units of $\frac{rad \cdot cm^2}{mCi \cdot hr}$?

Applicable Definitions:	Radioactivity	Page 7-12
	Radioactive Decay	Page 7-12
	Dose	Page 7-15
Applicable Formula:	Equation **#7-11**	Page 7-22
Solution to this Problem:	Page 7-64	

Workspace for Problem **#7.17**

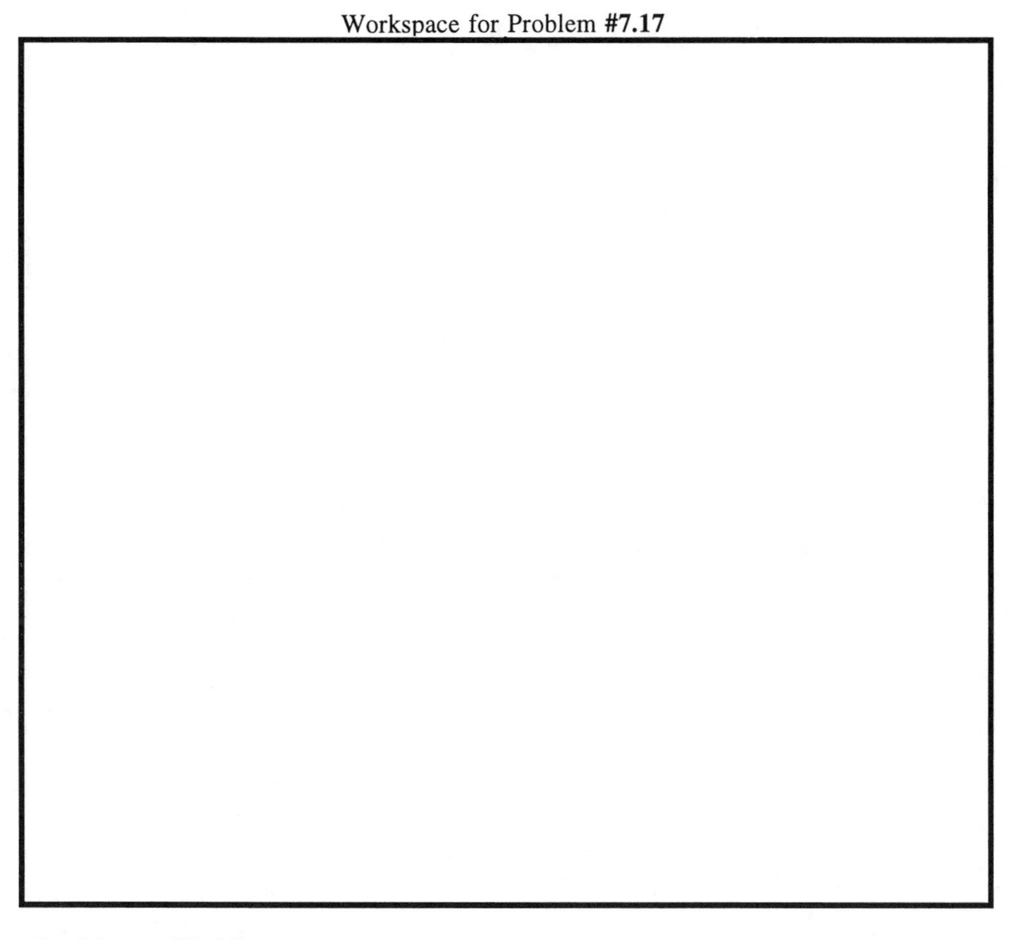

Problem #7.18:

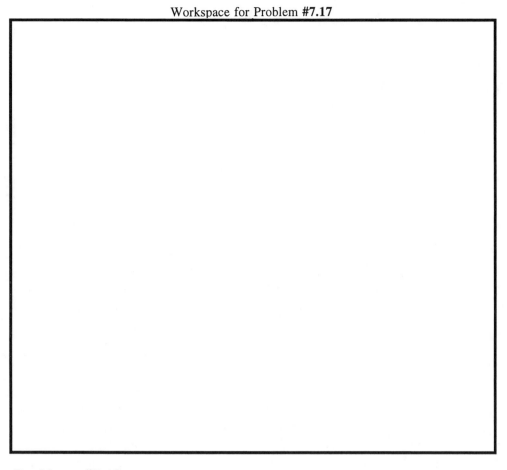

$^{60}_{27}Co$ decays, in part, by emitting relatively high energy γ-rays. It is widely used as a radiation source in the treatment of certain cancerous tumors. The Radiation Technician who operates the **C**obalt **R**adiation **S**ource **T**umor **T**reatment **A**pparatus [the CRSTTA] at a major hospital accumulates a steady 0.09 mrad, as an absorbed dose, for each hour he is in the room with the CRSTTA when its aperture is closed [i.e., when no patient treatment is occurring]. This Technician's absorbed dose increases to 0.44 mrad/hr whenever the CRSTTA's aperture is opened and a patient is being treated — during this treatment period, the Technician occupies a shielded cell in which his radiation exposure is reduced. On a typical day, this facility will have 2.2 hours of open aperture treatment time, and 5.8 hours of closed aperture operations (i.e., set-up, dosimeter development, etc.). What would this Technician's Dose Equivalent be, expressed in rems/day?

Applicable Definitions:	Gamma Radiation	Page 7-4
	Radioactivity	Page 7-12
	Radioactive Decay	Page 7-12
	Dose	Page 7-15
	Dose Equivalent	Page 7-15
Applicable Formula:	Equation **#7-12**	Pages 7-22 & 7-23
Solution to this Problem:	Pages 7-64 & 7-65	

Workspace for Problem **#7.18**

Problem #7.19:

Among the numerous facilities that it makes available to its staff of Research Scientists — a large research facility has a pool type nuclear reactor, equipped with an externally accessible graphite "Thermal Column". The Technicians who work 8 hours each day, 5 days each week, around this nuclear reactor accumulate a background absorbed radiation dose at a rate of 0.12 mrads/hour [from thermal, or low energy, neutrons], when the Thermal Column's access port is closed. Whenever the access port to this Thermal Column is opened — in order to work on one of the experiments in it, etc. — each Technician's absorbed dose rate from neutron exposure increases to 0.83 mrads/hour. If this access port is opened only 2 hours each week, and remains closed for the balance of the time, what will be the Dose Equivalent, expressed in an appropriate "sieverts/week" type unit [i.e., mSv/week, μSv/week, etc.], for each of the Technicians who work in this area?

Applicable Definitions:	Neutron Radiation	Pages 7-5 & 7-6
	Absorbed Dose	Page 7-15
	Dose Equivalent	Page 7-15
Applicable Formula:	Equation **#7-12**	Pages 7-22 & 7-23
Solution to this Problem:	Pages 7-65 & 7-66	

Workspace for Problem **#7.19**

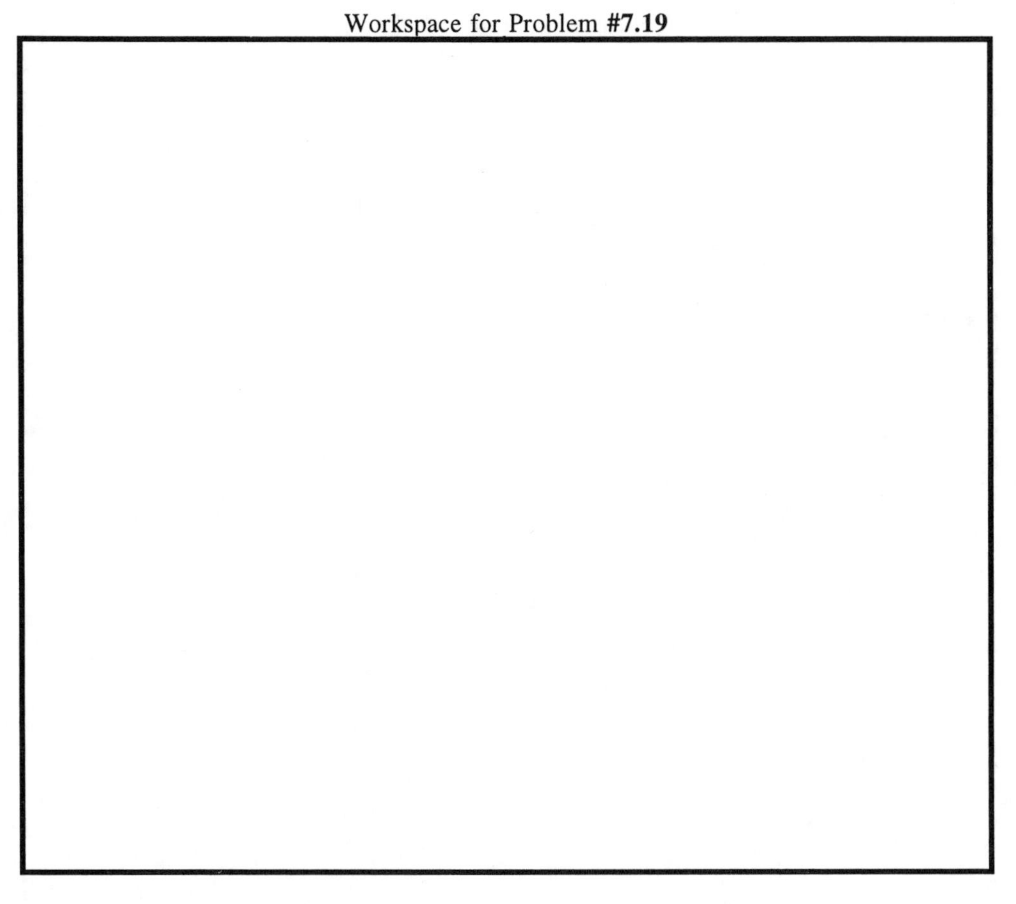

Problem #7.20:

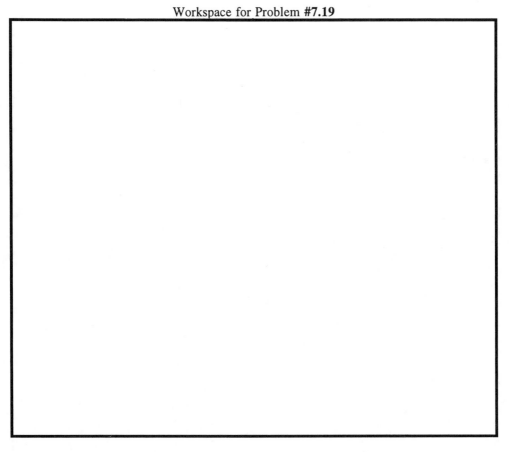

The facility described in Problem **#7.18**, on Page 7-43, was equipped with a new improved shielded cell from which the Radiation Technician could operate the CRSTTA when its aperture was open to provide treatment to a cancer patient. As a result of his operating this facility from his new improved cell, his radiation dose was reduced. The previous cell had been made from concrete, and was 18 inches thick. The improved cell shielding was fabricated from lead, and was 8 inches thick.

(1) If the Half-Value Thickness for concrete is 2.45 inches;
(2) If the radiation emission rate for the ${}^{60}_{27}Co$ Source in the CRSTTA is 40 rads/hour;
(3) If the Technician-to-Source geometry is unchanged [except for the improvements in the cell shielding]; and
(4) If the observed Radiation Emission Rate for this facility has been reduced by a factor of 775 when the Technician operates the CRSTTA from the new cell;

what then would you calculate the Half-Value Thickness of lead to be when used to shield the γ-rays from a ${}^{60}_{27}Co$ source?

Applicable Definitions:	Radioactivity	Page 7-12
	Radioactive Decay	Page 7-12
	Dose	Page 7-15
Applicable Formula:	Equation **#7-13**	Page 7-24
Solution to this Problem:	Pages 7-66 & 7-67	

Workspace for Problem **#7.20**

Problem #7.21:

A patient who has been injected with a quantity of ${}^{131}_{53}I$ for treatment of his thyroid cancer will, himself, become a radiation source for the γ-radiation emanating from this isotope as it accumulates in his cancerous thyroid gland. A doctor examining this patient from a distance of 15 cm would experience γ-radiation at an intensity of 2 mSv/hour. If this doctor were able to complete his examination at a distance of 25 cm, what would be the Dose Equivalent of radiation he would be experiencing at this increased distance?

Applicable Definitions:	Radioactivity	Page 7-12
	Radioactive Decay	Page 7-12
	Dose Equivalent	Page 7-15
Applicable Formula:	Equation **#7-14**	Page 7-25
Solution to this Problem:	Page 7-67	

Problem Workspace

Problem #7.22:

There is a commercially available therapeutic medical injectant incorporating a low level $^{226}_{88}Ra$ source [an α-emitter]. This material has been compounded so as to accumulate preferentially in a patient's liver. The adjusted radiation dose arising from this product is 40 μrem/hour, when this exposure is measured at a distance of 2.5 mm from the liver. An alternative form of this injectant is now also being marketed. This alternative has as its source, the radioisotope, $^{45}_{19}K$ [a β^--emitter]. This alternative injectant has been compounded so as to have the same low level of $^{45}_{19}K$ as is the case for the original form containing the $^{226}_{88}Ra$ source. What will be the effect on the adjusted radiation dose that would be experienced at the same 2.5 mm distance from a liver that contains this new injectant? At what distance from such a liver would the adjusted radiation dose be at the same 40 μrem/hour level as was the case for a point 2.5 mm from a liver containing the product with the radium isotope?

Applicable Definitions:	Radioactivity	Page 7-12
	Radioactive Decay	Page 7-12
	Dose Equivalent	Page 7-15
Applicable Formula:	Equation **#7-12** Table of "QFs"	Page 7-23
	Equation **#7-14**	Page 7-25
Solution to this Problem:	Pages 7-67 & 7-68	

Problem Workspace

Problem #7.23:

The Operator of a laser-based industrial metal trimmer [IR & visible light lasers] wears goggles that reduce the incident beam intensity at his eyes from 475 mW/cm^2 [at which level, he would not even be able to see the area he was trimming] to a more workable and safe level of 0.45 mW/cm^2, as the transmitted intensity. What is the effective Optical Density of the protective goggles he is wearing?

Applicable Definitions:	Visible Light	Page 7-7
	Infrared Radiation	Pages 7-7 & 7-8
Applicable Formula:	Equation **#7-15**	Page 7-26
Solution to this Problem:	Page 7-68	

Problem Workspace

Problem #7.24:

If the industrial metal trimmer identified in Problem **#7.23**, above on this page, were to be retrofitted with a new improved laser source that had 6.62 times the laser beam intensity of the original unit, and if it is hoped to reduce still further the transmitted beam intensity experienced by the Operator to a new lower level of 0.19 mW/cm^2 maximum, by how much must the Optical Density of the goggles be increased to satisfy these new requirements?

Applicable Definitions:	Visible Light	Page 7-7
	Infrared Radiation	Pages 7-7 & 7-8
Applicable Formula:	Equation **#7-15**	Page 7-26
Solution to this Problem:	Pages 7-68 & 7-69	

Workspace for Problem **#7.24**

Problem #7.25:

The UHF Microwave Systems that are used for transmitting a very large fraction of all the public telecommunications in the United States employ a circular antenna with a diameter of 40.5 inches. If these antennas transmit their information using UHF Microwaves that have a wavelength of 46 cm, what will be the distance from these antennas to the Far Field?

Applicable Definitions:	Microwave Radiation	Page 7-8
	Radiation vs. Field Characteristics	Pages 7-10 & 7-11
Applicable Formula:	Equation **#7-16**	Page 7-27
Solution to this Problem:	Page 7-69	

Problem Workspace

Problem #7.26:

If the average power output of the highly focused, very directional UHF Antennas described in Problem **#7.25**, on the previous page, is 0.05 kilowatts, what will be the approximate Power Density produced by this antenna at a point 12 inches directly in front of it? If the established 6-minute TLV-STEL for this frequency [$\nu \sim 600$ MHz] is 6.0 mW/cm^2, for what maximum time period can a Service Technician work, if his assigned task requires that he stand 12 inches away from and directly in front of this transmitting antenna? You may assume that his exposure must not exceed the established 6-minute TLV-STEL.

Applicable Definitions:	Time Weighted Averages	Page 3-2
	Microwave Radiation	Page 7-8
	Radiation vs. Field Characteristics	Pages 7-10 & 7-11
Applicable Formulae:	Equation **#3-1**	Page 3-8
	Equation **#7-17**	Page 7-28
Solution to this Problem:	Pages 7-69 & 7-70	

Problem Workspace

Problem #7.27:

In order never to exceed the established TLV-STEL provided in Problem **#7.26**, on Page 7-51, what is the closest distance (directly in front of one of these transmitting UHF Antennas) that a Service Technician may safely work for periods of time longer than 6 minutes? Would this position be in the Near or the Far Field for this UHF antenna?

Applicable Definitions:	Microwave Radiation	Page 7-8
	Radiation vs. Field Characteristics	Pages 7-10 & 7-11
Applicable Formula:	Equation **#7-19**	Pages 7-28 & 7-29
Solution to this Problem:	Pages 7-70 & 7-71	

Problem Workspace

Problem #7.28:

If these "line-of-sight" UHF Microwave Antenna Systems can successfully transmit voice or digital data over a distance 85 miles, what is the minimum Power Density Level at which the system's receiving antenna can still be expected to operate successfully?

Applicable Definitions:	Microwave Radiation	Page 7-8
	Radiation vs. Field Characteristics	Pages 7-10 & 7-11
Applicable Formula:	Equation **#7-18**	Pages 7-28 & 7-29
Solution to this Problem:	Page 7-71	

Workspace for Problem **#7.28**

Problem #7.29:

Highway Patrol Officers frequently use a J-Band Radar system to measure the speed of vehicles on the highway. These J-Band Speed Radar Guns operate at a microwave frequency of 10.525 GHz, and they radiate from a 4.02-inch diameter antenna. How far is it to the Far Field for such a Radar Gun?

Applicable Definitions:	Microwave Radiation	Page 7-8
	Radiation vs. Field Characteristics	Pages 7-10 & 7-11
Applicable Formula:	Equation **#7-1**	Page 7-16
	Equation **#7-16**	Page 7-27
Solution to this Problem:	Pages 7-71 & 7-72	

Problem Workspace

Workspace Continued on the Next Page

Continuation of Workspace for Problem **#7.29**

Problem #7.30:

The J-Band Speed Radar Gun described in Problem **#7.29**, on Page 7-53, has an output power of 45 mW. What is the minimum distance in front of this Gun's antenna that a Highway Patrolman must be in order to ensure that his exposure will never exceed the 6-minute TLV-STEL, which has been established at 10.0 mW/cm^2 for this frequency of microwave radar?

Applicable Definitions:	Microwave Radiation	Page 7-8
	Radiation vs. Field Characteristics	Pages 7-10 & 7-11
Applicable Formula:	Equation **#7-1**	Page 7-16
	Equation **#7-18**	Pages 7-28 & 7-29
Solution to this Problem:	Pages 7-72 & 7-73	

Workspace for Problem **#7.30**

SOLUTIONS TO THE IONIZING AND NON-IONIZING RADIATION PROBLEM SET

Problem #7.1:

To solve this problem, we must use Equation **#7-1**, from Page 7-16; however, we must first convert the wavelength [given in microns] into its equivalent in meters:

$$\lambda = (3.35 \text{ microns})(10^{-6} \text{ meters}/\text{micron}) = 3.35\times10^{-6} \text{ meters}$$

$$c = \lambda\nu \qquad \text{[Eqn. \#7-1]}$$

$$\nu_{\text{C-H Stretch}} = \frac{c}{\lambda} = \frac{3.0\times10^{8}}{3.35\times10^{-6}} = 8.96\times10^{13} \text{ Hertz} = 8{,}960 \text{ GHz}$$

$$\therefore \quad \nu_{\text{C-H Stretch}} = 8{,}960 \text{ GHz}$$

Problem #7.2:

To solve this problem, we must again use **Equation #7-3**, from Page 7-17. To apply this relationship, we must use the form of the frequency as it was initially calculated in the previous problem, namely $\nu = 8.96\times10^{13}$ Hertz:

$$E = h\nu \qquad \text{[Eqn. \#7-3]}$$

$$E = (4.136\times10^{-15})(8.96\times10^{13}) = 0.37 \text{ electron volts}$$

$$\therefore \quad E_{\text{carbon-hydrogen stretch photon}} = 0.37 \text{ ev}$$

Problem #7.3:

We can use either Equation **#7-1** or **#7-2** to solve this problem. The most direct and simple approach is to use Equation **#7-2**, so that is the way it will be done:

$$\lambda = \frac{1}{k} \qquad \text{[Eqn. \#7-2]}$$

Rearranging to solve for the wavenumber, k, we get:

$$k = \frac{1}{\lambda} = \frac{1}{3.35\times10^{-6}} = 298{,}507 \text{ meters}^{-1}$$

$$\therefore \quad k_{\text{carbon-hydrogen stretch photon}} = 2.99\times10^{5} \text{ meters}^{-1}$$

$$\text{or} \quad k_{\text{carbon-hydrogen stretch photon}} = 2.99\times10^{5} \text{ cycles/meter}$$

$$\text{or} \quad k_{\text{carbon-hydrogen stretch photon}} = 2{,}990 \text{ cm}^{-1} = 2{,}990 \text{ wavenumbers}$$

Problem #7.4:

To solve this problem, we must use Equation **#7-1**, from Page 7-16, but first we must convert the frequency of this gamma photon, which has been provided in the problem statement in units of megahertz, into plain Hertz, as is required for this equation to apply:

$$\left(2.84\times10^{14}\ \text{MHz}\right)\left(10^{6}\ \text{Hz}/_{\text{MHz}}\right) = 2.86\times10^{20}\ \text{Hz, and}$$

$$c = \lambda\nu \qquad \text{[Eqn. \#7-1]}$$

Solving for the wavelength, λ, we get:

$$\lambda = \frac{c}{\nu} = \frac{3.0\times10^{8}}{2.86\times10^{20}} = 1.05\times10^{-12}\ \text{meters}$$

Now converting this wavelength, λ, in meters, to its equivalent in microns, we get:

$$\left(1.05\times10^{-12}\ \text{meters}\right)\left(10^{6}\ \text{microns}/_{\text{meter}}\right) = 1.05\times10^{-6}\ \text{microns} = 1.05\times10^{-6}\ \mu$$

$$\therefore\quad \lambda_{\gamma\text{-ray photon}} = 1.05\times10^{-6}\ \mu$$

Problem #7.5:

Again, we will use the second form of Equation **#7-1** to solve this problem. First we will have to convert the frequency, ν, given in MHz to its equivalent in Hertz:

$$\left(2.84\times10^{14}\ \text{MHz}\right)\left(10^{6}\ \text{Hz}/_{\text{MHz}}\right) = 2.86\times10^{20}\ \text{Hz, and}$$

$$c = \frac{\nu}{k} \qquad \text{[Eqn. \#7-1]}$$

Rearranging to solve for the wavenumber, k, we get:

$$k = \frac{\nu}{c} = \frac{2.86\times10^{20}}{3.0\times10^{8}} = 9.53\times10^{11}\ \text{meters}^{-1}$$

Now converting this wavenumber, k, in meters^{-1} to its equivalent in cm^{-1}, we get:

$$\left(9.53\times10^{11}\ \text{meters}^{-1}\right)\left(10^{-2}\ \text{cm}^{-1}/_{\text{meter}^{-1}}\right) = 9.53\times10^{9}\ \text{cm}^{-1} = 9.53\times10^{9}\ \text{wavenumbers}$$

$$\therefore\quad k_{\gamma\text{-ray photon}} = 9.53\times10^{9}\ \text{wavenumbers}$$

Problem #7.6:

To solve this problem, we must use Equation **#7-3**, from Page 7-17. First, we must convert the frequency, ν, from its listed value in MHz to its equivalent value in Hertz:

$$\left(2.84\times10^{14}\ \text{MHz}\right)\left(10^{6}\ \text{Hertz}/_{\text{MHz}}\right) = 2.84\times10^{20}\ \text{Hertz}$$

$$E = h\nu \qquad \text{[Eqn. \#7-3]}$$

$$E = \left(4.136\times10^{-15}\right)\left(2.84\times10^{20}\right) = 1{,}174{,}624\ \text{electron volts}$$

$$\therefore \quad E_{\gamma\text{-ray photon}} = 1.17 \text{ Mev}$$

Problem #7.7:

To solve this problem, we must calculate the energy that was delivered to the atom by the UV-B photon. As a result of its having absorbed this UV-B photon, the atom will clearly have entered an excited or metastable state. In this state, it "decayed" by emitting the photon of visible light. Clearly, however, since the emitted photon [visible light] is less energetic than the absorbed one [UV-B], there will be a net energy gain by the atom. This gain will simply be the difference between the energy levels of these two photons. The first step then will involve the use of Equation **#7-1**, from Page 7-16. We will use this relationship to calculate the frequency of each of the two photons involved in this situation. We must start by converting the two wavelengths, one of which has been listed in nanometers, and the other in microns to their corresponding values in meters:

$$\lambda_{\text{UV-B Photon}} = (274 \text{ nm})(10^{-9} \text{ meters}/_{\text{nm}}) = 2.74\times10^{-7} \text{ meters,}$$

$$\lambda_{\text{Visible Light Photon}} = (0.46\,\mu)(10^{-6} \text{ meters}/_{\mu}) = 4.6\times10^{-7} \text{ meters, \&}$$

$$c = \lambda\nu \qquad \text{[Eqn. \textbf{\#7-1}]}$$

Rewriting to solve for the wavelength, λ, we get:

$$\lambda = \frac{c}{\nu}$$

$$\nu_{\text{UV-B Photon}} = \frac{c}{\lambda_{\text{UV-B Photon}}} = \frac{3.0\times10^{8}}{2.74\times10^{-7}} = 1.09\times10^{15} \text{ meters}^{-1}$$

$$\nu_{\text{Visible Light Photon}} = \frac{c}{\lambda_{\text{Visible Light Photon}}} = \frac{3.0\times10^{8}}{4.6\times10^{-7}} = 6.52\times10^{-14} \text{ meters}^{-1}$$

Now we can apply the second required relationship, namely, Equation **#7-3**, from Page 7-17, to obtain the required answer:

$$E = h\nu \qquad \text{[Eqn. \textbf{\#7-3}]}$$

$$E_{\text{UV-B Photon}} = (4.136\times10^{-15})(1.09\times10^{15}) = 4.51 \text{ electron volts}$$

$$E_{\text{Visible Light Photon}} = (4.136\times10^{-15})(6.52\times10^{14}) = 2.70 \text{ electron volts}$$

Now, finally we can simply determine the energy absorbed by the atom by determining the difference between the energy levels of these two photons:

$$\text{Energy absorbed by the atom} = 4.51 - 2.70 = 1.81 \text{ ev}$$

Since the ionization energy for this atom is listed as 1.2 ev, and since the net energy absorbed by it was 1.81 ev, it is obvious that one of the results of this overall process was the ionization of the atom.

$\therefore$ The atom absorbed a total of 1.81 ev and was ionized in the process.

Problem #7.8:

To solve this problem, we must use Equation **#7-4**, from Page 7-18, and actually apply this relationship twice, once for each of the two subsequent times identified in the problem statement:

$$N_t = N_0 e^{-kt} \qquad \text{[Eqn. \#7-4]}$$

Consider first the quantity of $^{131}_{53}I$ that remained on January 20th of the same year. Clearly, this date involves a time interval after receiving this isotope of 19 days, thus, t = 19 days:

$$N_{t=19\text{ days}} = (2.0)e^{-(0.0862)(19)} = (2.0)e^{-1.6378} = (2.0)(0.1944) = 0.3888\ \mu g$$

Consider next the quantity of $^{131}_{53}I$ that remained on the one year anniversary of the arrival at the hospital of this isotope. For this case, t = 365 days:

$$N_{t=365\text{ days}} = (2.0)e^{-(0.0862)(365)} = (2.0)e^{-31.463} = (2.0)\left(2.167\times10^{-14}\right) = 4.33\times10^{-14}\ \mu g$$

Since the atomic weight of the $^{131}_{53}I$ isotope is ~ 130.91 amu, and since one mole of this isotope would weigh 130.91 grams and would contain 6.022×10^{23} atoms of $^{131}_{53}I$, we can calculate that the number of atoms remaining in 4.33×10^{-14} μg to be:

$$\text{Number of Atoms} = \left[\frac{4.33\times10^{-14}}{130.91}\right]\left[6.022\times10^{23}\right] = 1.99\times10^{10}\text{ atoms of } ^{131}_{53}I$$

∴ On January 20th, there will be only ~ 0.39 μg of this $^{131}_{53}I$ isotope remaining [19.4% of the material that was received on January 1st]. One year later, there will only be an almost certainly undetectable 4.33×10^{-14} μg — 1.99×10^{10} atoms — remaining.

Problem #7.9:

To solve this problem we must use, first, Equation **#7-5**, from Page 7-19, to determine the Half-Life of this isotope. Next we will use Equation **#7-6**, also from Page 7-19, to determine the Mean Life of an $^{131}_{53}I$ atom. Let us start with the Half-Life determination:

$$T_{1/2} = \frac{0.693}{k} \qquad \text{[Eqn. \#7-5]}$$

$$T_{1/2} = \frac{0.693}{0.0682} = 10.16\text{ days}$$

Next, let us determine the Mean Life of an $^{131}_{53}I$ atom:

$$\tau = \frac{1}{k} \qquad \text{[Eqn. \#7-6]}$$

$$\tau = \frac{1}{0.0682} = 14.66\text{ days}$$

∴ The Half-Life of the $^{131}_{53}I$ isotope = 10.16 days; & the Mean Life of an $^{131}_{53}I$ atom = 14.66 days.

Problem #7.10:

To solve this problem, we will have to employ a large number of relationships. To start with, we will have to determine the actual number of $^{131}_{53}I$ atoms that are present in each of the three time-based scenarios. This atom count is required for the application of either Equation **#7-7**, from Page 7-20, and/or Equation **#7-9**, from Pages 7-20 & 7-21.

To make these required determinations, we must have specific designations for the three times of interest; thus we shall assign: $t_0 = 0$ days — for the situation on January 1st, the date when the isotope was delivered to the hospital; $t_{19} = 19$ days — for the situation on January 20th of that same year; and $t_{365} = 365$ days — for the situation one year after the delivery of the isotope to the hospital. We must start this process, for each of the times required, by determining the number of moles of the isotope present at that time. This is done by using Equation **#1-10**, from Pages 1-19 & 1-20:

$$n = \frac{m}{MW} \qquad \text{[Eqn. #1-10]}$$

For the situation on January 1st, when the isotope was delivered to the hospital:

$$n = \frac{2.0 \times 10^{-6}}{130.9061} = 1.53 \times 10^{-8} \text{ moles of } ^{131}_{53}I$$

Next, we must apply Equation **#1-11**, from Page 1-20, to determine the atom count on January 1st, when the isotope was delivered to the hospital:

$$n = \frac{N}{6.022 \times 10^{23}} \qquad \text{[Eqn. #1-11]}$$

Transposing this relationship in order to solve for the atom count, Q, we get:

$$N_{t=0} = \left(6.022 \times 10^{23}\right)(n) = \left(6.022 \times 10^{23}\right)\left(1.53 \times 10^{-8}\right) = 9.20 \times 10^{15} \text{ atoms}$$

Now, we can address the determination of the required Activity on this date, this time using Equation **#7-7**, from Page 7-20:

$$A_b = kN \qquad \text{[Eqn. #7-7]}$$

Now, using this relationship, we can determine the Activity level required:

$$A_{b/t=0} = kN_{t=0} = (0.0862)\left(9.20 \times 10^{15}\right) = 7.93 \times 10^{14} \text{ becquerels}$$

or, expressed in curies, $A_{c/t=0} = \dfrac{7.93 \times 10^{14}}{3.70 \times 10^{10}} = 21{,}435$ curies — a very hot source!!

Next, we can use Equation #7-9, from Pages 7-20 & 7-21, to determine the Activities of this source at the other two times of interest:

$$A_t = kN_0 e^{-kt} \qquad \text{[Eqn. #7-10]}$$

First, for t = 19 days, we have:

$$A_{t=19\text{ days}} = (0.0862)\left(9.20 \times 10^{15}\right)e^{-(0.0862)(19)} = \left(7.93 \times 10^{14}\right)e^{-1.6378}$$

$$A_{t=19\text{ days}} = \left(7.93 \times 10^{14}\right)(0.1944) = 1.54 \times 10^{14} \text{ becquerels}$$

or again expressing this Activity in curies, we have:

$$A_{t=19\ \text{days}} = \frac{1.54\times10^{14}}{3.70\times10^{10}} = 4{,}167 \text{ curies — still a } \underline{\underline{\text{very}}}\ \underline{\underline{\text{hot}}}\ \underline{\underline{\text{source}}}!!$$

Finally, for t = 365 days, we have:

$$A_{t=365\ \text{days}} = (0.0862)\left(9.20\times10^{15}\right)e^{-(0.0862)(365)} = \left(7.93\times10^{14}\right)e^{-31.463}$$

$$A_{t=365\ \text{days}} = \left(7.93\times10^{14}\right)\left(2.167\times10^{-14}\right) = 17.2 \text{ becquerels}$$

∴ The three source Activity levels required for the solution of this problem are:

$$A_{t=0\ \text{days}} = 7.93\times10^{14} \text{ becquerels} = 21{,}435 \text{ curies}$$
$$A_{t=19\ \text{days}} = 1.54\times10^{14} \text{ becquerels} = 4{,}167 \text{ curies}$$
$$A_{t=365\ \text{days}} = 17.2 \text{ becquerels}$$

Problem #7.11:

To solve this problem, we must again use Equation **#7-4**, from Page 7-18:

$$N_t = N_0e^{-kt} \qquad \text{[Eqn. \#7-4]}$$

Since we know both the "starting" and the "ending" masses of the isotope of einsteinium that is to be used for an experiment, we know all of the factors, except for the time, that are required for the solution of this problem:

$$1.5\times10^{-10} = \left(8.8\times10^{-6}\right)e^{-(0.502)t}$$

$$e^{-(0.502)t} = \frac{1.5\times10^{-10}}{8.8\times10^{-6}} = 1.705\times10^{-5}$$

Next, we take the natural logarithm of both sides of this equation:

$$ln\left[e^{-(0.502)t}\right] = ln\left[1.705\times10^{-5}\right], \ \&$$

$$-(0.502)t = -10.980$$

$$t = \frac{-10.980}{-0.502} = 21.87 \text{ minutes}$$

∴ This Scientist must complete her experiment, exhausting her supply of $^{245}_{99}Es$, in the next 21.87 minutes. That is to say, this amount of $^{245}_{99}Es$ will have decayed to the "barely detectable" level in 21.87 minutes, or 21 minutes and 52.3 seconds.

Problem #7.12:

To solve this problem, we must use Equation **#7-5** from Page 7-19:

$$T_{1/2} = \frac{0.693}{k} \qquad \text{[Eqn. \#7-5]}$$

$$k = \frac{0.693}{12.26} = 0.0565 \text{ years}^{-1}$$

∴ The Radioactive Decay Constant for ${}^{3}_{1}H$ = $k_{tritium}$ = 0.0565 $years^{-1}$

Problem #7.13:

The solution to this problem also requires the use of Equation **#7-5** from Page 7-19:

$$T_{1/2} = \frac{0.693}{k} \qquad \text{[Eqn. \#7-5]}$$

First, for the iodine isotope, ${}^{131}_{53}I$, which has a Radioactive Decay Constant = 0.0862 $days^{-1}$:

$$T_{1/2} = \frac{0.693}{0.0862} = 8.039 \text{ days}$$

Next, for the einsteinium isotope, ${}^{245}_{99}Es$, for which the Radioactive Decay Constant = 0.502 $minutes^{-1}$:

$$T_{1/2} = \frac{0.693}{0.502} = 1.381 \text{ minutes}$$

∴ The requested Half-Lives are as follows:
For the iodine isotope, ${}^{131}_{53}I$, $T_{1/2}$ ~ 8.04 days
For the einsteinium isotope, ${}^{245}_{99}Es$, $T_{1/2}$ ~ 1.38 minutes

---

Problem #7.14:

The solution to this problem will require the use of Equation **#17-10**, from Pages 7-20 & 7-21:

$$A_t = \left[\frac{0.693}{T_{1/2}}\right] N_0 e^{-(0.693)t/T_{1/2}} \qquad \text{[Eqn. \#7-10]}$$

Before directly addressing the solution to this problem by using the foregoing Equation, we must first convert the Half-Life of americium from the form in which it has been given in the problem statement [that being in units of "years"] to units of "seconds", since the problem is to determine the number of disintegrations per second or becquerels:

$$T_{1/2} = (432.2\,\text{years})\left(365\,{}^{\text{days}}\!/_{\text{year}}\right)\left(24\,{}^{\text{hours}}\!/_{\text{day}}\right)\left(3,600\,{}^{\text{seconds}}\!/_{\text{hour}}\right) = 1.363 \times 10^{10} \text{ seconds}$$

We are interested in the specific number of disintegrations that are occurring in each second, namely in the Activity in becquerels, at that instant in time when the actual mass of americium present is exactly 1.75 μg; thus as we substitute numbers and values into Equation

#7-10, we shall use a time, t = 0 seconds, and the number of americium atoms that 1.75 μg of this isotope represents:

$$A_t = \left[\frac{(0.693)(1.75)\left(4.0\times10^{15}\right)}{1.363\times10^{10}}\right]e^{-\left[\frac{(0.693)(0)}{1.363\times10^{10}}\right]}$$

& since the exponent of e = $-\left[\frac{(0.693)(0)}{1.363\times10^{10}}\right]$ = 0, and since $e^0 = 1$, we have:

$$A_t = \frac{(0.693)(1.75)\left(4.0\times10^{15}\right)(1)}{1.363\times10^{10}} = \frac{4.851\times10^{15}}{1.363\times10^{10}} = 3.559\times10^{5}\text{ becquerels}$$

∴ A Smoke Detector will operate properly if there is a minimum of 355,900 $^{241}_{95}Am$ atoms disintegrating, or decaying, each second.

Problem #7.15:

To solve this problem, we must first use Equation **#7-5**, from Page 7-19, to determine the Radioactive Decay Constant of this americium isotope, and then use Equation **#7-4**, from Page 7-18, to develop the answer that has been requested in the problem statement:

$$k = \frac{0.693}{T_{1/2}} \qquad \text{[Eqn. #7-5]}$$

$$k = \frac{0.693}{432.2} = 1.60\times10^{-3}\text{ years}^{-1}$$

Knowing this value, we can apply Equation **#7-4**, from Page 7-18, to obtain the solution:

$$N_t = N_0e^{-kt} \qquad \text{[Eqn. #7-4]}$$

$$1.75\times10^{-6} = \left(1.80\times10^{-6}\right)e^{-\left(1.60\times10^{-3}\right)t}$$

$$\frac{1.75\times10^{-6}}{1.80\times10^{-6}} = e^{-\left(1.60\times10^{-3}\right)t} = 0.972$$

Now, taking the natural logarithm of both sides of this equation, we get:

$$ln[0.972] = ln\left[e^{-\left(1.60\times10^{-3}\right)t}\right], \&$$

$$-0.282 = -\left(1.60\times10^{-3}\right)t, \&$$

$$t = \frac{-0.282}{-\left(1.60\times10^{-3}\right)} = 17.607\text{ years}$$

∴ It will take ~17.6 years for the amount of the $^{241}_{95}Am$ provided in a new Smoke Detector to decay to the extent that there will be fewer than the minimum required 355,900 disintegrations/second occurring — i.e., the Activity will have fallen below 355,900 becquerels. It is unlikely that we can consider that the manufacturer has "built in" product obsolescence as a result of this "lifetime", since this time period [17.6 years] is simply too long. I cannot imagine that the average Smoke Detector owner would note that he will have to purchase a new unit 17+ years after the date of his current purchase; rather, after 17+ years, that person's Smoke Detector would simply cease to function properly without any obvious clue that it had done so!

Problem #7.16:

To solve this problem, we must employ Equation **#7-11**, from Page 7-22:

$$E = \frac{\Gamma A}{d^2} \qquad \text{[Eqn. #7-11]}$$

For this relationship to apply, we must have the distance between the target and the source expressed in units of centimeters; therefore, we must recognize that 1.0 meters = 100 cm:

$$E = \frac{(18.81)(550)}{(100)^2} = \frac{10,345.5}{10,000} = 1.035 \text{ rads/hour}$$

∴ The requested Dose Exposure Rate ~ 1.04 rads/hour.

Problem #7.17:

The solution to this problem also employs Equation **#7-11**, from Page 7-22:

$$E = \frac{\Gamma A}{d^2} \qquad \text{[Eqn. #7-11]}$$

$$48.0 = \frac{(440)\Gamma}{(30)^2}$$

$$\Gamma = \frac{(48.0)(30)^2}{440} = \frac{43,200}{440} = 98.18$$

∴ The Radiation Constant for $^{226}_{88}Ra$ is 98.2 $\frac{\text{Rad}\cdot\text{cm}^2}{\text{mCi}\cdot\text{hr}}$.

Problem #7.18:

To solve this problem, we must use Equation **#7-12**, on Pages 7-22 & 7-23:

$$D_{Rem} = D_{Rad}[QF] \qquad \text{[Eqn. #7-12]}$$

From the Tabulation of Quality Factors by Radiation Type, on Page 7-23, we can see that the Quality Factor is QF = 1.0, for γ-rays. Let us first consider the situation for this Technician when he is operating the CRSTTA with its aperture closed:

$$D_{Rem/Closed\ Aperture} = (0.09)(1.0) = 0.09 \text{ mrems/hour}$$

Next, let us deal with the open aperture situation:

$$D_{Rem/Open\ Aperture} = (0.44)(1.0) = 0.44 \text{ mrems/hour}$$

Now, since we know that this Technician experiences 2.2 hours/day of open aperture operations and the balance of the day, or 5.8 hours, of closed aperture operations, and since further we know the time rate of Equivalent Dose exposure, we can calculate directly the total Dose Equivalent this Technician will experience:

$$[\text{Dose Equivalent}]_{total} = (2.2)(0.44) + (5.8)(0.09) = 0.968 + 0.522 = 1.49 \text{ mrems}$$

∴ This Technician will experience a daily Dose Equivalent Exposure of 1.49 mrems.

Problem #7.19:

To solve this problem we must first understand the relationship between the units in which each Technician's Absorbed Dose numbers have been provided, and the units in which the required Dose Equivalent values have been requested in the problem statement. Specifically, remember that, for Absorbed Dose values: 1.0 Gy = 100 rad, and for Dose Equivalent values: 1.0 Sv = 100 rem. In the problem statement we have been given the Absorbed Dose data for each Technician in units of mrads/hour; therefore, it is easy and direct then to convert these data into their corresponding Dose Equivalent values in mrems/hour, using Equation **#7-12**, from Pages 7-22 & 7-23.

Note, too, that we are dealing in an "external" radiation situation — in contrast to an "internal" one, wherein the radiation source is located in some part of a person's body. We know that Quality Factors for all energy states of neutrons are valid for both internal and external radiation source situations; therefore, we can simply proceed using a QF = 5.0 to develop the solution to this problem.

As stated on the previous page, for thermal neutrons, the Quality Factor = QF = 5.0. We can assume that these Technicians are employed for a 5-day work week, with each workday consisting of 8 hours, thus a 40-hour work week. Let us consider, first, the situation that exists when the Thermal Column's access port is closed:

$$H_{Rem} = D_{Rad}[QF] \qquad \text{[Eqn. \#7-12]}$$

$$H_{rem\text{-}closed} = (0.12)(5.0) = 0.60 \text{ mrem/hr}$$

Let us consider next the situation when this access port is open:

$$H_{rem\text{-}open} = (0.83)(5.0) = 4.15 \text{ mrem/hr}$$

Next, we can combine these items of specific hourly Dose Equivalent information to the stated weekly time intervals for the two scenarios under which these Technicians routinely work. Consider first the closed port situation, which involves 38 hours per week:

$$H_{closed} = (0.60 \text{ mrem/hour})(38 \text{ hours}) = 22.80 \text{ mrem}$$

Next, we consider the situation that exists when the access port is open:

$$H_{open} = (4.15 \text{ mrem/hour})(2 \text{ hours}) = 8.30 \text{ mrem}$$

For the next step, we need only combine these two data items to determine the Dose Equivalent, in units of mrem, that each Technician experiences each week:

$$H_{\text{weekly total}} = 22.80 + 8.30 = 31.10 \text{ mrem}$$

And, since we know that 100 mrem = 1 mSv, we can obtain the answer in the form asked for in the problem statement:

$$(31.10_{\text{ mrem}})(10^{-2}{}_{\text{ mSv/mrem}}) = 0.311 \text{ mSv} = 311 \text{ μSv}$$

∴ These Technicians experience a Dose Equivalent of 311 μSv/week for thermal neutrons.

Problem #7.20:

To solve this problem, we must use Equation **#7-13**, from Page 7-24:

$$ER_{\text{goal}} = \frac{ER_{\text{source}}}{2^{x/HVL}} \qquad \text{[Eqn. \#7-13]}$$

We must consider two cases, namely: (1) a shielded cell fabricated from concrete; and (2) a shielded cell made of lead. We shall assume "effective relative" target Radiation Emission Rates for these two situations as follows:

775 for the concrete shielded cell, and

1.0 for the lead shielded cell.

Consider first the concrete shielded cell:

$$ER_{\text{goal}_{\text{ concrete}}} = \frac{ER_{\text{source}}}{2^{\left[\frac{x_{\text{concrete}}}{HVL_{\text{concrete}}}\right]}}$$, and substituting in the appropriate "effective relative" values:

$$775 = \frac{ER_{\text{source}}}{2^{\left[\frac{18}{2.45}\right]}} = \frac{ER_{\text{source}}}{2^{7.35}} = \frac{ER_{\text{source}}}{162.80}$$

Next, rearranging this equation to solve for the observed Radiation Emission Rate of the source, we get:

$$ER_{\text{source}} = (775)(162.80) = 126{,}168.42 \text{ mrads/hour} \sim 126.2 \text{ rads/hr}$$

Next we must consider the lead shielded cell:

$$ER_{\text{goal}_{\text{ lead}}} = \frac{ER_{\text{source}}}{2^{\left[\frac{x_{\text{lead}}}{HVL_{\text{lead}}}\right]}}$$, and now substituting in the lead "effective relative" values:

$$1.0 = \frac{126{,}168.42}{2^{\left[\frac{8}{HVL_{\text{lead}}}\right]}}$$, and rearranging:

$$2^{\left[\frac{8}{HVL_{\text{lead}}}\right]} = 126{,}158.42 \text{ mrads/hour} \sim 126.2 \text{ rads/hr}$$

Next, we must take the common logarithm of both sides of this equation:

$$(log[2])\left(\frac{8}{HVL_{\text{lead}}}\right) = log[126{,}168.42]$$

$$(0.301)\left(\frac{8}{HVL_{lead}}\right) = 5.101\text{, and rearranging:}$$

$$\frac{8}{HVL_{lead}} = \frac{5.101}{0.301} = 16.945$$

Finally, now, solving for the Half-Value Thickness of lead, we get:

$$HVL_{lead} = \frac{8}{16.945} = 0.472 \text{ inches}$$

∴ Thus, we see that the Half-Value Thickness of lead is ~ 0.47 inches ~ 1.20 cm.

Problem #7.21:

The solution to this problem requires the use of Equation **#7-14**, from Page 7-25:

$$\frac{ER_a}{ER_b} = \frac{S_b^2}{S_a^2} \qquad \text{[Eqn. \#7-14]}$$

$$\frac{ER_a}{2.0} = \frac{(15)^2}{(25)^2} = \frac{225}{625} = 0.36$$

$$ER_a = (0.36)(2.0) = 0.72 \text{ mSv/hr}$$

∴ The Doctor would experience a radiation intensity of 0.72 mSv/hour at the greater distance from the Patient.

Problem #7.22:

The solution to this problem will require, initially, the use of the Quality Factor Table, from Page 7-23; this tabulation is associated with Equation **#7-12**, which is described on Pages 7-22 & 7-23. Note, in this problem, we are dealing with: (1) an "internal" radiation situation, and (2) both α- and/or β-radiation sources — for each of which, the listed Quality Factors apply only to "internal" radiation situations. In order, the Quality Factors for each of these classes of radiation are, respectively: $QF_\alpha = 20.0$, and $QF_\beta = 1.0$.

Knowing these data points, we must next use Equation **#7-14**, from Page 7-25. We will first determine what the impact on the adjusted radiation Dose Equivalent, expressed in μrem, would be as a result of the change in the character of the radiation source (i.e., the source changing from being an alpha emitter to being a beta emitter):

$$\frac{QF_{\text{Beta Source}}}{QF_{\text{Alpha Source}}} = \frac{1.0}{20.0} = 0.05$$

Thus we see that by simply changing the type of radiation being emitted [by changing the radioactive source], we will diminish the experienced radiation Dose Equivalent by a factor of 20, thus:

$$(40_{\mu\text{rem/hour}})(0.05) = 2.0 \ \mu\text{rem/hour}$$

Next, we must apply Equation **#7-14**, from Page 7-25, to determine the distance at which the Dose Equivalent of the new beta source would equal 40 μrem/hour:

$$\frac{ER_a}{ER_b} = \frac{S_b^2}{S_a^2} \qquad \text{[Eqn. #7-14]}$$

Substituting in, we get:

$$\frac{2}{40} = \frac{S_b^2}{(2.5)^2}$$

Solving now for S_b, we get:

$$S_b^2 = \frac{(2.5)^2(2)}{40} = \frac{12.5}{40} = 0.313, \text{ and}$$

$$S_b = \sqrt{0.313} = 0.559 \text{ mm}$$

∴ The adjusted hourly radiation Dose Equivalent produced by the injectant containing the ${}^{45}_{19}K$ source vs. the value for the injectant containing the ${}^{226}_{88}Ra$ source would be 2.0 μrem vs. 40.0 μrem. A Dose Equivalent value of 40.0 μrem/hour could be measured for the injectant containing the ${}^{45}_{19}K$ source at a distance of approximately 0.56 mm from the liver in which this injectant has accumulated.

Problem #7-23:

To solve this problem, we must use Equation **#7-15**, from Page 7-26:

$$OD = log\left[\frac{I_{incident}}{I_{transmitted}}\right] \qquad \text{[Eqn. #7-15]}$$

$$OD = log\left[\frac{475}{.45}\right] = log[1,055.56]$$

$$OD = log\left[\frac{475}{.45}\right] = log[1,055.56] = 3.023$$

∴ The Optical Density of the Operator's Goggles = OD = 3.02.

Problem #7.24:

This problem also requires the use of Equation **#7-15**, from Page 7-26:

$$OD = log\left[\frac{I_{incident}}{I_{transmitted}}\right] \qquad \text{[Eqn. #7-15]}$$

$$OD = log\left[\frac{(475)(6.62)}{0.19}\right] = log\left[\frac{3,144.5}{0.19}\right] = log[16,550.0] = 4.219$$

We have been asked, in the statement of the problem, not simply for the new Optical Density, but rather what increase in Optical Density would be required under the conditions that prevail for the new; thus we shall proceed as follows:

$$\Delta OD = \frac{OD_{new}}{OD_{former}}, \text{ and substituting the two known Optical Densities:}$$

$$\Delta OD = \frac{4.219}{3.023} = 1.395 = 139.5\%$$

> ∴ The Optical Density of the goggles will have to be increased by approximately 140%, to a new Optical Density value of 4.22.

Problem #7.25:

This problem requires the use of Equation **#7-16**, from Page 7-27:

$$r_{FF} = \frac{\pi D^2}{8\lambda} \qquad \text{[Eqn. \#7-16]}$$

Remembering that since the antenna diameter term, "D", for this expression must be in units of "centimeters", rather than "inches", we must convert the dimension provided in the problem statement from inches into centimeters:

$D = (40.5 \text{ inches})(2.54 \text{ cm/inch}) = 102.87$ cm, and now substituting in:

$$r_{FF} = \frac{\pi(102.87)^2}{(8)(46)} = \frac{33{,}245.08}{368} = 90.34 \text{ cm}$$

> ∴ The Far Field is ~ 90.3 cm ~ 35.6 inches out in front of this UHF antenna.

Problem #7.26:

The first part of this problem can be solved by the application of Equation **#7-17**, from Page 7-28. From the previous problem, we see that the distance to the Far Field is just under 3 feet; therefore, we are dealing with a situation in the Near Field, which means that this relationship is the valid way to obtain the solution to this problem. We must now convert all the units provided in the problem statement into the units required for use in this equation:

$$P = (0.05 \text{ kilowatts})(10^6 \text{ milliwatts/kilowatt}) = 5\times10^4 \text{ mW}$$

$D = (40.5 \text{ inches})(2.54 \text{ cm/inch}) = 102.87$ cm, and now substituting in:

$$W_{NF} = \frac{16P}{\pi D^2} \qquad \text{[Eqn. \#7-17]}$$

$$W_{NF} = \frac{(16)(5\times10^4)}{\pi(102.87)^2} = \frac{8\times10^5}{\pi(10{,}582.24)} = \frac{8\times10^5}{33{,}245.08} = 24.06 \text{ mW/cm}^2$$

We must next determine the time interval during which the Service Technician can safely work in a position in the Near Field in front of this radiating antenna. To obtain this answer, we shall use Equation **#3-1**, from Page 3-8:

$$TWA = \frac{\sum_{i=1}^{n} T_i C_i}{\sum_{i=1}^{n} T_i} = \frac{T_1C_1 + T_2C_2 + \ldots + T_nC_n}{T_1 + T_2 + \ldots + T_n} \qquad \text{[Eqn. \#3-1]}$$

We shall assume that this Service Technician will spend "t" minutes [out of every 6-minute time intervals] working at a point in the Near Field in front of this antenna, and the remainder of the time during this 6-minute period, namely, "(6 – t)" minutes, this Technician will be well away from this antenna — in a position where the microwave Power Density = 0.0 mW/cm^2. We shall further assume that the 6-minute TWA exposure that this Service Technician experienced was at the TLV-STEL level of 6.0 mW/cm^2:

$$6 = \frac{24.06t + 0.0(6-t)}{t + (6-t)} = \frac{24.06}{6}t = 4.011t, \text{ and solving for "t", we get:}$$

$$t = \frac{6}{4.011} = 1.496 \text{ minutes} \sim 1.50 \text{ minutes}$$

∴ The Power Density in the Near Field directly in front of this UHF antenna will be ~ 24 mW/cm^2.

As to the second part of this problem, this Service Technician will only be able to work for very brief periods — each being 1.5 minutes [1 minute, 30 seconds] or less out of every 6 minutes — at the designated point in the Near Field, directly in front of this antenna, whenever it is transmitting. In essence, he will only be able to complete productive work for ~ 25% of the time; he must spend the remaining ~ 4.5 minutes out of every 6 away from this antenna — i.e., ~ 75% of his time. If he exceeds this duty cycle, he will exceed the TLV-STEL, which is 6.0 mW/cm^2.

Problem #7.27:

We do not know for certain that the answer to this problem will be a distance that will be great enough to put it in the Far Field for this antenna — i.e., a distance that is greater than 90.3 cm = 35.6 inches. For the moment, however, we will assume that the location is, in fact, in the Far Field. Based on this assumption, this problem will require the use of Equation **#7-19**, from Pages 7-28 & 7-29. Recall the following data and factors that apply to this antenna:

$D = 40.5$ inches $= 102.87$ cm

$\lambda = 46$ cm

$P = 5 \times 10^4$ mW

$W_{FF} = 6.0$ mW/cm^2 — a power level equal to the TLV-STEL for this frequency; therefore, an electrical field power density that ensures that an exposed Technician will never exceed the established TLV-STEL.

$$r = \frac{D}{2\lambda}\sqrt{\frac{\pi P}{W_{FF}}} \qquad \text{[Eqn. \textbf{\#7-19}]}$$

Now plugging these values into this equation, we get:

$$r = \frac{102.87}{(2)(46)}\sqrt{\frac{\pi(5\times10^4)}{6.0}} = \frac{102.87}{92}\sqrt{\frac{157,079.63}{6}}$$

$$r = 1.118\sqrt{26,179.94} = (1.118)(161.8) = 180.92 \text{ cm} = 71.23 \text{ inches}$$

Although we did not know if the maximum required electrical field Power Density Level of 6.0 mW/cm^2 would be in the Far Field at the start of this problem, we can now see that it truly is a Far Field location [recall that the Far Field starts 90.3 cm from the antenna]. Had the calculated value for this problem been less than this distance, we would have been compelled to use the Near Field Power Density Level relationship, namely, Equation **#7-17**, from Page 7-28, to obtain the solution to this problem.

> ∴ Although it is difficult to determine how it would actually be done at the calculated distance, the closest that a Service Technician could safely work in front of this UHF antenna [and be certain never to exceed the 6-minute TLV-STEL of 6.0 mW/cm^2] would be at a distance of ~ 180.9 cm or ~ 71.2 inches. He would have to have VERY long arms!!

Problem #7.28:

This problem will also require the use of Equation **#7-18**, from Pages 7-28 & 7-29. Clearly, this point is in the Far Field. In fact, we will have to convert this very large distance into the units required for this Equation; thus:

$$(85 \text{ miles})(5,280 \text{ feet/mile})(12 \text{ inches/foot})(2.54 \text{ cm/inch}) = 1.368 \times 10^7 \text{ cm}$$

$$W_{FF} = \frac{\pi D^2 P}{4\lambda^2 r^2} \qquad \text{[Eqn. \#7-18]}$$

Now substituting in the appropriate values, we get:

$$W_{FF} = \frac{\pi(102.87)^2(5\times10^4)}{(4)(46)^2(1.368\times10^7)^2} = \frac{\pi(10,582.24)(5\times10^4)}{(4)(2,116)(1.871\times10^{14})}$$

$$W_{FF} = \frac{1.662\times10^9}{1.584\times10^{18}} = 1.05\times10^{-9} \text{ mW/cm}^2$$

> ∴ The minimum Power Density required for this UHF receiving antenna to be able to operate successfully appears to be 1.05×10^{-9} mW/cm^2 = 1.05×10^{-6} μW/cm^2.

Problem #7.29:

This problem must be solved by first applying Equation **#7-1**, from Page 7-16, to determine the wavelength of this J-Band microwave radar [often referred to in police circles as "X-Band" Speed Radar; since during World War II, this particular frequency was in a band that was identified as the "X-Band"]. When we have this wavelength information, we can then utilize Equation **#7-16**, from Page 7-27, to develop the result that was requested in the problem statement:

$$\lambda = \frac{c}{\nu} \qquad \text{[Eqn. \#7-1]}$$

Remembering that the frequency term, n, must be in Hertz rather than gigahertz, we can substitute in the values provided:

$$\lambda = \frac{3\times10^8}{10.525\times10^9} = 2.85\times10^{-2} \text{ meters} = 2.85 \text{ cm}$$

Now since we now know the wavelength of this J-Band Speed Radar, we can now determine the distance to the Far Field, using Equation **#7-16**, from Page 7-27:

$$r_{FF} = \frac{\pi D^2}{8\lambda} \qquad \text{[Eqn. \#7-16]}$$

To apply this relationship, we must have the diameter of the radar antenna in units of centimeters:

$$(4.02 \text{ in})(2.54 \text{ cm/in}) = 10.21 \text{ cm}$$

Now we can substitute numerical values and determine the distance to the Far Field for this Speed Radar unit:

$$r_{FF} = \frac{\pi(10.21)^2}{(8)(2.85)} = \frac{327.49}{22.80} = 14.36 \text{ cm}$$

∴ It is ~ 14.36 cm or ~ 5.66 inches to the Far Field for this J-Band Speed Radar Gun.

Problem #7.30:

This problem will require the use of Equation **#7-1**, from Page 7-16, to calculate the wavelength of this J-Band Radar. When this value is known, we use Equation **#7-18**, from Pages 7-28 & 7-29, to obtain the solution requested in the problem statement:

$$\lambda = \frac{c}{\nu} \qquad \text{[Eqn. \#7-1]}$$

Remembering that the frequency term, n, must be in Hertz rather than gigahertz, we can substitute in the values provided:

$$\lambda = \frac{3 \times 10^8}{10.525 \times 10^9} = 2.85 \times 10^{-2} \text{ meters} = 2.85 \text{ cm}$$

Since we now know the wavelength of this J-Band Speed Radar, we can determine the distance at which the required maximum Power Density level occurs; for this step, as stated above, we will use Equation **#7-18**, from Pages 7-28 & 7-29:

$$W_{FF} = \frac{\pi D^2 P}{4\lambda^2 r^2} \qquad \text{[Eqn. \#7-18]}$$

Again, remembering that the diameter of this J-Band radar is 4.02 inches = 10.21 cm, we can substitute in the numeric values that we have. Note, this relationship is most commonly used to calculate the Power Density in the Far Field; however, the distance term, "r", is a factor in the denominator. Thus we will use this relationship — rearranged — to calculate the distance required in the problem statement:

$$10.0 = \frac{\pi(10.21)^2(45)}{(4)(2.85)^2 r^2}$$

$$r^2 = \frac{\pi(10.21)^2(45)}{(4)(2.85)^2(10.0)} = \frac{14,739.47}{324.9} = 45.37, \text{ \& taking the square root of both sides:}$$

$$r = \sqrt{45.37} = 6.74 \text{ cm}$$

∴ The distance — in front of this J-Band Speed Radar Gun's antenna — at which the Power Density Level will be $\leq$ 10 mW/cm^2 will be ~ 6.74 cm, which is ~ 2.65 inches. For any distance greater than this, there will never be any danger of an individual exceeding the established TLV-STEL of 10 mW/cm^2; however, at distances less than this, exposures — especially relatively long duration exposures — may well exceed this Standard.

Chapter 8
Statistics and Probability

This chapter will discuss the broad areas of statistics and probability, as these disciplines can be applied to the routine practice of occupational safety and health. Decision making on matters of employee safety frequently involves the evaluation of statistical data, and the subsequent development from these data of the probabilities of the occurrence of future events. These evaluations and the subsequent projections are important because the events being considered may involve workplace hazards. These two subjects: (1) the statistical aspects and (2) the probability considerations will be considered separately.

RELEVANT DEFINITIONS

Populations

A **Population** is any set of values of some variable measure of interest — for example, a listing of *the orthodontia bills of every person living on the island of Guam*, or a tabulation showing the count of *the number of Letters to the Editor that were received by the Washington Post newspaper each day during 1996,* would each make up a **Population**. A **Population** is the entire set of those values, the entire family of objects, data, measurements, events, etc. being considered from a statistical, probabilistic, or combinatorial perspective. A **Population** may consist of "events" that are either random or deterministic. For reference, a deterministic event is one that can be characterized as "cause-and-effect" related — i.e., when a person loses his grip on a baseball [the "cause"], the ball will fall to the ground [the "effect" event that was deterministically produced in a totally predictable manner by the identified "cause"]. **Populations** may also consist of "members" whose values are themselves functions of a second, or a third, or even some higher number of random variables. The two example **Populations** listed above are most likely random [and therefore, not deterministic] — i.e., in each case, the values in either of these **Populations** are not obviously related to, or functions of, any other identifiable random factor or variable.

Distributions

A **Distribution** is a special type or subset of a population. It is a population, the values of whose "members" are related or a function of some identifiable and quantifiable random variable. A **Distribution** is virtually always spoken of or characterized as being "a function of some random variable"; the most common mathematical way to represent such a **Distribution** is to speak of it as a function of "x" — i.e., **f(x)**, where "x" is the random variable. Examples of Distributions might be *the per acre yield of soybeans as a function of such things as: (1) the amount of fertilizer applied to the crop, (2) the volume of irrigation water used, (3) the average daytime temperature during the growing season, (4) the acidity of the soil*, etc. Any **Distribution** that is characterized as being an **f(x)**, for "x", some continuous random variable, can be and is also frequently described as being:

(1) a Probability Density Function,
(2) a Probability Distribution,
(3) a Frequency Function, and/or
(4) a Frequency Distribution, etc.

Specific Types of Distributions

Uniform Distribution

A **Uniform Distribution** is one in which the value of every member is the same as the value of every other member. An example of a **Uniform Distribution** would be *the situation where the Safety Manager of a manufacturing plant had to complete safety inspections of various production areas at random times during the 8-hour workday.* If this workday is thought of as being divided up into 480 one-minute intervals, the probability of the Safety Manager visiting during any one of these intervals will be equally likely. Clearly — if the Safety Manager actually makes his visits on a random basis — each of these intervals will be equally likely to be selected; thus the "value" for each of these intervals will be equal [i.e., the probability of a visit during any specific interval will be 1/480, or 0.00208], and the population of these values can be said to constitute a **Uniform Distribution**.

Normal Distribution

A **Normal Distribution** is one of the most familiar types in this overall category of distributions — its applications apply to virtually any naturally occurring event. The "graphical" representation of a **Normal Distribution** is the well-known and widely understood "bell-shaped curve", or "normal probability distribution curve". The **Normal Distribution** is almost certainly the most important and widely used foundation block in the science of statistical inference, which is the process of evaluating data for the purpose of making predictions of future events. This type of distribution is always perfectly symmetrical about its Mean [described on Page 8-4]. Examples of **Normal Distributions** are: (1) *the number of tomatoes harvested during one growing season from each plant in a one-acre field of this crop*; (2) *the annual rainfall at some specific location on the island of Kauai, HI*; (3) *the magnitude of the errors that arise in the process of reading a dial oven thermometer*, etc.

Binomial Distribution

A **Binomial Distribution** is one in which every included event will have only two possible outcomes. It is a distribution made up of members whose values depend upon a binomial random variable. This category of variable can be most easily understood by considering one of its most familiar members, namely, the result of flipping a coin — a process for which there are only two possible outcomes, "HEADS" and/or "TAILS" [here we assume that the coin cannot land on and remain on its edge]. An example of a Binomial Distribution would be *the genders of all the individuals standing in the Ticket Line for the musical, Phantom of the Opera.* **Binomial Distributions** in general, and particularly those with a large number of members, can be considered and handled, for any necessary computational effort, as Normal Distributions.

Exponential Distribution

An **Exponential Distribution** is frequently described as the Waiting Time Distribution, since many populations in this category involve considerations of variable time intervals. This class of distribution is relatively easy to understand by considering a couple of examples. A first might be *the lengths of time between Magnitude 7.5+ earthquakes on the San Andreas Fault in California.* Another example might be *the distances traveled by a municipal bus between major mechanical breakdowns, etc.* Both of these populations would be characterized as **Exponential Distributions**.

Characteristics of Populations and/or Distributions

Member

A **Member** of any population or distribution is simply one item from the set that makes up the whole. The **Member** can be any quantifiable characteristic — i.e., the height of any individual who belongs to some social group; the number of shrimp caught each day by any member of the Freeport, TX, fishing fleet; the number of times that the dice total 12 in a game of Craps, etc.

Variable

A **Variable** is a characteristic or property of any individual member of a population or distribution. The name, "**Variable**", derives from the fact that any particular characteristic of interest may assume different values among the individual members of the population or distribution being considered. If one was considering the distribution of the weights of elephant calves born in captivity throughout the world, one might evaluate such data from a variety of different random perspectives, or from the relationship of these birth weights to a variety of **Variables**. Among such **Variables** might be: (1) the country in which the birth occurred, (2) whether or not the birth occurred in a zoo, (3) a situation where the calf was the offspring of a "work elephant", or (4) the age of the mother elephant, etc.

Sample

A **Sample** is a subset of the members of an entire population. **Samples**, per se, are employed whenever one must evaluate some measurable characteristic of the members of an entire population in a situation where it is simply not feasible to consider or measure every member of that population. For example, one might have to answer a question of the following type:

1. Does the average digital clock produced in a clock factory actually keep correct time? or
2. Is the butterfat content of the daily output of homogenized milk from a dairy at or above an established standard for this factor?

In order to make any of these types of determinations, it is not usually considered necessary to sample and test every member of the population — rather such a determination can usually be made by obtaining and testing a **Sample** from the population of interest. For the two questions asked above, one might sample and test one of every 10 clocks, or one of every 1,000 gallons of milk, etc.

Parameter

A **Parameter** is a *calculated quantitative measure* that provides a useful description or characterization of a population or distribution of interest. **Parameters** are calculated directly from observations, the summary tabulation of which make up the population or distribution being considered. For any population or distribution of interest, an example of a **Parameter** would be that population's or distribution's Mean or Median [i.e., see Page 8-4 for complete descriptions of these terms].

Sample Statistic

A **Sample Statistic** is a specific numeric descriptive measure of a sample. It is calculated directly from observations made on the sample itself. Basically, a **Sample Statistic** is a *parameter* that is determined for a sample — i.e., the sample standard deviation [see Page 8-5 for a complete description of this term]. It is very common for a measured **Sam-**

ple Statistic to be thought of as representative of or applicable to the entire population or distribution of interest.

Parameters of Populations and/or Distributions

Frequency Distribution

A **Frequency Distribution** is a tabulation of any of variable characteristics of any population that can be measured, counted, tabulated, or correlated. For example, from the **Frequency Distribution** that represents the results of the performance of high school seniors on the Scholastic Aptitude Test, it can be predicted that a score of 1,290 will place the student in the top 5% of all similar students taking this test.

Range

The **Range** of any set of variable data — taken from some population or distribution of interest — will be the calculated result that is obtained when the value of the numerically smallest member of the set is subtracted from the value of the numerically largest member of that same set — see Equation **#8-1**, from Page 8-10.

Mean

The **Mean** of any set of variable data — from some population or distribution of interest — is the sum of the individual values of the items of that data set, divided by the total number of items that make up the set. The **Mean** is the average value for the set of data being considered, and, in fact, the word "Average" is almost always used synonymously with **Mean**. The **Mean** is the first important measure of the "central tendency" of that set of variables — see Equation **#8-3**, from Page 8-11.

Geometric Mean

The **Geometric Mean** is a common alternative measure of the "central tendency" of any set of variable data — from some population or distribution of interest. It is a somewhat more useful measure than the simple Mean for any situation where the population or distribution being evaluated has a very large range of values among its members — i.e., a range of values varying over several orders of magnitude. Specifically, for any set of data, for which the ratio $R \geq 200$ or $\log R \geq 2.30$ — where R is defined as follows:

$$R = \frac{\text{the numeric value of the largest member of a population or distribution of interest}}{\text{the numeric value of the smallest member of a population or distribution of interest}} \text{ —}$$

the **Geometric Mean** may be a better measure of this population's or distribution's central tendency — See Equation **#8-4**, from Pages 8-11 & 8-12.

Median

The **Median** of any set of variable data — taken from some population or distribution of interest — is the middlemost value of that data set. When all the individual variable members of the set have been arranged either in ascending or descending order, the **Median** will be either:

(1) the data point that is exactly in the center position, or

(2) if there are a number of same value data points at, near, or around the center position, then this parameter will be the value of the data point that is centermost.

It can be regarded as the "Midpoint" value in any Normal Distribution containing "n" different numeric values, x_i. For such a set, it is that specific value of $x_{n/2}$, for which there are as many values in the distribution greater than this number, as there are values in the distribution less than this number. It is the second important measure of the "central tendency" of the set of variables being considered — see Equation **#8-5**, from Pages 8-12 & 8-13.

Mode

The **Mode** of any set of variable data points — taken from some population or distribution of interest — is the value of the most frequently occurring member of that set. The **Mode** is the "most populous" value in any Normal Distribution containing "n" different numeric values, x_i. For such a set, it is that specific x_i which is the most frequently occurring value in the entire distribution. The **Mode** is the third most important measure of the "central tendency" of the set of variables being considered; however, it does not have to be a value that is close to the center of that population. It can be numerically the smallest, or the largest, or any other value in the set, so long as it appears more frequently than any other value — see Equation **#8-6**, from Page 8-13.

Sample Variance

The **Sample Variance** of any set of "n" data points — taken from some population or distribution of interest — is equal to the sum of the squared distances of each member of that set from the set's Mean. This squared "distance" must then be divided by one less than "n", the number of members of that set — i.e., the denominator in this process is the quantity, "(n – 1)" — see Equation **#8-7**, from Pages 8-13 & 8-14.

This parameter looks at the absolute "distance" between each value in the set and the value of the set's Mean. If one were simply to obtain a simple "average" of these distances, the result would be zero, since some of these values would be negative, while a compensating number would be positive. To correct for this in the computation of the **Sample Variance**, each of these "distances" is squared; thus the result for each of these operations will always be positive, and a measure of the absolute "value-to-mean distance" will thereby be obtained.

The **Sample Variance** is always designated by the term, "s^2", and its dimensions will always be the square of the dimensions of the values of the members of the population or distribution being considered — i.e., if the population is a set of values measured in U.S. Dollars, then s^2 will be in units of $[\text{U.S. Dollars}]^2$.

For a Normal Distribution, the **Sample Variance** will probably be the best and least biased [i.e., the most unbiased] estimator of the true Population Variance.

Sample Standard Deviation

The **Sample Standard Deviation** of any set of variable data points — taken from some population or distribution of interest — is equal to the positive square root of the Sample Variance, as defined above on this page. For the relationship that defines this parameter, see Equation **#8-9**, on Pages 8-14 & 8-15.

The **Sample Standard Deviation** is always designated by the term, "s", and its dimensions will always be the same as the dimensions of each member in the population or distribution being considered — i.e., if the population is a set of values measured in U.S. Dollars, then "s" [unlike the Sample Variance, "s^2", of which "s" is the square root] will also be in units of U.S. Dollars.

For a Normal Distribution, the **Sample Standard Deviation** will be a better, less biased estimator of the true and most useful Population Standard Deviation.

Sample Coefficient of Variation

The **Sample Coefficient of Variation** is simply the ratio of the Sample Standard Deviation to the Mean of or for the population or distribution being considered — see Equation **#8-11**, from Pages 8-15 & 8-16. This parameter is also commonly described as the **Relative Standard Deviation.**

For any Normal Distribution, the **Sample Coefficient of Variation** is thought to be a good to very good measure of the specific dispersion of the values that make up the set being examined. This coefficient is most commonly designated as "CV_{sample}", and it is a dimensionless number. Since the **Sample Coefficient of Variation** is regarded as a less biased, and therefore better estimator of the dispersion that characterizes the data in the distribution being considered, and does so more effectively than does its more biased counterpart, the Population Coefficient of Variation, this parameter tends to be the much more widely used of the two.

Population Variance

The **Population Variance** of any set of "n" data points — taken from some population or distribution of interest — is equal to the average of the squared distances of each member of that set from the Mean of the set — see Equation **#8-8**, from Page 8-14.

This parameter, like its Sample Variance counterpart, also looks at the absolute "distance" between each value in the set and the value of the set's Mean. Again, if one were simply to obtain a simple "average" of these distances, the summation result would always be zero, since roughly half of these distances are negative, while the remainder are positive. To correct for this in this computation and thereby obtain a true measure of the absolute distance, each of these "distances" is squared; thus the result will always be a positive number, and a very effective measure of the absolute "value-to-mean distance" will thereby be obtained.

The **Population Variance** is always designated by the term, "σ^2", and its dimensions will always be the square of the dimensions of each member in the population being considered — i.e., if the population is a set of values measured in units of "lost time injuries/1,000 work days", then σ^2 will be in units of [lost time injuries/1,000 work days]^2.

For a Normal Distribution, the **Population Variance** will usually be slightly more biased in determining a useful and precise value for this parameter than will its Sample Variance counterpart, and for this reason, it is used less frequently than the Sample Variance.

Population Standard Deviation

The **Population Standard Deviation** of any set of variable data points — taken from some population or distribution of interest — is equal to the positive square root of the Population Variance, as defined above — see Equation **#8-10**, from Page 8-15, for the mathematical relationship for the **Population Standard Deviation.**

The **Population Standard Deviation** is always designated by the term, "σ", and its dimensions will always be the same as the dimensions of each value in the population being considered — i.e., if the population is a set of values measured in "lost time injuries/1,000 work days", then "σ" [unlike the Population Variance, of which "σ" is the square root] will also be in units of "lost time injuries/1,000 work days".

For a Normal Distribution, the **Population Standard Deviation** will be slightly more biased as an estimator; thus, it is used less frequently in these determinations than the Sample Standard Deviation.

Population Coefficient of Variation

The **Population Coefficient of Variation** is simply the ratio of the Population Standard Deviation to the Mean of or for the population or distribution being considered — see Equation **#8-12**, from Page 8-16.

For any Normal Distribution, the **Population Coefficient of Variation** is thought to be a slightly biased measure of the specific dispersion of the values that make up the set being examined. This coefficient is most commonly designated as "$CV_{population}$", and it is a dimensionless number. Since the **Population Coefficient of Variation** is regarded as a slightly more biased, and therefore poorer estimator of the dispersion that characterizes the data in the distribution being considered, its counterpart, the Sample Coefficient of Variation, tends to be much more widely used.

Probability Factors and Terms

Experiment

An **Experiment** is a procedure or activity that will ultimately lead to some identifiable outcome that cannot be predicted with certainty. A good example of an **Experiment** might be the result of throwing a fair die and observing the number of dots that appear on the up-face. There are six possible result outcomes for such an **Experiment**; in order they are: one dot, two dots, three dots, four dots, five dots, and six dots. Each of these outcomes is equally likely; however, the specific result of any single **Experiment** can never be predicted with certainty.

Result

A **Result** is the most basic and simple outcome of any Experiment — i.e., for the Experiment of throwing of a fair die, there are a total of six possible **Results**, as described above.

Sample Space

The **Sample Space** of any Experiment is the totality of all the possible Results of that Experiment. For the Experiment of throwing a fair die described above, the **Sample Space** would be: one, two, three, four, five, and six. This **Sample Space** is most frequently represented symbolically in the following way:

$$S: \{1, 2, 3, 4, 5, 6\}$$

Event

An **Event** is a sub-set of specific Results from some well-defined overall Sample Space — i.e., for the fair die throwing Experiment described above, a specific **Event** might be the occurrence of an even number on the up-face of the die. From the totality of the Sample Space for this Experiment, the even number on the up-face of the die **Event** would be the following sub-set: two, four, and six — or listing this **Event** as a sort of Sub-Sample Space, the following would be its symbolic representation:

$$S_{even}: \{2, 4, 6\}$$

Compound Event

A **Compound Event** is some useful or meaningful combination of two or more different Events. Compound Events are structured in two very specific ways. In order, these structures are shown below:

1. The UNION of two Events — say, M & N — is the first type of a **Compound Event**. A UNION is said to have taken place whenever either M or N, or both M & N occur as the outcome of a single execution of the Experiment. Symbolically, a UNION, as the first category of a **Compound Event**, is represented in the following way — again assume we are dealing with the two Events, M & N:

   $$M \cup N$$

   Considering again the Experiment of throwing a fair die and observing its up-face, we might have an interest in the following two events: (1) M = the Result is an even number, and (2) N = the Result is a number greater than three. The Sub-Sample Space that makes up the UNION of these two Events would be:

   $$S_{M \cup N}: \{2, 4, 5, 6\}$$

2. The INTERSECTION of two Events — again, say, M & N — is the second type of **Compound Event**. An INTERSECTION is said to have taken place whenever both M & N occur as the outcome of a single execution of the Experiment. Symbolically, an INTERSECTION, as the second category of a **Compound Event**, is represented in the following way — again assume we are dealing with the two Events, M & N:

   $$M \cap N$$

   Considering again the die throwing Experiment, and the same two events described above in the section on the UNION, the Sub-Sample Space that makes up the INTERSECTION of these two events would be:

   $$S_{M \cap N}: \{4, 6\}$$

Complementary Event

A **Complementary Event** is the totality of all the alternatives to some specific Event of interest. Within any Sample Space, the **Complement** to some Event of interest — say, M — will be every other possible Result that is not included within M. That is to say, whenever M has not occurred, its **Complement** — designated symbolically as M′ — will have occurred.

Considering again the Experiment of throwing a fair die and observing its resultant up-face, we might have an interest in the event: M = the Result is an even number. For this event, its **Complement**, M′ = the Result, is an odd number. The Sub-Sample Spaces for the Event, M, would be shown symbolically as:

$$S_M: \{2, 4, 6\}$$

The Sub-Sample Space for the **Complement** to M, again designated as M′, would be:

$$S_{M'}: \{1, 3, 5\}$$

Probabilities Associated with Results

The **Probability of the Occurrence of a Result** must always lie between 0 and 100% [or as a decimal, between 0.00 and 1.00]. This probability is a measure of the relative frequency of occurrence of the Result of interest. It is the outcome frequency that would be expected to occur if the Experiment were repeated over and over and over — i.e., a very large number of repetitions.

For example, in the Experiment of throwing and observing the up-face of a fair die, the probability of observing a "two" would be 1/6. This 1/6 factor would also be the probability associated with each one of the other five Results that exist within this Experiment's Sample Space.

It is important to note in this context that the probabilities of all the Results within any Sample Space must always equal 100%, or 1.00.

Probability of the Occurrence of Any Type of Event

The **Probability of the Occurrence of any Type of Event** can be determined by following the following five-step process:

1. Define as completely as possible the **Experiment** — i.e., describe the process involved, the methodology of making observations, the way these observations will be documented, etc.
2. Identify and list all the possible individual experimental **Results**.
3. Assign a probability of occurrence to each of these **Results.**
4. Identify and document the specific **Results** that will make up or are contained in the **Event**, the **Compound Event**, or the **Complementary Event** of interest.
5. Sum up the **Result** probabilities to obtain the **Probability of the Occurrence of the Event**, the **Compound Event**, or the **Complementary Event** of interest.

RELEVANT FORMULAE & RELATIONSHIPS

Parameters Relating to Any Population or Distribution

Equation #8-1:

The following Equation, **#8-1**, defines the **Range** for any data set, population, or distribution of interest. It is determined by subtracting the **Value of the Numerically Smallest Member** of the set from the **Value of the Numerically Largest Member**.

$$R = \left[x_{i_{maximum}} - x_{i_{minimum}} \right]$$

Where:

R = the **Range** of the data set, population, or distribution consisting of "**n**" different members designated as "x_i";

x_i = any of the "**n**" members of the data set, population, or distribution being considered;

$i_{maximum}$ = the subscript index of the numerically largest member of the data set, population, or distribution being considered — indicating in Equation **#8-1** the numerically largest member of the set by the term: $x_{i_{maximum}}$; &

$i_{minimum}$ = the subscript index of the numerically largest member of the data set, population, or distribution being considered — indicating in Equation **#8-1** the numerically smallest member of the set by the term: $x_{i_{minimum}}$.

Equation #8-2:

The relationship that is used to characterize the relative magnitude of the range for any data set, distribution, or population under consideration is given by Equation **#8-2**. This expression is simply the ratio of the numerically largest member of any data set to its smallest member. This ratio is used to characterize the magnitude of the range for any distribution, population, or data set. Whenever a distribution, population, or data produces a value for **R** that is greater than 200, that distribution, population, or data set is said to have a relatively large range.

$$R = \frac{x_{i_{maximum}}}{x_{i_{minimum}}}$$

Where:

R = the ratio of the largest member of any distribution or population to the smallest member of the same distribution or population;

$x_{i_{maximum}}$	=	is the Value of the largest member of the distribution or population under consideration; &
$x_{i_{minimum}}$	=	is the Value of the smallest member of the distribution or population under consideration.

Equation #8-3:

The following Equation, **#8-3**, defines the first, and the most important and, almost certainly the most widely used measure of location — or "central tendency" — for any type of population, distribution, or data set. This measure has been identified under a variety of names, among which are: **Mean**, Average, Arithmetic Mean, Arithmetic Average, etc. For the purpose of discussion in this text from this point forward, this parameter will always be identified as the **Mean**. In general, the **Mean** is designated either by the Greek letter, "μ", or by "$\overline{x}$".

$$\mu = \overline{x} = \frac{1}{n}\sum_{i=1}^{n} x_i$$

Where:

$\mu = \overline{x}$	=	the **Mean** of the population, distribution, or data set of "**n**" different values of x_i — the dimensions of the **Mean** and the individual members in the population, distribution, or data set will always be identical;
x_i	=	the value of the "**ith**" member of the total of "**n**" members in the overall population, distribution, or data set;
n	=	the number of members in the overall population, distribution, or data set being considered; &
i	=	the "index" of the population, distribution, or data set being considered, this term will always appear as a subscript on the term representing a variable member of the overall population, distribution, or data set; this index will identify the position of the subscripted member within the overall population, distribution, or data set.

Equation #8-4:

The following Equation, **#8-4**, characterizes and defines a second measure of location — or "central tendency" — for any measurable or quantifiable parameter, for any distribution (normal or otherwise). This measure is called the **Geometric Mean** of the distribution. It is somewhat more useful than the simple Mean — at least as a measure of this "central tendency" — whenever the distribution being examined or analyzed has a very large range, which might be defined as one with values varying over several orders of magnitude [i.e., a range for which **R** ≥ 200, or *log***R** ≥ 2.30 — see Equation **#8-2**, on Pages 8-10 & 8-11].

Whenever a distribution has such a large range, the **Geometric Mean** will probably be a better indicator of its "central tendency" than will the simple Mean. It must be noted, however, that one can determine a **Geometric Mean** value for any distribution, population, or data set regardless of the magnitude of its range.

The relationships that are used to calculate this parameter are given below in two forms: the first is simply the direct mathematical relationship representing the definition of the **Geometric Mean**, while the second is presented in a format that will probably prove to be slightly easier to use in any case where the value of this parameter must be determined — particularly, for any distribution that has a relatively large to very large range.

$$M_{geometric} = \sqrt[n]{(x_1)(x_2)(x_3)\ldots(x_{n-1})(x_n)}$$

$$M_{geometric} = 10^{\left[\frac{1}{n}\sum_{i=1}^{n} \log x_i\right]}$$

Where: $M_{geometric}$ = the **Geometric Mean** of the distribution, population, or data set under consideration;

x_i = is the value of the "**ith**" of "**n**" members of the overall distribution, population, or data set under consideration;

n = the number of members in the distribution, population, or data set under consideration.

Equation #8-5:

The following Equation, **#8-5**, is actually more of a definition. It characterizes the third measure of location, or "central tendency", for any quantifiable parameter, preferably for the situation in which the information being analyzed makes up a normal distribution. This parameter is called the **Median**. Although it is considered to be most applicable to normal distributions, a **Median** value can be determined for any other type of distribution, population, or data set.

M_e = the **Median** or "midpoint" value [principally for a normal distribution] of "**n**" different numeric values of "x_i" — i.e., when all the members of the distribution, population, or data set have been arranged in an increasing or a decreasing order by their numeric values, the **Median** will be in the middle position of the resultant ordered set. If "**n**" is odd, then the **Median** will be the actual middle number in the data set. If "**n**" is even, then the **Median** will be the numeric average, or mean, of the two members of the ordered data set that jointly occupy the middle position of that set.

Where: M_e = the **Median** of the distribution, population, or data set consisting of "**n**" different values of x_i;

x_i = is the value of the "**ith**" of "**n**" members of the overall distribution, population, or data set under consideration;

n = the number of members in the overall distribution, population, or data set under consideration.

Equation #8-6:

The following Equation, **#8-6**, is also more of a definition. It characterizes the fourth measure of location, or "central tendency", for any quantifiable parameter, again preferably for a situation in which the resultant distribution is normal. This parameter is called the **Mode**. Although it is considered to apply most effectively to normal distributions, the **Mode** can also be determined for any other type of distribution, population, or data set.

> M_o = the **Mode** or "most populous" value in any distribution, population, or data set consisting of "**n**" different numeric values of "x_i", i.e., that specific numeric value of "x_i" which is the most frequently occurring value in the entire distribution, population, or data set. Although the **Mode** is considered to be an important measure of location or "central tendency", this value can occur at any position in the data set — i.e., it could be the smallest value, or the largest, or any other value. In a normal distribution, the **Mode** will usually be fairly close in value to the Median, and therefore, this parameter will provide its most useful information when applied to this important class of distribution.

Where:

M_o = the **Mode** of the distribution, population, or data set of "**n**" different Values of "x_i";

x_i = is the value of the "**ith**" of "**n**" members of the overall distribution, population, or data set under consideration;

n = the number of members in the overall distribution, population, or data set under consideration.

Equation #8-7:

The following Equation, **#8-7** is shown in two equivalent forms, and defines the **Sample Variance**, which is the first and most widely used measure of variability, or dispersion, of the data in any distribution, population, or data set of interest.

$$s^2 = \frac{\sum_{i=1}^{n}[x_i - \mu]^2}{n - 1} = \frac{\sum_{i=1}^{n}[x_i - \bar{x}]^2}{n - 1}$$

Where:

s^2 = the **Sample Variance** for the entire distribution, population, or data set of "**n**" different values of "x_i";

x_i = is the value of the "**ith**" of "**n**" members of the overall distribution, population, or data set under consideration;

n = the number of members in the overall distribution, population, or data set under consideration; &

$\mu = \bar{x}$ = the Mean of the distribution, population, or data set.

Equation #8-8:

The following Equation, **#8-8**, is shown in two equivalent forms, and defines the **Population Variance**, which is the second measure of variability, or dispersion, of the data in any distribution, population, or data set of interest.

$$\sigma^2 = \frac{\sum_{i=1}^{n}[x_i - \mu]^2}{n} = \frac{\sum_{i=1}^{n}[x_i - \bar{x}]^2}{n}$$

Where:

σ^2 = the **Population Variance** for the entire distribution, population, or data set of "**n**" different values of "x_i";

x_i = is the value of the "**ith**" of "**n**" members of the overall distribution, population, or data set under consideration;

n = the number of members in the overall distribution, population, or data set under consideration; &

$\mu = \bar{x}$ = the Mean of the distribution, population, or data set.

Equation #8-9:

The following Equation, **#8-9**, which like its two predecessors is shown in two equivalent forms, defines the **Sample Standard Deviation**, which is the third — and probably most important — measure of variability, or dispersion, of the data in any distribution, population, or data set of interest. In general, the **Sample Standard Deviation** is believed to be most applicable to normal distributions; however it can be and is applied to any type of data set.

$$s = \sqrt{s^2} = \sqrt{\frac{\sum_{i=1}^{n}[x_i - \mu]^2}{n - 1}} = \sqrt{\frac{\sum_{i=1}^{n}[x_i - \bar{x}]^2}{n - 1}}$$

Where:

s = the **Sample Standard Deviation** for the entire distribution, population, or data set of "**n**" different values of "x_i";

s^2 = the **Sample Variance** for the entire distribution, population, or data set of "**n**" different values of "x_i";

x_i = is the value of the "**i**th" of "**n**" members of the overall distribution, population, or data set under consideration;

n = the number of members in the overall distribution, population, or data set under consideration; &

$\mu = \bar{x}$ = the Mean of the distribution, population, or data set.

Equation #8-10:

The following Equation, **#8-10**, which like its three predecessors is shown in two equivalent forms, defines the **Population Standard Deviation**, which is the fourth measure of variability, or dispersion, of the data in any distribution, population, or data set of interest. In general, the **Population Standard Deviation** is believed to be the least important of the variability or dispersion quantifying parameters.

$$\sigma = \sqrt{\sigma^2} = \sqrt{\frac{\sum_{i=1}^{n}[x_i - \mu]^2}{n}} = \sqrt{\frac{\sum_{i=1}^{n}[x_i - \bar{x}]^2}{n}}$$

Where:

σ = the **Population Standard Deviation** for the entire distribution, population, or data set of "**n**" different values of "x_i";

σ^2 = the **Population Variance** for the entire distribution, population, or data set of "**n**" different values of "x_i";

x_i = is the value of the "**i**th" of "**n**" members of the overall distribution, population, or data set under consideration;

n = the number of members in the overall distribution, population, or data set under consideration; &

$\mu = \bar{x}$ = the Mean of the distribution, population, or data set.

Equation #8-11:

The following Equation, **#8-11**, defines the **Sample Coefficient of Variation** or **Relative Standard Deviation**, which is the first measure of the specific dispersion of all the data in any population, distribution, or data set being considered. This expression is shown in two identical forms below:

$$CV_{sample} = \frac{s}{\mu} = \frac{s}{\bar{x}}$$

Where: CV_{sample} = the **Sample Coefficient of Variation** for any population, distribution, or data set of "**n**" different values of "x_i";

s = the Sample Standard Deviation for the entire distribution, population, or data set of "**n**" different values of "x_i"; &

$\mu = \bar{x}$ = the Mean of the distribution, population, or data set.

Equation #8-12:

The following Equation, **#8-12**, defines the **Population Coefficient of Variation**, which is the second measure of the specific dispersion of all the data in any population, distribution, or data set being considered. Proceeding logically from the previous relationship — i.e., Equation **#8-11** — this one has been provided below in two useful formats:

$$CV_{population} = \frac{\sigma}{\mu} = \frac{\sigma}{\bar{x}}$$

Where: $CV_{population}$ = the **Population Coefficient of Variation** for the population, distribution, or data set of "**n**" different values of "x_i";

σ = the Population Standard Deviation for the entire distribution, population, or data set of "**n**" different values of "x_i";

$\mu = \bar{x}$ = the Mean of the distribution, population, or data set.

STATISTICS & PROBABILITY PROBLEM SET

Data Set for Problem #s 8.1 through 8.11:

The following data set lists — for a large metal foundry — the "Workdays Without a Lost-Time Accident" experience — i.e., the **WDWLTA** experience — for each of this company's fifteen different functional departments. Every previous analysis of this foundry's Lost-Time Accident information has produced data that were normally distributed; you may, therefore, assume that the data below also will be normally distributed.

Although it is not a specific requirement of any part of the several problems that have been developed for this data set, a space has been provided to be used for the retabulation of the data provided below. A retabulation in an ordered sequence, plus calculations of the three derived values [also listed below], should greatly facilitate the determination of the answers that have been requested in the eleven problem statements that are based on this data set.

Dept. #	WDWLTA	Dept. #	WDWLTA	Dept. #	WDWLTA
1	85	2	71	3	102
4	43	5	90	6	87
7	55	8	118	9	63
10	62	11	77	12	62
13	95	14	82	15	69

The following space has been provided for the data retabulation to which reference was made above.

1	2	3	4	5
Dept. #	x_i = WDWLTA	$log\ x_i$	$x_i - \mu$	$[x_i - \mu]^2$

Column Summations

——			——	

Problem #8.1:

What is the Range of these data?

Applicable Definitions:	Normal Distribution	Page 8-2
	Range	Page 8-4
Applicable Formula:	Equation **#8-1**	Page 8-10
Solution to this Problem:	Page 8-30	

Problem Workspace

Problem #8.2:

What is the Mean of these data?

Applicable Definitions:	Normal Distribution	Page 8-2
	Mean	Page 8-4
Applicable Formula:	Equation **#8-3**	Page 8-11
Solution to this Problem:	Page 8-30	

Problem Workspace

Problem #8.3:

What is the Geometric Mean of these data?

Applicable Definitions:	Normal Distribution	Page 8-2
	Geometric Mean	Page 8-4
Applicable Formula:	Equation **#8-2**	Pages 8-10 & 8-11
	Equation **#8-4**	Pages 8-11 & 8-12
Solution to this Problem:	Pages 8-30 & 8-31	

Problem Workspace

Problem #8.4:

What is the Median of these data?

Applicable Definitions:	Normal Distribution	Page 8-2
	Median	Pages 8-4 & 8-5
Applicable Formula:	Equation **#8-5**	Pages 8-12 & 8-13
Solution to this Problem:	Page 8-31	

Problem Workspace

Problem #8.5:

What is the Mode of these data?

Applicable Definitions:	Normal Distribution	Page 8-2
	Mode	Page 8-5
Applicable Formula:	Equation **#8-6**	Page 8-13
Solution to this Problem:	Page 8-31	

Problem Workspace

Problem #8.6:

What is the Sample Variance for these data?

Applicable Definitions:	Normal Distribution	Page 8-2
	Sample Variance	Page 8-5
Applicable Formula:	Equation **#8-7**	Pages 8-13 & 8-14
Solution to this Problem:	Pages 8-31 & 8-32	

Problem Workspace

Problem #8.7:

What is the Sample Standard Deviation for these data?

Applicable Definitions:	Normal Distribution	Page 8-2
	Sample Standard Deviation	Page 8-5
Applicable Formula:	Equation **#8-9**	Pages 8-14 & 8-15
Solution to this Problem:	Page 8-32	

Problem Workspace

Problem #8.8:

What is the Sample Coefficient of Variation for these data?

Applicable Definitions:	Normal Distribution	Page 8-2
	Sample Coefficient of Variation	Page 8-6
Applicable Formula:	Equation **#8-11**	Pages 8-15 & 8-16
Solution to this Problem:	Page 8-32	

Problem Workspace

Problem #8.9:

What is the Population Variance for these data?

Applicable Definitions:	Normal Distribution	Page 8-2
	Population Variance	Page 8-6
Applicable Formula:	Equation **#8-8**	Page 8-14
Solution to this Problem:	Pages 8-32 & 8-33	

Problem Workspace

Problem #8.10:

What is the Population Standard Deviation for these data?

Applicable Definitions:	Normal Distribution	Page 8-2
	Population Standard Deviation	Page 8-6
Applicable Formula:	Equation **#8-10**	Page 8-15
Solution to this Problem:	Page 8-33	

Problem Workspace

Problem #8.11:

What is the Population Coefficient of Variation for these data?

Applicable Definitions:	Normal Distribution	Page 8-2
	Population Coefficient of Variation	Page 8-7
Applicable Formula:	Equation **#8-12**	Page 8-16
Solution to this Problem:	Page 8-33	

Problem Workspace

Data Set for Problem #s 8.12 through **8.17**:

A petrochemical company has a total of 1,784 refinery employees at its refinery location in a large gulf coast city. This company's employee rolls at this location can be characterized according to: (1) the age; (2) the gender; and (3) the compensation category of each employee, according to the following listing:

Hourly Employees

————	< 25 years	25 to 34 years	35 to 44 years	> 44 years
Male	56	259	309	191
Female	48	206	341	168

Salaried Employees

————	< 25 years	25 to 34 years	35 to 44 years	> 44 years
Male	8	31	29	42
Female	5	43	29	19

The company has decided that all of these employees must "attend" a 1-hour duration, interactive, computer-based, safety orientation program. To implement this program, the company will have available a total of 10 computer terminals. Every one of the company's 1,784 employees will be required to complete this course. A maximum of 60 employees can complete this training on any one day — made up of 6 sessions of 10 employees each.

The employees who will be involved in this safety orientation program will be selected randomly, and no more than two salaried employees will ever be permitted to be simultaneously involved in any one 10-person session of this course.

Problem #8.12:

What is the probability that the very first person selected for the very first 10-person session of this course will be female?

Applicable Definitions:	Experiment	Page 8-7
	Result	Page 8-7
	Sample Space	Page 8-7
	Event	Page 8-7
	Probabilities Associated with Results	Pages 8-8 & 8-9
	Probabilities Associated with Events	Page 8-9
Solution to this Problem:	Pages 8-33 & 8-34	

Problem Workspace

Problem #8.13:

What is the probability that the very first person selected for the very first 10-person session of this course will be a salaried female?

Applicable Definitions:	Experiment	Page 8-7
	Result	Page 8-7
	Sample Space	Page 8-7
	Event	Page 8-7
	Probabilities Associated with Results	Pages 8-8 & 8-9
	Probabilities Associated with Events	Page 8-9
Solution to this Problem:	Page 8-34	

Problem Workspace

Problem #8.14:

What is the probability that the very first person selected for the very first 10-person session of this course will be male and over 35 years of age?

Applicable Definitions:	Experiment	Page 8-7
	Result	Page 8-7
	Sample Space	Page 8-7
	Event	Page 8-7
	Probabilities Associated with Results	Pages 8-8 & 8-9
	Probabilities Associated with Events	Page 8-9
Solution to this Problem:	Pages 8-34 & 8-35	

Problem Workspace

Problem #8.15:

What is the probability that the very first person selected for the very first 10-person session of this course will either be a man or be over 44 years of age?

Applicable Definitions:	Experiment	Page 8-7
	Result	Page 8-7
	Sample Space	Page 8-7
	Event	Page 8-7
	Compound Event	Pages 8-7 & 8-8
	Probabilities Associated with Results	Pages 8-8 & 8-9
	Probabilities Associated with Events	Page 8-9
Solution to this Problem:	Pages 8-35 & 8-36	

Workspace for Problem **#8.15**

Problem #8.16:

What is the probability that the very first person selected for the very first 10-person session of this course will neither be salaried, nor over 44 years of age, nor female?

Applicable Definitions:	Experiment	Page 8-7
	Result	Page 8-7
	Sample Space	Page 8-7
	Event	Page 8-7
	Compound Event	Pages 8-7 & 8-8
	Complementary Event	Page 8-8
	Probabilities Associated with Results	Pages 8-8 & 8-9
	Probabilities Associated with Events	Page 8-9
Solution to this Problem:	Pages 8-36 & 8-37	

Workspace for Problem **#8.16**

Problem #8.17:

What is the probability that the very first 10-person session of this course will include two salaried males younger than 25 years of age?

Applicable Definitions:	Experiment	Page 8-7
	Result	Page 8-7
	Sample Space	Page 8-7
	Event	Page 8-7
	Compound Event	Pages 8-7 & 8-8
	Complementary Event	Page 8-8
	Probabilities Associated with Results	Pages 8-8 & 8-9
	Probabilities Associated with Events	Page 8-9
Solution to this Problem:	Pages 8-37 & 8-38	

Workspace for Problem **#8.17**

SOLUTIONS TO THE STATISTICS & PROBABILITY PROBLEM SET

The solutions to the first eleven problems [i.e., **#s 8.1** through **8.11**] require the application of all but one of the twelve equations documented in this chapter.

The Reader is advised that the mathematical operations that will be used to determine the value of each of the required parameters included in the solutions section of this chapter will rely on, and be carried out using, a calculator that has a full library of statistical functions. As a result of this, there may well be some differences between the solutions listed in this section and those that the Reader calculates — i.e., the Reader may not have available a calculator with full statistical capabilities. Most calculators that have statistical functions maintain intermediate results in their internal memories, carrying out each of these determinations to many more decimal places than would be practical for a simple hand held calculating process. Because of this, the step-by-step hand-held calculators that use numbers accurate to only two or three decimal places will likely produce slightly different results than those that will be documented here. Again, these results were obtained by the more precise statistical sub-routines in a calculator that has a full range of statistical capabilities.

At this point, it will probably be useful to reproduce the reorganized/restructured data that were developed from the raw data provided in the Problem Statement. The following tabulation shows an ordered listing in which the "Workdays Without a Lost-Time Accident" experience values — i.e., the **WDWLTA** or "x_is" — appear in an increasing value order:

1	2	3	4	5
Dept.	x_i = WDWLTA	*log* x_i	$x_i - \mu$	$[x_i - \mu]^2$
4	43	1.633	– 34.40	1,183.36
7	55	1.740	– 22.40	501.76
10	62	1.792	– 15.40	237.16
12	62	1.792	– 15.40	237.16
9	63	1.799	– 14.40	207.36
15	69	1.839	– 8.40	70.56
2	71	1.851	– 6.40	40.96
11	77	1.886	– 0.40	0.16
14	82	1.914	4.60	21.16
1	85	1.929	7.60	57.76
6	87	1.940	9.60	92.16
5	90	1.954	12.60	158.76
13	95	1.978	17.60	309.76
3	102	2.009	24.60	605.16
8	118	2.072	40.60	1,648.36
		Column Summations		
——	1,161	28.130	——	5,371.6

The values that appear on the line immediately above are obviously summations of all the individual values listed in each of the second, third, and fifth columns. These summation values will be useful in obtaining several of the results required in the various problem statements.

For reference, these summation values are shown on the following page.

Column #2: $\sum_{i=1}^{15} x_i = 1{,}161$

Column #3: $\sum_{i=1}^{15} \log x_i = 28.130$

Column #5: $\sum_{i=1}^{15} [x_i - \mu]^2 = 5{,}371.6$

We can now determine the statistical parameters required by the first eleven problem statements in this section.

Problem #8.1:

The **Range** of these data can be determined by using Equation **#8-1**, from Page 8-10:

$$R = \left[x_{i_{maximum}} - x_{i_{minimum}}\right] \qquad \text{[Eqn. \#8-1]}$$

$$R = [118 - 43] = 75$$

∴ The Range of these data = 75 Workdays Without a Lost-Time Accident.

Problem #8.2:

The **Mean** of these data can be determined by using Equation **#8-3**, from Page 8-11:

$$\mu = \bar{x} = \frac{1}{n}\sum_{i=1}^{n} x_i \qquad \text{[Eqn. \#8-3]}$$

Utilizing the summation total determined for all of the WDWLTA data included in Column #2 of the reordered data set for this problem, we have:

$$\sum_{i=1}^{n} x_i = 1{,}161$$

Therefore: $\mu = \bar{x} = \frac{1}{15}(1{,}161) = 77.4$

∴ The Mean = 77.4 Workdays Without a Lost-Time Accident.

Problem #8.3:

The **Geometric Mean** of these data can be determined by using Equation **#8-4**, from Pages 8-11 & 8-12:

$$M_{geometric} = 10^{\left[\frac{1}{n}\sum_{i=1}^{n} \log x_i\right]} \qquad \text{[Eqn. \#8-4]}$$

Utilizing the summation total determined for the *log* x_i data included in Column #3 of the reordered data set for this problem, we have:

$$\sum_{i=1}^{15} \log x_i = 28.13$$

Therefore: $M_{geometric} = 10^{\left[\frac{1}{15}(28.13)\right]} = 10^{\frac{28.13}{15}} = 10^{1.875} = 75.05$

And just to check on the relative magnitude of the Range of these data, we can apply Equation **#8-2**, from Pages 8-10 & 8-11:

$$R = \frac{x_{i_{maximum}}}{x_{i_{minimum}}} \qquad \text{[Eqn. \#8-2]}$$

$$\& \quad R = \frac{118}{43} = 2.74 \ll 200$$

Therefore, the Range of these data is not considered to be relatively large [i.e., the value of the ratio, R, is not greater than 200]; therefore, the Geometric Mean is probably no better an estimator of the central tendency of these data than is the simple Mean.

∴ The Geometric Mean = 75.05 Workdays Without a Lost-Time Accident.

Problem #8.4:

The **Median** of these data can be determined by using Equation **#8-5**, from Pages 8-12 & 8-13:

The midpoint value from the listing of WDWLTA data as shown in Column #2 of the reordered data set for this problem, on Page 8-29, is 77.

∴ The Median = 77 Workdays Without a Lost-Time Accident.

Problem #8.5:

The **Mode** of these data can be determined by using Equation **#8-6**, from Page 8-13:

The most populous value from the listing of WDWLTA data as shown in Column #2 of the reordered data set for this problem, on Page 8-29, is 62.

∴ The Mode = 62 Workdays Without a Lost-Time Accident.

Problem #8.6:

The **Sample Variance** for these data can be determined by using Equation **#8-7**, from Pages 8-13 & 8-14:

$$s^2 = \frac{\sum_{i=1}^{n}[x_i - \mu]^2}{n-1} = \frac{\sum_{i=1}^{n}[x_i - \bar{x}]^2}{n-1} \qquad \text{[Eqn. \#8-7]}$$

Now utilizing the summation total determined for the $\sum_{i=1}^{15}[x_i - \mu]^2$ data included in Column #5 of the reordered data set for this problem, on Page 8-29:

$$\sum_{i=1}^{15}[x_i - \mu]^2 = 5{,}371.6$$

& knowing that n = 15 — therefore, (n – 1) = 14 — we can calculate the value of s^2:

$$s^2 = \frac{5{,}371.6}{15 - 1} = \frac{5{,}371.6}{14} = 383.69$$

∴ The Sample Variance = 383.69 [Workdays Without a Lost-Time Accident]^2.

Problem #8.7:

The **Sample Standard Deviation** for these data can be determined by using Equation **#8-9**, from Pages 8-14 & 8-15:

$$s = \sqrt{s^2} = \sqrt{\frac{\sum_{i=1}^{n}[x_i - \mu]^2}{n - 1}} = \sqrt{\frac{\sum_{i=1}^{n}[x_i - \bar{x}]^2}{n - 1}} \qquad \text{[Eqn. #8-9]}$$

This parameter is simply the square root of the Sample Variance obtained in the previous problem and listed immediately above on this page.

$$s = \sqrt{\frac{5{,}371.6}{14}} = \sqrt{383.69} = 19.59$$

∴ The Sample Standard Deviation = 19.59 Workdays Without a Lost-Time Accident.

Problem #8.8:

The **Sample Coefficient of Variation** for these data can be determined by using Equation **#8-11**, from Pages 8-15 & 8-16:

$$CV_{sample} = \frac{s}{\mu} = \frac{s}{\bar{x}} \qquad \text{[Eqn. #8-11]}$$

Using the results obtained for Problem **#s 8.2** [the Mean], and **8.7** [the Sample Standard Deviation], we get:

$$CV_{sample} = \frac{19.59}{77.4} = 0.253$$

∴ The Sample Coefficient of Variation = 0.253.

Problem #8.9:

The **Population Variance** for these data can be determined by using Equation **#8-8**, from Page 8-14:

$$\sigma^2 = \frac{\sum_{i=1}^{n}[x_i - \mu]^2}{n} = \frac{\sum_{i=1}^{n}[x_i - \bar{x}]^2}{n} \qquad \text{[Eqn. #8-8]}$$

Now again utilizing the summation total determined for the $\sum_{i=1}^{15}[x_i - \mu]^2$ data included in Column #5 of the reordered data set for this problem, on Page 8-29:

$$\sum_{i=1}^{15}[x_i - \mu]^2 = 5{,}371.6$$

and knowing that n = 15, we can calculate the value of σ^2:

$$\sigma^2 = \frac{5{,}371.6}{15} = 358.11$$

∴ The Population Variance = 358.11 [Workdays Without a Lost-Time Accident]^2.

Problem #8.10:

The **Population Standard Deviation** for these data can be determined by using Equation **#8-10**, from Page 8-15:

$$\sigma = \sqrt{\sigma^2} = \sqrt{\frac{\sum_{i=1}^{n}[x_i - \mu]^2}{n}} = \sqrt{\frac{\sum_{i=1}^{n}[x_i - \bar{x}]^2}{n}} \qquad \text{[Eqn. \#8-10]}$$

This parameter is simply the square root of the Population Variance obtained in the previous problem and listed immediately above on this page.

$$\sigma = \sqrt{\frac{5{,}371.6}{15}} = \sqrt{358.11} = 18.92$$

∴ The Population Standard Deviation = 18.92 Workdays Without a Lost-Time Accident.

Problem #8.11:

The **Population Coefficient of Variation** for these data can be determined by using Equation **#8-12**, from Page 8-16:

$$CV_{population} = \frac{\sigma}{\mu} = \frac{\sigma}{\bar{x}} \qquad \text{[Eqn. \#8-12]}$$

Using the results obtained for Problem **#s 8.2** [the Mean], and **8.10** [the Population Standard Deviation], we get:

$$CV_{population} = \frac{18.92}{77.4} = 0.245$$

∴ The Population Coefficient of Variation = 0.245.

Problem #8.12:

The Event that is the subject of this problem is the random selection of a woman to fill the first position in the first 10-person interactive, computer-based, safety orientation program.

There are a number of specific Results that make up the sub-set of possibilities for this Event. As an example, the selection of a 30-year-old, hourly, female employee would be a Result that complies fully with the requirement of having a female employee be the first randomly selected participant in this program.

What must be done then is to determine the total number of possible favorable Results [i.e., the only possible favorable Result for this situation would be a female employee being the first person randomly selected to participate in this program] that make up this Event. Clearly the total number of favorable results is equal to the number of female employees of any age and any type of compensation. This sum is given by:

$$N_{\text{female employees}} = 48 + 206 + 341 + 168 + 5 + 43 + 29 + 19 = 859 \text{ females}$$

Thus, there are 859 possible successful outcomes to this Experiment.

Next, we must determine the total number of Results — of any type — that are possible outcomes for this experiment. Simply, this number is the total of the employees who work for this company in this location, and that number has been given as 1,784. Therefore, the requested probability is simply the ratio of these two numbers.

$$\text{P[first person selected is a woman]} = \frac{859}{1,784} = 0.482$$

∴ The probability that the first person randomly selected to participate in this safety orientation program will be a woman = P[first person selected is a woman] = 0.482.

Problem #8.13:

The number of specific Results in the sub-set that makes up this Event will be the total number of salaried Female employees. This sum is given by:

$$N_{\text{salaried female employees}} = 5 + 43 + 29 + 19 = 96 \text{ salaried females}$$

Thus, there are 96 possible successful outcomes to this Experiment.

Since the total number of Results — of any type — that are possible outcomes for this Experiment is simply the total of the employees who work for this company in this location, and since we know that number to be 1,784, we can proceed directly to the answer:

$$\text{P[first person selected is a salaried woman]} = \frac{96}{1,784} = 0.054$$

∴ The probability that the first person randomly selected to participate in this safety orientation program will be a salaried woman = P[first person selected is a salaried woman] = 0.054.

Problem #8.14:

The number of specific Results in the sub-set that makes up this Event will be the total number of Male employees who are 35 years old or older. This sum is given by:

$$N_{\text{male employees 35+ years old}} = 309 + 191 + 29 + 42 = 571 \text{ >35-year-old males}$$

Thus, there are 571 possible successful outcomes for this Experiment.

Since the total number of Results — of any type — that are possible outcomes for this Experiment is simply the total of the employees who work for this company in this location, and since we know that number to be 1,784, we can proceed directly to the answer:

$$P[\text{first person selected is a >35-year-old man}] = \frac{571}{1,784} = 0.320$$

∴ The probability that the first person randomly selected to participate in this safety orientation program will be a 35+ year old man = P[first person selected is a >35-year-old man] = 0.320.

Problem #8.15:

In this situation, we are dealing with a Compound Event, namely, the Union of two specific Events. We shall designate these Events as "A" & "B", as follows:

A: The Event in which the first person randomly selected is a man.

B: The Event in which the first person randomly selected is more than 44 years old.

This Compound Event or Union is indicated schematically as $A \cup B$.

Let us now consider the total number of Results that make up each of these two Events.

For Event "A", we have:

$$N_{\text{male employees}} = 56 + 259 + 309 + 191 + 8 + 31 + 29 + 42 = 925 \text{ male employees}$$

Next, we must consider the total number of Results that make up Event "B".

For Event "B", we have:

$$N_{\text{44+ year old employees}} = 191 + 168 + 42 + 19 = 420 \text{ each > 44-year-old employees}$$

Finally, we must determine if there are any specific Results that are members, simultaneously, of both Events "A" & "B". In essence, in the problem statement we have been asked for the probability of an either/or situation — i.e., the Union of two Events. To evaluate the probability of such a Compound Event, we must consider the possibility of there being a Result that is simultaneously a member of the two sub-set Events. For example, clearly, a 50-year-old salaried male would be a member of both sub-sets.

To determine the correct number of successful Results for the Union of two Events, we must be sure not to count the same Result twice; thus we must subtract from the sum of these two sub-sets, those Results that also are a part of or included in both sub-sets. Only in this way will we be certain not to double count some specific Result. Clearly, there are two groups that satisfy the condition of being members of both Events "A" & "B". In order, these two groups are:

the 191 hourly male employees over 44 years of age, &

the 42 salaried male employees over 44 years of age.

Therefore, we can now determine the total number of successful Results which will be required for the solution of the probability asked for in the problem statement:

$$N_{A \cup B} = 925 + 420 - 191 - 42 = 1,112 \text{ male and/or >44-year-old employees}$$

Since the total number of Results — of any type — that are possible outcomes for this Experiment is still simply the total of the employees who work for this company in this location, and since we know that number to be 1,784, we can proceed directly to the answer.

$$P[\text{first person selected is either male or >44 years old}] = \frac{1{,}112}{1{,}784} = 0.623$$

∴	The probability that the first person randomly selected to participate in this safety orientation program will be either male or 44+ years old = P[first person selected is either male or >44 years old] = 0.623.

Problem #8.16:

For this situation we are dealing with a type of Compound Event that is known as a Complementary Event. Specifically, we have been asked to determine the probability of an overall Compound Event that is comprised of a set of Results from which several clearly identified individual Results, which make up three other specific Events, have been excluded. Let us first consider these "excluded" Events, designating them in order as "A", "B", & "C":

A: The Event in which the first person randomly selected is salaried.

B: The Event in which the first person randomly selected is more than 44 years old.

C: The Event in which the first person randomly selected is a woman.

Let us now consider the total number of Results that make up each of these three Events.

For Event "A", we have:

$$N_{\text{salaried employees}} = 8 + 5 + 31 + 43 + 29 + 29 + 42 + 19 = 206 \text{ salaried employees}$$

Next, we must consider the total number of Results that make up Event "B".

For Event "B", we have:

$$N_{>44\text{-year-old employees}} = 191 + 168 + 42 + 19 = 420 \text{ >44-year-old employees}$$

We must next consider the total number of Results that make up Event "C".

For Event "C", we have:

$$N_{\text{female employees}} = 48 + 206 + 341 + 168 + 5 + 43 + 29 + 19 = 859 \text{ females}$$

In essence, to obtain information about the Complementary Event asked for in the problem statement, we have two avenues to pursue in order to reach the desired answer. The first is to evaluate the Union of the three Events designated above and recognize that the sum of the probability of this Union and its Complement, which is what we are seeking, is 1.00.

Therefore, we must determine if there are any specific Results that are members, simultaneously, of any two or possibly even all three of Events "A", "B", & "C", as listed above. Consider first those groups that are simultaneously members of Events "A" & "B":

the 42 salaried male employees over 44 years of age

the 19 salaried female employees over 44 years of age

Next, consider those groups that are simultaneously members of Events "B" & "C":

the 168 hourly female employees over 44 years of age

the 19 salaried female employees over 44 years of age

Next, consider those groups that are simultaneously members of Events "A" & "C":

the 5 salaried female employees under 25 years of age

the 43 salaried female employees 25 to 34 years old

the 29 salaried female employees 35 to 44 years old

the 19 salaried female employees over 44 years of age

At this point, it should be obvious that that single group, namely, the 19 salaried female employees over 44 years of age, is included in all three of the paired sub-sets listed on this and the previous page. Clearly this group must not be counted more than once to obtain the totality of the Results that make up this three Event Union. For this Compound Event, then we have:

$$N_{A \cup B \cup C} = 206 + 420 + 859 - 42 - 19 - 168 - 5 - 43 - 29 - 19 = 1{,}160$$

Therefore, the correct number of successful Results for the Union of these three Events is 1,160 employees, and the probability of the occurrence of this union would be given by the following expression:

$$P[A \cup B \cup C] = \frac{1{,}160}{1{,}784} = 0.650$$

Since the probability of the Complement to this Compound Event is simply the difference between this probability and 1.00, we can determine this probability very simply:

P[first person selected is not female, not 44+ years old, & not salaried] = 1.000 – 0.650

P[first person selected is not female, not 44+ years old, & not salaried] = 0.350

As stated earlier, this problem could have been approached from the perspective of seeking directly those groups that satisfy the required condition. Clearly, those groups would be:

the 56 hourly male employees under 25 years of age

the 259 male employees between 25 & 34 years of age

the 309 male employees between 35 & 44 years of age

The total number of employees in these three groups clearly is:

$$N_{\text{hourly male employees less than 44 years old}} = 56 + 259 + 309 = 624$$

The selection of any person from any of these three groups would be a Result that satisfies the requirement requested in the problem statement; thus, the probability would be given by the following expression:

$$P[\text{first person selected is not female, not 44+ years old, \& not salaried}] = \frac{624}{1{,}784} = 0.350$$

Therefore, we see that we obtain the same result from both perspectives.

∴ The probability that the first person randomly selected to participate in this safety orientation program will be neither female, nor 44+ years old, nor salaried =
P[first person selected is not female, not 44+ years old, & not salaried] = 0.350.

Problem #8.17:

The solution to this problem will require the sequential consideration of two separate situations. Specifically, we must first consider the probability of having a salaried male less than 25 years of age be the first person selected randomly for this course. This is a very simple and direct problem. For the second problem, we must consider a new altered sample space, namely, one in which both of the following conditions exist:

1. There is one fewer salaried male less than 25 years of age — i.e., the number of men in this category has been reduced from eight to seven, since one member of this group was the first person selected to take this course.
2. There is one fewer person in the overall employee pool — again, since one person, a salaried male less than 25 years of age, has been selected for the course.

Now to the first part of this problem. The Event that is the subject here is the random selection of a salaried male less than 25 years of age to fill the first position in the first 10-person interactive, computer-based, safety orientation program. There are exactly eight specific Results that make up the sub-set of possibilities for this Event, each one corresponding to a member of this group.

$$\text{P[first person selected is a salaried male less than 25 years old]} = \frac{8}{1,784} = 0.0045$$

Next, we must consider the probability of having a salaried male less than 25 years of age be the second person selected for this course. As before, this is a very simple and direct problem; however, for this situation we must consider the new altered sample space, alluded to and described above.

The Event that is the subject here is the random selection of a salaried male less than 25 years of age from a pool that now contains only seven members; thus, there are exactly seven specific Results that make up the sub-set of possibilities for this second Event. We must also note that the overall pool has now shrunk to 1,783 persons. Therefore, this probability is given by:

$$\text{P[second person selected is a salaried male less than 25 years old]} = \frac{7}{1,783} = 0.0039$$

Now the overall probability of having the first two employees selected for this course is simply the product of these two individual, sequential probabilities:

$$\text{P[first 2 selections are salaried males} < \text{25 years old]} = \left(\frac{8}{1,784}\right)\left(\frac{7}{1,783}\right) = 1.76 \times 10^{-5}$$

> ∴ The probability that the first two people randomly selected to participate in this safety orientation program will be salaried males less than 25 years of age =
> P[first 2 selections are salaried males < 25 years old] = 1.76×10^{-5}

Appendix A

The Atmosphere

Components of the Atmosphere

Because virtually all of the measurements that are made by professionals working in the fields of occupational safety and health, industrial hygiene, and/or the environment are completed in the earth's atmosphere, it is important to understand the nature of the negative impacts that this atmospheric matrix might have on some of these measurements [i.e., negative impacts = factors, circumstances, atmospheric components, etc. that will tend to cause a particular measurement to be incorrect]. Certainly many of the measurements that are made in the atmosphere are totally unaffected by any of the elements or compounds that make it up; however, there are quite a few for which this is not true. To understand the relationships between the atmospheric matrix and the measurements that are made in it, we must know its major components, as well as the concentration range of each.

The most common way to tabulate these components is to consider DRY AIR — i.e., air in which there is absolutely no water vapor. Clearly, nowhere on earth is there ever any air mass that is completely free of water vapor. In fact, for most situations, water vapor will exist at a concentration that would place it in the third highest position, trailing only nitrogen and oxygen. The range of the ambient concentrations of water vapor varies over more than 2.5 orders of magnitude — i.e., ~100 ppm(vol) at points in the Antarctic in winter, to 35,000+ in an equatorial rain forest. Because water vapor can have such a wide concentration range, as stated above, the most common way to list the components that make up the air is to consider them on a DRY, or water vapor free, basis. The following tabulation, Table #1A, lists the seventeen most common atmospheric components in the order of decreasing concentrations, again on a water vapor free or DRY basis:

Table #1A

No.	Component	Concentration – in ppm(vol)
1.	Nitrogen	780,840
2.	Oxygen	209,459
3.	Argon	9,303
4.	Carbon Dioxide	average = 370; range: 350 to 580
5.	Neon	18.2
6.	Helium	5.2
7.	Methane	average = 2.8; range: 2.5 to 4.5
8.	Krypton	1.1
9.	Hydrogen	0.4
10.	Nitrous Oxide	average = 0.3; range: 0.3 to 0.6
11.	Sulfur Dioxide	average = 0.2; range: 0.2 to 0.3
12.	Xenon	0.07
13.	Ozone	average = 0.02; range: 0.02 to 0.8
14.	Nitrogen Dioxide	average = 0.02; range: 0.01 to 0.7
15.	Carbon Monoxide	average = 0.01; range: 0.01 to 0.7
16.	Iodine	< 0.01
17.	Ammonia	< 0.01
	SUMMATION	1,000,000.32

Of the components listed on the previous page, seven have been shown both in terms of their average atmospheric concentration, as well as the range of ambient concentrations common for each. For the remaining ten components, the generally accepted, ambient, dry basis concentration value has been provided.

For the seven components having both an average concentration and a concentration range, there are a variety of activities, situations, and naturally occurring circumstances that are responsible for producing the observed concentration variations. Included among these factors are such things as: industrial production, changes in rain forest acreage, volcanic eruptions, etc. Clearly any of these factors can and do occur in a wide variety of locations on the earth, at any time of the day or night, etc. For the seven variable concentration components, some of the specific factors, activities, situations, and naturally occurring circumstances that are responsible for producing the specific concentration changes are listed below:

Carbon Dioxide: combustion of organic fuels — i.e., coal, gasoline, etc.
Methane: decay of organic matter in the soil, mammalian flatulence, etc.
Nitrous Oxide: industrial production, electrical discharges — i.e., lightning
Sulfur Dioxide: volcanic activity, combustion of high sulfur coal
Ozone: solar activity, industrial activity, electrical discharges
Nitrogen Dioxide: combustion of organic fuels in the presence of excess air
Carbon Monoxide: incomplete combustion of organic fuels

There are several common atmospheric concentration measurements that experience interferences — some positive, some negative — from one or more of the seventeen components of the air. A listing of some of the frequently made ambient concentration measurements that experience interferences from the various atmospheric components is shown in the following tabulation, Table #2A. In this tabulation, both the specific measurement and its most likely measurement technology have been listed against the interfering component and the direction of the anticipated interference — with this interference direction shown following the component in brackets as "+", for positive or additive interferences, and "–" for negative or subtractive interferences.

Table #2A

No.	Measurement	Monitoring Technology	Interfering Component
1	Ambient Hydrocarbons	Flame Ionization Detectors	Methane [+]
2	Organic Vapors	Photo Ionization Detectors	Methane [+ or –]
3	Mercury Vapors	UV Photometry	Ozone [+]
4	SF_6 Tracer Gas	ECD Chromatography	Oxygen [++]

This listing is by no means complete; however, it does show four of the common ambient measurements that are impacted by various atmospheric components. The first two of these measurements are associated with routine air pollution monitoring. The measurement of ambient mercury vapor concentrations is a common industrial hygiene measurement made in any situation where mercury is used — i.e., mining, high voltage rectification, etc. The measurement of the ambient levels of the tracer, sulfur hexafluoride [SF_6], is virtually always accomplished by using a gas chromatograph equipped with an Electron Capture Detector [ECD]. This methodology is used because of the extremely low concentration level of SF_6 (i.e., ppb and even ppt by volume) that is common in atmospheric tracer work.

On balance, as a result of the earth's gravitational field, one would expect that the heavier molecules or atoms from the Table #1A listing would tend to concentrate at lower elevations, while the reverse would be true for the lighter components. This seems logical because in indoor situations, it has long been recognized that heavy molecules such as carbon dioxide tend to concentrate at the floor level. On this basis, the heaviest and the lightest three or four components listed on the following page, as Table #3A [Heavies] and Table #4A [Lights], would logically be expected to exhibit this sort of behavior — i.e., the heavier members, down; and lighter ones, up.

Table #3A: Heavy Components

No.	Component	Atomic or Molecular Weights
12.	Xenon	131.3 amu
16.	Iodine	126.9 amu
8.	Krypton	83.8 amu
11.	Sulfur Dioxide	64.0 amu

Note, the "effective" molecular weight of air is 28.964 amu; thus each of the following five listed atmospheric components — including WATER, which heretofore has not been discussed — is "effectively" lighter than the air mass in which each is a component. Again, the basic laws of physics would seem to apply, with the result being a situation where each of these components should experience a buoyant force, from the "heavier" air matrix in which they exist, and therefore tend to lift or loft up to steadily higher altitudes. That this effect does not occur results from the highly efficient meteorological mixing of the atmosphere that occurs constantly at all points and altitudes on the Earth.

Table #4A: Light Components

No.	Component	Atomic or Molecular Weights
9.	Hydrogen	2.0 amu
6.	Helium	4.0 amu
7.	Methane	16.0 amu
17.	Ammonia	17.0 amu
–	WATER	18.0 amu

Although the earth's atmosphere is very well mixed, the volume-based concentrations listed in Table #1A on Page A-1 are accurate for ALL terrestrial altitudes and/or barometric pressures — i.e., from the level of the Dead Sea to the top of Mount Everest. Of course, if the concentrations of these components were to have been expressed on a mass per unit volume basis — i.e., in units such as mg/m^3 — then there would have been a wide variation in the listed concentrations, heavily dependent on the altitude where the measurements were made.

As an example, the ambient concentration of Argon, expressed both in volumetric- and mass-based units, at five identifiable locations on the Earth (including the altitude of each location) is given in the following listing, Table #5A:

Table #5A

Location	Altitude, relative to Mean Sea Level	Ambient Argon Concentration ppm(vol)	mg/m^3
At the Dead Sea, Israel	– 396 meters	9,303 ppm	17,377 mg/m^3
At Sea Level	0 meters	9,303 ppm	16,581 mg/m^3
In Denver, CO, USA	+ 1,609 meters	9,303 ppm	13,623 mg/m^3
At Lake Titicaca, Peru	+ 3,806 meters	9,303 ppm	10,343 mg/m^3
Atop Mt. Everest, Nepal	+ 8,848 meters	9,303 ppm	5,065 mg/m^3

Atmospheric Humidity — Ambient Water Vapor Concentrations

The concentration of water vapor in the atmosphere can and does vary over a wide range of values, as has been stated earlier. Because of these variations in concentration, water vapor is usually omitted from any tabulation of the component materials that make up the air — i.e., see Table #1A on Page A-1.

Whenever one thinks about the ambient water vapor concentration, the perspective is of a measurement expressed in one of the two common "water vapor specific" sets of concentration units, units that are NEVER applied to the measurement of the concentration of any other component of the air. These two widely used units of water vapor concentration are:

(1) the Relative Humidity at some average Ambient Temperature, &

(2) the Dew Point.

Rarely, if ever, does one hear about the water vapor concentration in what might be a more useful format, namely, in some unit of its "absolute humidity", rather than its "relative humidity" cousin. It is important here to note the specific differences between these concentration units, and the most effective way to accomplish this is to reproduce the definitions of each (Note: these Definitions are also provided on Page 4-4, in Chapter 4 — Ventilation).

Absolute Humidity: The specific amount of vaporous water in the atmosphere, measured on a basis of its mass per unit volume. Absolute humidity is most frequently expressed in units of lbs/ft^3 or mg/m^3.

Relative Humidity: The RATIO of the actual or measured vapor pressure of water vapor in an air mass of some known temperature, to the saturated vapor pressure of pure water, at the same temperature as that of the air mass. Relative Humidity can be quantified as the value of the following ratio [determined at some specific known ambient temperature].

$$RH = 100\left[\frac{\text{Measured Absolute Humidity, in mg/m}^3}{\text{Maximum Possible Absolute Humidity, in mg/m}^3}\right]$$

Note the ratio must be multiplied by 100 to convert it into a percent.

Dew Point: The Dew Point of an air mass is also a measure of relative humidity. It is the temperature at which the water vapor present in that air mass would start to condense — or in the context of the term, Dew Point, it is the temperature at which that water vapor would turn to "dew". It can be determined experimentally simply by cooling a mass of air until "dew", or condensate, forms; the temperature at which this event occurs is defined to be the Dew Point of the air mass that had been cooled.

Let us next consider two tabulations of the actual concentration of water in ambient air [i.e., the absolute humidity] vs. various specific values of relative humidity. In the first of these two tabulations, namely, Table #6A on the following page, the water vapor concentrations will be volumetric, namely, ppm(vol): (1) at five specific relative humidities, and (2) at the Ambient Temperature at which each of the five relative humidities was determined.

The water vapor concentrations tabulated in Table #6A at the 100% relative humidity level are — by definition — the concentrations that would exist in any atmosphere at its Dew Point. Condensation always forms at the Dew Point because the air is saturated with water vapor under those conditions — i.e., its relative humidity equals 100%. Thus, whenever one wants the ambient water vapor volume-based concentration at Sea Level for some identifiable dew point, this concentration can either be (1) taken directly from the last column of Table #6A, or (2) determined by linear interpolation between adjacent pairs of values taken from this column for those situations in which the temperature involved is not among those tabulated.

In every listing in Table #6A on the following page, it has been assumed that the prevailing ambient barometric pressure is 760 mm Hg, or 1.0 atmosphere — i.e., the barometric pressure at Sea Level. At different (notably lesser) barometric pressures, the values in the last two columns of Table #6A will likely be incorrect; however the values in all the other columns will still be accurate.

Finally, if it is ever necessary to determine a Sea Level volume-based water vapor concentration at some untabulated ambient temperature and/or relative humidity [i.e., the water vapor concentration at a temperature of 13.2°C, and a relative humidity of 67%], a simple horizontal and/or vertical linear interpolation from the data in Table #6A will provide a reasonably accurate result, certainly to within ±1 or 2% of the true value.

For reference, the water vapor concentration under the example conditions listed in the final paragraph on the previous page — obtained by a combination of horizontal and vertical linear interpolation — would be 9,997.3 ppm(vol).

Table #6A

Volume-Based Water Vapor Concentrations, in ppm(vol) at Various Levels of Relative Humidity, & at 760 mm Hg

Temperatures °C	°F	0% RH	20% RH	40% RH	60% RH	80% RH	100% RH
– 20	– 4.0	0	214	428	643	857	1,071
– 18	– 0.4	0	258	517	775	1,034	1,292
– 16	3.2	0	315	631	946	1,261	1,576
– 14	6.8	0	384	768	1,153	1,537	1,921
– 12	10.4	0	466	932	1,397	1,863	2,329
– 10	14.0	0	560	1,120	1,680	2,240	2,800
– 8	17.6	0	666	1,333	1,999	2,665	3,332
– 6	21.2	0	785	1,571	2,356	3,141	3,926
– 4	24.8	0	917	1,833	2,750	3,666	4,853
– 2	28.4	0	1,061	2,121	3,182	4,242	5,303
0	32.0	0	1,237	2,474	3,710	4,947	6,184
2	35.6	0	1,447	2,895	4,342	5,790	7,237
4	39.2	0	1,658	3,316	4,973	6,631	8,289
6	42.8	0	1,895	3,790	5,684	7,579	9,474
8	46.4	0	2,158	4,316	6,473	8,631	10,789
10	50.0	0	2,447	4,895	7,342	9,790	12,237
12	53.6	0	2,763	5,526	8,290	11,053	13,816
14	57.2	0	3,132	6,263	9,395	12,526	15,658
16	60.8	0	3,553	7,105	10,658	14,210	17,763
18	64.4	0	4,026	8,053	12,079	16,106	20,132
20	68.0	0	4,553	9,105	13,658	18,210	22,763
21.1	70.0	0	4,895	9,789	14,684	19,579	24,474
22	71.6	0	5,158	10,316	15,473	20,631	25,789
23.9	75.0	0	5,804	11,608	17,413	23,217	29,021
24	75.2	0	5,842	11,684	17,527	23,369	29,211
26	78.8	0	6,605	13,210	19,816	26,421	33,026
28	82.4	0	7,447	14,895	22,342	29,790	37,237
30	86.0	0	8,368	16,737	25,105	33,474	41,842
32	89.6	0	9,421	18,842	28,263	37,684	47,105
34	93.2	0	10,553	21,105	31,658	42,210	52,763
36	96.8	0	11,816	23,632	35,447	47,263	59,079
38	100.4	0	13,184	26,368	39,553	52,737	65,921
40	104.0	0	14,684	29,368	44,053	58,737	73,421
42	107.6	0	16,316	32,632	48,947	65,263	81,579
46	111.2	0	18,079	36,158	54,237	72,316	90,395
48	114.8	0	20,000	40,000	60,000	80,000	100,000

Table #7A on the following page is analogous to Table #6A. For this second tabulation, the water vapor concentrations have been listed in mass-based units of concentration, namely, mg/m^3. For most situations that require or involve knowing the ambient water vapor concentration on a true or absolute humidity basis, the mass-based concentration units in Table #7A will be the most useful reference.

As was true for Table #6A, the water vapor concentrations tabulated in Table #7A at the 100% relative humidity level are again — by definition — the ambient concentrations, or absolute humidity levels, that would exist in the listed atmosphere at its Dew Point. Thus, whenever one wants the ambient Sea Level water vapor concentration in mg/m^3 at some identifiable Dew Point, this concentration can either be: (1) taken directly from the last column of Table #7A, or (2) determined by linear interpolation between adjacent pairs of values taken from this column.

As was true for Table #6A, for every listing in Table #7A on the following page, it has been assumed that the prevailing ambient barometric pressure is 760 mm Hg, or 1.0 atmosphere, the barometric pressure at Sea Level. At different barometric pressures, the values in every column of Table #7A will be incorrect. For each mass-based concentration shown in this table, the listed value will be correct, ±1%, at the temperature tabulated in the first two columns, and at the 1.0 atmosphere or 760 mm Hg barometric pressure.

Finally, if it is ever necessary to determine a mass-based absolute water vapor concentration at Sea Level for some untabulated ambient temperature and/or relative humidity [i.e., the water vapor concentration at a temperature of 62°F, and a relative humidity of 45%], a simple horizontal and/or vertical linear interpolation from the data in Table #7A will provide a reasonably accurate result, certainly to within ±1 to 2% of the true value.

For reference, the water vapor concentration under the example conditions listed immediately above in the previous paragraph — obtained by a combination of horizontal and vertical linear interpolation — would be 6,317.9 mg/m^3.

In the event that it is ever necessary to determine an absolute, mass-based water vapor concentration at (1) an unlisted ambient temperature, (2) an unlisted relative humidity, and (3) a barometric pressure other than 1.0 atmosphere or 760 mm Hg, the procedure required to accomplish this is a bit more complicated. Basically the following two step process must be followed. The first step involves using a simple horizontal and/or vertical linear interpolation from the data in Table #6A — NOT Table #7A — in order to obtain the volume-based water vapor concentration at the required ambient temperature and relative humidity. Note, this volume-based concentration is applicable at a barometric pressure of 760 mm Hg or 1.0 atmosphere. Once this value has been determined, the second step requires the use of Equation **#3-9**, from Page 3-13, which will then provide the mass-based water vapor concentration under any prevailing ambient conditions. The required Equation is reproduced below:

$$C_{mass} = \frac{P[MW]}{RT}[C_{volume}]$$

For determinations involving water, using as a value for R, the Universal Gas Constant, 62.36 $^{(liter)(mm\ Hg)}/_{(K)(mole)}$, and since water has a molecular weight of 18.0153 amu, we get:

$$C_{mass} = [0.2886]\frac{P}{T}[C_{volume}]$$

This overall two step process, which obviously uses this specific relationship, will give the required mass-based absolute water vapor concentration at any ambient temperature, barometric pressure, and level of relative humidity.

Table #7A

Mass-Based Water Vapor Concentrations, in mg/m^3 at Various Levels of Relative Humidity, & at 760 mm Hg

Temperatures °C	°F	0% RH	20% RH	40% RH	60% RH	80% RH	100% RH
– 20	– 4.0	0	185.8	371.5	557.3	743.1	928.8
– 18	– 0.4	0	222.3	444.7	667.0	889.4	1,111.7
– 16	3.2	0	269.1	538.2	807.3	1,076.4	1,345.6
– 14	6.8	0	325.5	651.0	976.5	1,302.0	1,627.5
– 12	10.4	0	391.6	783.2	1,174.8	1,566.4	1,958.0
– 10	14.0	0	467.2	934.4	1,401.6	1,868.9	2,336.1
– 8	17.6	0	551.8	1,103.6	1,655.4	2,207.2	2,759.0
– 6	21.2	0	645.3	1,290.6	1,935.9	2,581.2	3,226.5
– 4	24.8	0	791.7	1,583.5	2,375.2	3,166.9	3,958.7
– 2	28.4	0	858.8	1,717.5	2,576.3	3,435.1	4,293.8
0	32.0	0	993.4	1,986.8	2,979.4	3,972.8	4,966.2
2	35.6	0	1,153.9	2,307.8	3,461.8	4,615.7	5,769.6
4	39.2	0	1,312.1	2,624.2	3,936.3	5,248.4	6,560.6
6	42.8	0	1,489.0	2,977.9	4,466.9	5,955.8	7,444.8
8	46.4	0	1,683.6	3,367.1	5,050.7	6,734.2	8,417.8
10	50.0	0	1,896.0	3,792.0	5,688.1	7,584.1	9,480.1
12	53.6	0	2,125.7	4,251.3	6,377.0	8,502.6	10,628.3
14	57.2	0	2,392.3	4,784.6	7,176.8	9,569.1	11,961.4
16	60.8	0	2,695.1	5,390.2	8,085.4	10,780.5	13,475.6
18	64.4	0	3,033.6	6,067.2	9,100.7	12,134.3	15,167.9
20	68.0	0	3,406.6	6,813.3	10,219.9	13,626.6	17,033.2
21.1	70.0	0	3,649.0	7,298.0	10,947.0	14,596.0	18,245.0
22	71.6	0	3,833.3	7,666.7	11,500.0	15,333.4	19,166.7
23.9	75.0	0	4,289.9	8,579.8	12,869.7	17,159.6	21,449.1
24	75.2	0	4,312.8	8,625.6	12,938.3	17,251.1	21,563.9
26	78.8	0	4,843.4	9,686.8	14,530.3	19,373.7	24,217.1
28	82.4	0	5,424.7	10,849.4	16,274.2	21,698.9	27,123.6
30	86.0	0	6,055.4	12,110.8	18,166.1	24,221.5	30,276.9
32	89.6	0	6,772.4	13,544.7	20,317.1	27,089.4	33,861.8
34	93.2	0	7,536.4	15,072.8	22,609.3	30,145.7	37,682.1
36	96.8	0	8,384.0	16,768.0	25,159.9	33,535.9	41,919.9
38	100.4	0	9,294.8	18,589.6	27,884.4	37,179.2	46,474.0
40	104.0	0	10,286.2	20,572.4	30,858.5	41,144.7	51,430.9
42	107.6	0	11,356.6	22,713.2	34,069.7	45,426.3	56,782.9
46	111.2	0	12,426.1	24,852.3	37,278.4	49,704.6	62,130.7
48	114.8	0	13,660.9	27,321.8	40,982.6	54,643.5	68,304.4

Water vapor can cause problems for a variety of atmospheric measurements, mostly those designed to determine the ambient concentration of some chemical or particulate of interest.

Water vapor concentrations by themselves — unless they are extremely high [i.e., > 25,000 ppm(vol) or > 18,000 mg/m^3] — will usually not be the source of too many problems; the principal exception to this rule is the quantification, by particle diameter, of the size distributions of certain dusts. For this specific analytical category, high ambient water vapor concentration levels will have a tendency to cause various smaller particles to agglomerate together, thereby forming steadily larger sized particles. Such a situation will obviously skew the size distribution results, shifting the entire population upward as expressed in effective diameters, producing what might appear to be a less hazardous distribution [i.e., particles having the smallest diameters — namely, the respirable, the alveolic, and/or the thoracic fractions — are removed from the population simply as a result of being agglomerated together, thereby becoming particles that are larger, and correspondingly, less hazardous].

For determinations of the ambient concentration levels of certain vapors, water vapor concentration changes are the source of numerous problems. In this area, the measurements that are made must almost always be compensated for by the level of water vapor in the ambient air. Such compensations are usually fairly easy. The problems arise because of the manner in which ambient water vapor concentrations can vary both: (1) over short time intervals, and (2) from point-to-point.

Changes in water vapor concentration levels over short time intervals can occur indoors as a result of a variety of factors, among which are the following:

1. the reduction (or the increase, in certain situations) of indoor humidity levels produced by the operation of air conditioning equipment;
2. the equilibration of the indoor and outdoor humidity levels — to the extent that these levels differ — by the opening of windows, etc.; and
3. the operation of an evaporative cooler (i.e., a swamp cooler) during the summer months in desert areas having relatively dry climates.

This listing is by no means complete; however, it does show some of the common ways in which short time interval very large changes in indoor humidity levels can be produced. As to outdoor situations where humidity levels experience major shifts over a short time period, the most obvious circumstance that could cause such an event would be the passage of a weather front. In any of these man or nature caused situations, humidity changes of 30 to 50% can and do occur — i.e., from an 80% RH level @ 75°F [~ 23,217 ppm(vol) or ~ 17,139 mg/m^3] to a 50% RH level @ 68°F [~ 11,382 ppm(vol) or ~8 8,517 mg/m^3]. Such a shift would be a change of approximately 50%. Short time interval humidity shifts of these magnitudes can be produced by efficient air conditioning systems, swamp coolers, or weather fronts.

Point-to-point changes in the humidity level occur principally indoors. Typically, an individual would likely experience large changes in these levels if he were to pass from an area where the temperature and humidity had been set for optimum human comfort to an area where these parameters were established in order to provide perfect operating conditions for a computer or other sensitive machinery, equipment, or system. It will always be the large scale humidity changes — like those listed in the previous paragraph — that will likely prove to have been responsible for producing problems for an individual who must make any of a variety of ambient measurements.

Examples of some but not all of the specific types of measurements that might or might not be affected by INCREASES in humidity, along with the measurement technology most likely to have been used for each determination, are tabulated on the following page in Table #8A.

Table #8A — Increasing Humidity Levels

No.	Measurement	Monitoring Technology	Direction of Interference
1	Organic Vapors	IR Spectrophotometry	plus
2	Organic Vapors	NDIR	mostly plus
3	Organic Vapors	Photo Ionization Detectors	minus
4	Organic Vapors	Flame Ionization Detectors	none
5	Any Measurable Vapor	Electrochemical Detectors	some plus, some minus
6	Any Measurable Vapor	Solid State Detectors	some plus, some minus
7	Combustible Gases	Catalytic Bead Detectors	none
8	Oxygen	Electrochemical Detectors	minus

This tabulation is by no means complete; however, it does identify some of the most important measurements that can be affected by increases in humidity.

Examples of some of the specific types of measurements that might or might not be affected by DECREASES in humidity, along with the measurement technology most likely to have been used for each determination, are tabulated below in Table #9A.

Table #9A — Decreasing Humidity Levels

No.	Measurement	Monitoring Technology	Direction of Interference
1	Organic Vapors	IR Spectrophotometry	minus
2	Organic Vapors	NDIR	none
3	Organic Vapors	Photo Ionization Detectors	plus
4	Organic Vapors	Flame Ionization Detectors	none
5	Any Measurable Vapor	Electrochemical Detectors	some minus, some plus
6	Any Measurable Vapor	Solid State Detectors	some minus, some plus
7	Combustible Gases	Catalytic Bead Detectors	none
8	Oxygen	Electrochemical Detectors	plus

For virtually all of the measurements that experience problems arising from changes in the ambient water vapor concentration, the mechanism of the problem — expressed in the form of a sequence of events — will be as follows:

1. The analyzer involved is zeroed, or zero checked, at a location that is known to be free of whatever analyte is to be quantified. Care is not exercised in this process to note the prevailing ambient humidity level at the location where this process is completed.
2. The individual doing the analysis then proceeds to the location where the measurements are to be made, an area where the humidity level differs from that at the location where the analytical system was zeroed, or zero checked. No notice of this change in humidity level is made by the analyst.
3. Measurements are then made in the desired location. If the direction of the interference is negative (minus), then the individual is fairly likely to recognize the problem, since he or she may see the analytical result appear as a negative concentration, something that cannot possibly exist. If the direction of the interference is positive (plus), it is unlikely that the analyst will be able to determine that his or her readings are in error on the high side. The readings will simply appear to be indications of possibly higher than expected concentrations of whatever analyte is being measured.

It is interesting to note that in virtually all of the problem areas identified in either Table #s 8A or 9A, the magnitude of the interference produced by changes in the humidity will turn out to have been only a very small percentage of the absolute change in the water vapor concentration. The problem is simply that the magnitude of the absolute changes in the water vapor concentration can be extremely large — i.e., ±5,000 ppm(vol), or even more.

Consider an example. Assume that the magnitude of the error caused by a change in the ambient water vapor concentration was determined — for some specific analysis — to be only ±0.2% of the actual change in water vapor concentration. In most cases such an error, particularly expressed as a percentage, would appear to be unimportant and could be ignored. The actual error produced for the example suggested on the previous page, namely, one produced by a change in the absolute humidity level of, say, ±5,000 ppm(vol) — calculated as ±0.2% of this change in the water vapor concentration level — would be in the range of ±10 ppm(vol). If the measurement being made was for the ambient concentration acetone, for which the OSHA PEL-TWA is 1,000 ppm(vol), an error of ±10 ppm(vol) would, in actual fact, be sufficiently unimportant that it could be ignored. In contrast, if the ambient concentration level of formaldehyde [OSHA PEL-TWA = 0.75 ppm(vol), & OSHA PEL-STEL = 2.0 ppm(vol)] was being sought, an error of ±10 ppm(vol) would be totally unacceptable.

The result of all this is that very considerable attention must be paid to the prevailing humidity levels whenever the goal is the accurate determination of the ambient concentration of some important volatile compound or chemical. Very frequently, as stated earlier, a potential for humidity related problems does exist — problems that are both a function of analyte involved and the analytical method being used to make the determination.

On rare occasions, it may be necessary to know the make-up — on a volume-based concentration basis — of an air mass on an actual, or NON-DRY, basis. In such a situation, the accurate composition of air mass must be calculated. The true atmospheric composition will clearly differ from the tabulated DRY basis concentrations shown in Table #1A on Page A-1 since the air mass is now no longer DRY. The determination of the actual make-up of the air mass of interest will require adjustments to each of the volume-based concentrations of the "normal" DRY basis components, adjustments that are obviously required in order to compensate for the amount of water vapor present. In essence, the air mass must be recognized as being no longer DRY, or water vapor free.

To make the required corrections, the following formula should be used to determine the factor, $\mathbf{f}_{\mathbf{volume}}$, by which each of the DRY volume-based concentrations shown in Table #1A on Page A-1 must be multiplied in order to produce the true wet-basis concentrations for each of these components.

$$\mathbf{f}_{\mathbf{volume}} = \left[1 - \frac{\text{Water Vapor Concentration, in ppm(vol)}}{10^6}\right]$$

Thus, if we are in a situation where the relative humidity is 80% at 70°F, and the water vapor concentration — see Table #6A, on Page A-5 — is 19,579 ppm(vol), we can determine the required multiplication factor as follows:

$$\mathbf{f}_{\mathbf{volume}} = \left[1 - \frac{19{,}579}{10^6}\right] = [1 - 0.0196] = 0.9804$$

Applying this factor to the seven highest concentration members of a DRY air mass, as shown in Table #1A on Page A-1, we would obtain the true composition of this atmosphere, on a wet-basis, as shown on the following page:

No.	Component	True Concentration – in ppm(vol)
1.	Nitrogen	765,552
2.	Oxygen	205,258
–	WATER	19,579
3.	Argon	9,121
4.	Carbon Dioxide	average = 363; range: 343 to 569
5.	Neon	17.8
6.	Helium	5.1
7.	Methane	average = 2.7; range: 2.5 to 4.4

Note, for this air mass, water vapor is the member with the third highest concentration. This is a very common situation — ambient water vapor concentrations usually fall into the third position on the basis of decreasing component concentration for any air mass.

Flow Measurements vs. the Make-Up of the Gas Being Measured

One of the most common devices used to measure the flow rate of a gas is the rotameter, which is typically a transparent, circular cross section tube in which the inside diameter increases steadily as one travels upward in the tube. This rotameter tube will usually have graduated markings on its exterior surface, markings that permit a clear identification of any vertical position on that tube. In the center of the interior of this tube will be a "float" of some suitable material. This float is situated such that it can move freely upward and downward in its tube, in response to the passage of whatever gas is flowing in the rotameter. A rotameter then, as described here, is simply a variable area flow meter.

Rotameters are typically calibrated by their manufacturer. That is to say, the rotameter in question will be challenged under precisely known conditions of pressure and temperature with well-documented flows of some gas and the resulting position of the float noted. It is this tabulation, or the graphical representation of this tabulation, that constitutes the calibration of a rotameter. The most common conditions under which rotameter manufactures calibrate their rotameters are listed below — namely, STP conditions for either air or nitrogen as the calibrant gas.

Gas being Measured:	Air or Nitrogen
Barometric Pressure:	1 atmosphere or 760 mm Hg
Atmospheric Temperature:	0°C or 32°F

Whenever one must measure a gas other than air or nitrogen [for which the effective and actual molecular weights are, respectively: $MW_{air} = 28.964$, and $MW_{nitrogen} = 28.013$ amu] a correction must be made — calibrations with either air or nitrogen are considered to be equivalent, since the effective and actual molecular weights [and densities] of these two fluids are virtually identical. The corrections that are required are based on the differences in the densities of the gases — i.e., the density of the "different fluid" or alternative gas that will be flowing in the rotameter vs. the density of the gas — either air or nitrogen — that was used to develop the initial rotameter calibration.

To understand the nature of this alternative gas correction, it is important to recognize that the flow rate that a typical user will want to achieve in his operations will be what the equivalent flow rate of the alternative gas would be if the operating conditions were actually to be Standard Temperature and Pressure. For the purposes of the formula on the following page, it will be assumed that whatever this alternative gas is, the conditions of the actual measurement under which the rotameter will be used will be Standard Temperature and Pressure. The goal of the correction formula, then, will be to identify the required rotameter *scale reading* that must be established to produce some specific STP equivalent flow rate.

The quantitative relationship is listed below:

$$X = X_{calibrated}\sqrt{\frac{\rho}{\rho_{standard}}}$$

Where:

X = the required *scale reading* on a rotameter carrying a gas other than air or nitrogen, i.e., the *scale reading* that will be required in order to indicate that the flow rate of the alternative gas is numerically equivalent to some particular STP calibrated flow rate of either air or nitrogen;

$X_{calibrated}$ = a specific *scale reading* obtained during the calibration of this rotameter — a reading that corresponded to a particular specific flow rate of the calibration gas, air or nitrogen, under conditions of STP;

ρ = the density of the gas currently flowing through the rotameter, in some suitable set of units; &

$\rho_{calibrated}$ = the density of the calibration gas — either air or nitrogen — in the same set of units as was the case for ρ.

To understand the specifics of this situation, let us consider an example.

Assume we are dealing with a rotameter that is to carry helium. Assume further that an investigator <u>must know</u> the specific rotameter *scale reading* that is required to produce an STP flow rate of 100.0 ml/min of helium in a rotameter that was calibrated using nitrogen under conditions of STP. Assume finally that the manufacturer's calibration of this rotameter showed that a flow rate of 100 ml/min of nitrogen occurred at a rotameter *scale reading* of 4.50 cm. To apply this formula we must know the values of the densities, at STP, for both the alternative gas, helium, and the calibration gas, nitrogen. In order, these densities are:

$$\rho_{helium} = 1.79 \times 10^{-4} \text{ grams/ml} = 0.179 \text{ grams/liter}$$

$$\rho_{nitrogen} = 1.25 \times 10^{-3} \text{ grams/ml} = 1.25 \text{ grams/liter}$$

Now applying the required formula we get:

$$X = X_{calibrated}\sqrt{\frac{\rho}{\rho_{standard}}} = (4.50)\sqrt{\frac{1.79 \times 10^{-4}}{1.25 \times 10^{-3}}}$$

$$X = (4.50)\sqrt{0.143} = (4.50)(0.378) = 1.70 \text{ cm}$$

What this means is that the rotameter float must be set at a *scale reading* of 1.70 cm if it is to deliver the required 100.0 ml/min STP equivalent flow rate of helium. Although everything stated in this example is completely correct and accurate, in actual practice, the situation is slightly different. To understand this, it is necessary to consider a typical rotameter.

Rotameters vary in overall length from less than 5.0 cm to more than 25 cm. They are always fabricated from some type of transparent material [i.e., glass, polycarbonate, etc.]. This is obviously necessary in order to be able to observe the position of the rotameter float. Rotameters will always have their walls marked with labeled and uniformly spaced graduations — i.e., *scale readings* — in order to provide clear references as to the position of the float in any situation or circumstance. The major graduations on a typical rotameter will usually be measured in centimeters, with minor divisions at half or tenth centimeter distances. Obviously, the greeter the overall length of the rotameter, the lesser will be the distance between its minor divisions. Positioning a rotameter float at either one of its minor or its major divisions is a relatively easy process. Interpolating between minor *scale readings* to establish a float position at some non-graduated location is considerably more difficult.

In the example on the previous page, the desired *scale reading* was 1.70 cm. It is very likely that the rotameter being used would have 1/10 cm distances as its minor divisions. For such a rotameter setting the float at a *scale reading* of 1.70 cm would be a relatively simple matter. If the result of the calculation on the previous page had indicated that the *scale reading* had to be 1.72 cm, the situation would have proved to be very much more difficult. It is extremely challenging to center a rotameter float at a position that is <u>not</u> directly on one of the graduated marks on the rotameter body. Visual interpolation and setting of a rotameter float position to *scale readings* that are not directly on graduated major or minor division is process that is fraught with potential errors. Because of this, the actual process of setting the flow in a rotameter that is to carry a gas other than the one used to obtain the device's calibration is different than was described on the previous page.

Specifically, the process usually involves carefully setting the float of the rotameter that is carrying an alternative gas on an easy-to-read *scale reading*, namely, one that is centered directly on one of the rotameter's graduated major and/or minor divisions. Once this has been accomplished, the STP flow rate of the alternative gas can be determined by a two step process. The first of these steps involves using the quantitative relationship from the previous page [slightly rearranged, as shown below] to determine what the equivalent calibrated *scale reading* would have been for the gas that was used to obtain the calibration of that specific rotameter. The rotameter's calibration data set is then used, in conjunction with the just calculated equivalent *scale reading* to determine the actual STP flow rate of alternative gas. In this second step, if interpolation is necessary [which is usually the case], the interpolation is mathematical rather than visual, and the potential for errors significantly reduced.

Again, let us consider an example.

Assume we are dealing with a rotameter that was calibrated using air but is to carry oxygen in the application being considered. Assume further that the manufacturer's air-based calibration data set for this rotameter — at STP — is as shown by the following table.

Rotameter Calibration

Scale reading, cm	Flow Rate, ml/min
0.0	0
1.0	58
2.0	193
3.0	446
4.0	803
5.0	1,220
6.0	1,618
7.0	2,015
8.0	2,398
9.0	2,790
10.0	3,193

Assume, finally that this oxygen carrying rotameter is set at a *scale reading* of 4.0 cm. The actual flow rate of oxygen is determined as follows — using, as stated above, a rearranged form of the relationship from the previous page. This rearranged form is as follows:

$$X_{calibrated} = X\sqrt{\frac{\rho_{standard}}{\rho}}$$

Again to apply this formula we must know the values of the densities, at STP, for both the alternative gas, oxygen, and the calibration gas, air. In order, these densities are shown on the following page.

$$\rho_{oxygen} = 1.428 \times 10^{-3} \text{ grams/ml} = 1.428 \text{ grams/liter}$$

$$\rho_{air} = 1.292 \times 10^{-3} \text{ grams/ml} = 1.292 \text{ grams/liter}$$

Although these densities are fairly close together in value, there is enough difference between them to end up producing a different STP flow rate for the oxygen passing through the rotameter. Substituting into the listed formula, we get:

$$X_{calibrated} = X\sqrt{\frac{\rho_{standard}}{\rho}} = (4.0)\sqrt{\frac{1.292 \times 10^{-3}}{1.428 \times 10^{-3}}}$$

$$X_{calibrated} = (4.0)\sqrt{0.905} = (4.0)(0.951) = 3.81 \text{ cm}$$

Now, to determine the equivalent STP flow represented by an equivalent *scale reading* of 3.81 cm, we need only perform a simple linear interpolation between the calibrated flow rates for *scale readings* of 3.0 and 4.0 cm. From the calibration tabulation on the previous page, these flow rates are as follows:

at a scale reading of 3.0, the flow rate was 446 ml/min
at a scale reading of 4.0, the flow rate was 803 ml/min

The linear interpolation would be as follows:

$$\text{True Flow Rate} = 446 + [(0.81)(803 - 446)]$$

$$\text{True Flow Rate} = 446 + (0.81)(357) = 446 + 289.2 \approx 735 \text{ ml/min}$$

Clearly, the difference between the calibrated STP flow rate of air at a *scale reading* of 4.0 cm [803 ml/min], and the calculated true STP flow rate of oxygen at the same *scale reading* of 4.0 [735 ml/min] is significant — a actual difference of 68 ml/min or 9.25%.

Flow Measurements vs. Barometric Pressure & Temperature

The potential problem caused by making rotameter-based flow rate measurements at terrestrial altitudes other than Sea Level — i.e., at barometric pressures other than 760 mm Hg, at which pressure the rotameter was probably calibrated — is that *scale readings* at different altitudes will not correspond to the flow rates that were determined during the calibration. The same can be said for measurements that are made at temperatures other than 0°C. Basically the differences are directly analogous to the foregoing discussion on rotameters when they are carrying alternative gases.

The single factor that produces variations from the calibrated standard values in these situations will always be the density of the gas being transported through the rotameter vs. the density of the gas that was used to develop the calibration. In the discussion of the impact of having alternative gases flowing through the rotameter, the density factor was simply a function of the molecular weights of the materials involved, since the densities were always determined under the same ambient conditions, namely, STP.

Now, if a rotameter is used to determine the flow rate of air or nitrogen [the same gases most commonly used to develop the rotameter's basic calibration] at an altitude other than Sea Level, and/or at a temperature other than 0°C, there may be a difference in the density of the air or the nitrogen flowing through the rotameter vs. the corresponding density at Sea Level. This density shift will have been caused either by the change in the prevailing barometric pressure, or by changes in the ambient temperature, or by both. This is a situation for which an alert analyst must always be ready to make adjustments.

The formula that permits these readings to be corrected has been produced on the following page. It is directly analogous to the previous relationship, namely, the one that involves alternative gases. It is based on accounting for the different gas density values; however, this time the determination of these differences is made under two different sets of ambient conditions rather than from two different gases. The mathematical relationship accomplishes these corrections directly from the ambient pressure and temperature, without requiring any additional density calculations, since we are dealing here with either air or nitrogen, the gases that were used to develop the STP calibration of the rotameter:

$$X = [0.5995][X_{standard}]\sqrt{\frac{P_{ambient}}{T_{ambient}}}$$

Where: X = the required *scale reading* on a rotameter carrying air or nitrogen under temperature and pressure conditions different from those at STP;

$X_{standard}$ = a specific *scale reading* obtained during the calibration of this rotameter — a reading that corresponded to a particular specific flow rate of the calibration gas, air or nitrogen, under conditions of STP;

$P_{ambient}$ = the ambient barometric pressure, in mm Hg; &

$T_{ambient}$ = the ambient temperature, in K.

Again, this relationship can probably be most easily understood by considering a specific example.

Suppose there is available a rotameter that has been calibrated using nitrogen at STP. For this rotameter, a calibrated float position of 23.0 mm corresponded to a nitrogen flow rate of 1,626 ml/minute. If this rotameter is to be used to identify the flow rate of air in Denver, CO, where: (1) the altitude is 5,280 feet above mean Sea Level, (2) the barometric pressure is 642.5 mm Hg, and (3) the ambient temperature is 32°C, what will be the required *scale reading* on this rotameter — the reading at which the flow rate of air at the Denver location would equal 1,626 ml/minute?

The solution is as follows:

We have been given the values of almost all of the parameters we will require. The only calculation we must make is to convert the ambient temperature, which has been given in °C, to its corresponding value in K. This is done simply by adding 273.16° to the ambient value of 32°C, thus:

$$T = 32° + 273.16° = 305.16 \text{ K}$$

Now, substituting into the formula, we can develop the required answer:

$$X = (0.5995)(23.0)\sqrt{\frac{642.5}{305.16}}$$

$$X = (13.79)\sqrt{2.106} = (13.79)(1.451) = 20.0 \text{ mm}$$

Therefore, the required *scale reading* on the rotameter located in Denver, CO — i.e., the reading that will correspond to a 1,626 ml/minute flow rate of air at that location — would be 20.0 mm.

Other Effects of Barometric Pressure & Temperature

Whenever one must make ambient concentration measurements of some chemical of interest at ambient temperatures and/or pressures other than those characteristic of STP, certain problems may develop. For the most part, these problems — like those related to the measurement of flow rates — will turn out to have been related to the density of the atmosphere in which the measurements are being made.

Most manufacturers calibrate their gas analyzers under conditions of Normal Temperature and Pressure [NTP], namely, 25°C and 760 mm Hg. Whenever an analyzer must be used under different ambient conditions, problems for certain analytical approaches may occur. To understand these potential problem areas, it will be instructive to consider the basic detection methodology of an infrared spectrophotometric gas analyzer. To that end, let us consider an example in which this type of analyzer is tasked to perform an analysis of the ambient concentration of acetone.

An IR spectrophotometer's operation is based on the relationship between the absorbance produced by some analyte of interest (as a component in the gas mixture being analyzed), to that analyte's concentration in the gas mixture. Simply, for any analyte, it is fairly easy to select an infrared wavelength where that analyte will absorb — i.e., for acetone, a wavelength of 1,208 cm^{-1} [8.28 μ], in the mid-infrared band, works very well. For an acetone analysis conducted under conditions of NTP, an infrared spectrophotometric analyzer, using the previously listed acetone absorbing wavelength, will be able to provide excellent results. The calibration on this type of analyzer, having been done at NTP, ensures that it will provide precise and accurate analyses of ambient acetone concentrations at NTP — i.e., it will respond perfectly to the population of acetone molecules that are characteristic of the acetone concentration as this population would exist in a matrix at NTP. Because this type of analyzer responds to a specific population of absorbing molecules, there will always be a potential problem for an analysis conducted under conditions other than those at which the analyzer was calibrated!

To understand this process, we must start by identifying the acetone molecule population, say at a concentration of 100 ppm(vol), in an air matrix at NTP. For such a situation, there will be a total of $2.46 \times 10^{21} = N_{acetone}$ acetone molecules/m^3. Against its calibration at NTP, an infrared spectrophotometer will identify the absorbance produced by this number of molecules as corresponding to a concentration of 100 ppm(vol).

Assume next that we are dealing with an ambient acetone analysis in Denver, CO, on June 14, 1997, when the temperature was 88°F. Assume further that the acetone concentration in this Denver, CO, matrix was exactly 100 ppm(vol). Clearly, the ambient conditions in this situation were significantly different from those of NTP, specifically, the two sets of conditions, including data on the acetone molecule concentrations [both volume and mass-based], and the acetone molecule populations, are tabulated side-by-side below to emphasize the comparative differences:

Parameter	NTP Conditions	Conditions in Denver, CO, on 6/14/97
Altitude	Sea Level	5,280 feet above Sea Level
Ambient Temperature	25°C = 298.16 K	88°F = 31.11°C = 304.27 K
Barometric Pressure	760 mm Hg	642.5 mm Hg
True Acetone Conc.	100 ppm(vol)	100 ppm(vol)
Indicated Acetone Conc.	100 ppm(vol)	82.9 ppm(vol)
True Acetone Conc.	237.5 mg/m^3	196.9 mg/m^3
Acetone Molecule Count	2.46×10^{21} molecules/m^3	2.04×10^{21} molecules/m^3

For this matrix, with its reduced number of acetone molecules per unit volume, but with the same volumetric acetone concentration, an infrared spectrophotometer, calibrated under conditions of NTP, would produce a measured acetone concentration of 82.9 ppm(vol)! Clearly there is a problem!

The basic nature of the problem in this situation again relates to the density of the gas in the matrix where the measurements are being made. The formula that will permit the correction of this problem is listed immediately below, with descriptions of the factors involved in it listed on the following page:

$$C_{true} = 2.549 C_{indicated} \left[\frac{T_{ambient}}{P_{ambient}} \right]$$

Where: C_{true} = the true concentration of the analyte of interest under the prevailing, non-NTP conditions of ambient temperature and pressure;

$C_{indicated}$ = the concentration indicated by an analyzer for which the principle of detection is dependent upon the analyte's actual molecular population per unit volume in the matrix being monitored;

$P_{ambient}$ = the ambient barometric pressure, in mm Hg; &

$T_{ambient}$ = the ambient temperature, in K.

It remains now only to list some of the analytical methods for which this corrective formula will be applicable. Seven of the common ones are listed below in Table #10A:

Table #10A

No.	Monitoring Technology	Correction Formula Applicable
1	IR Spectrophotometry	YES
2	NDIR	YES
3	Photo Ionization Detectors	YES
4	Flame Ionization Detectors	YES
5	Electrochemical Detectors	YES
6	Solid State Detectors	YES
7	Catalytic Bead Detectors	YES

Appendix B

Conversion Factors Alphabetical Listing

1 atm (atmosphere) =	1.013 bars
	10.133 newtons/cm^2 (newtons/square centimeter)
	33.90 ft. of H_2O (feet of water)
	101.325 kp (kilopascals)
	1,013.25 mb (millibars)
	14.70 psia (pounds/square inch - absolute)
	760 torr
	760 mm Hg (millimeters of mercury)
1 bar =	0.987 atm (atmospheres)
	1×10^6 dynes/cm^2 (dynes/square centimeter)
	33.45 ft. of H_2O (feet of water)
	1×10^5 pascals [nt/m^2] (newtons/square meter)
	750.06 torr
	750.06 mm Hg (millimeters of mercury)
1 Bq (becquerel) =	1 radioactive disintegration/second
	2.7×10^{-11} Ci (curie)
	2.7×10^{-8} mCi (millicurie)
1 BTU (British Thermal Unit) =	252 cal (calories)
	1,055.06 j (joules)
	10.41 liter-atmospheres
	0.293 watt-hours
1 cal (calorie) =	3.97×10^{-3} BTUs (British Thermal Units)
	4.18 j (joules)
	0.0413 liter-atmospheres
	1.163×10^{-3} watt-hours
1 cm (centimeter) =	0.0328 ft (feet)
	0.394 in (inches)
	10,000 microns (micrometers)
	100,000,000 Å = 10^8 Å (Ångstroms)
1 cc (cubic centimeter) =	3.53×10^{-5} ft^3 (cubic feet)
	0.061 in^3 (cubic inches)
	2.64×10^{-4} gal (gallons)
	0.001 ℓ (liters)
	1.00 ml (milliliters)

DEFINITIONS, CONVERSIONS, AND CALCULATIONS

1 ft^3 (cubic foot) =	28,317 cc (cubic centimeters)
	1,728 in^3 (cubic inches)
	0.0283 m^3 (cubic meters)
	7.48 gal (gallons)
	28.32 ℓ (liters)
	29.92 qts (quarts)
1 in^3 (cubic inch) =	16.39 cc (cubic centimeters)
	16.39 ml (milliliters)
	5.79×10^{-4} ft^3 (cubic feet)
	1.64×10^{-5} m^3 (cubic meters)
	4.33×10^{-3} gal (gallons)
	0.0164 ℓ (liters)
	0.55 fl oz (fluid ounces)
1 m^3 (cubic meter) =	1,000,000 cc = 10^6 cc (cubic centimeters)
	35.31 ft^3 (cubic feet)
	61,023 in^3 (cubic inches)
	264.17 gal (gallons)
	1,000 ℓ (liters)
1 yd^3 (cubic yard) =	201.97 gal (gallons)
	764.55 ℓ (liters)
1 Ci (curie) =	3.7×10^{10} radioactive disintegrations/second
	3.7×10^{10} Bq (becquerel)
	1,000 mCi (millicurie)
1 day =	24 hrs (hours)
	1,440 min (minutes)
	86,400 sec (seconds)
	0.143 weeks
	2.738×10^{-3} yrs (years)
1°C (expressed as an interval) =	1.8°F = $\left[\frac{9}{5}\right]$°F (degrees Fahrenheit)
	1.8°R (degrees Rankine)
	1.0 K (degrees Kelvin)
°C (degree Celsius) =	$\left[\left(\frac{5}{9}\right)(°F - 32°)\right]$
1°F (expressed as an interval) =	0.556°C = $\left[\frac{5}{9}\right]$°C (degrees Celsius)
	1.0°R (degrees Rankine)
	0.556 K (degrees Kelvin)
°F (degree Fahrenheit) =	$\left[\left(\frac{9}{5}\right)(°C) + 32°\right]$

1 dyne =	1×10^{-5} nt (newtons)
1 ev (electron volt) =	1.602×10^{-12} ergs
	1.602×10^{-19} j (joules)
1 erg =	1 dyne-centimeters
	1×10^{-7} j (joules)
	2.78×10^{-11} watt-hours
1 fps (feet/second) =	1.097 kmph (kilometers/hour)
	0.305 mps (meters/second)
	0.01136 mph (miles/hour)
1 ft (foot) =	30.48 cm (centimeters)
	12 in (inches)
	0.3048 m (meters)
	1.65×10^{-4} nt (nautical miles)
	1.89×10^{-4} mi (statute miles)
1 gal (gallon) =	3,785 cc (cubic centimeters)
	0.134 ft^3 (cubic feet)
	231 in^3 (cubic inches)
	3.785 ℓ (liters)
1 gm (gram) =	0.001 kg (kilograms)
	1,000 mg (milligrams)
	1,000,000 ng = 10^6 ng (nanograms)
	2.205×10^{-3} lbs (pounds)
1 gm/cc (grams/cubic centimeter) =	62.43 lbs/ft^3 (pounds/cubic foot)
	0.0361 lbs/in^3 (pounds/cubic inch)
	8.345 lbs/gal (pounds/gallon)
1 Gy (gray) =	1 j/kg (joules/kilogram)
	100 rad
	1 Sv (sievert) — [unless modified through division by an appropriate factor, such as Q and/or N]
1 hp (horsepower) =	745.7 j/sec (joules/sec)
1 hr (hour) =	0.0417 days
	60 min (minutes)
	3,600 sec (seconds)
	5.95×10^{-3} weeks
	1.14×10^{-4} yrs (years)
1 in (inch) =	2.54 cm (centimeters)
	1,000 mils

DEFINITIONS, CONVERSIONS, AND CALCULATIONS

1 inch of water =	1.86 mm Hg (millimeters of mercury)
	249.09 pascals
	0.0361 psi (lbs/in^2)
1 j (joule) =	9.48×10^{-4} BTUs (British Thermal Units)
	0.239 cal (calories)
	10,000,000 ergs = 1×10^{7} ergs
	9.87×10^{-3} liter-atmospheres
	1.00 nt-m (newton-meters)
1 kcal (kilocalorie) =	3.97 BTUs (British Thermal Units)
	1,000 cal (calories)
	4,186.8 j (joules)
1 kg (kilogram) =	1,000 gms (grams)
	2.205 lbs (pounds)
1 km (kilometer) =	3,280.8 ft (feet)
	0.54 nt (nautical miles)
	0.6214 mi (statute miles)
1 kw (kilowatt) =	56.87 BTU/min (British Thermal Units/minute)
	1.341 hp (horsepower)
	1,000 j/sec (joules/sec)
1 kw-hr (kilowatt-hour) =	3,412.14 BTU (British Thermal Units)
	3.6×10^{6} j (joules)
	859.8 kcal (kilocalories)
1 ℓ (liter) =	1,000 cc (cubic centimeters)
	1 dm^3 (cubic decimeters)
	0.0353 ft^3 (cubic feet)
	61.02 in^3 (cubic inches)
	0.264 gal (gallons)
	1,000 ml (milliliters)
	1.057 qts (quarts)
1 m (meter) =	1×10^{10} Å (Ångstroms)
	100 cm (centimeters)
	3.28 ft (feet)
	39.37 in (inches)
	1×10^{-3} km (kilometers)
	1,000 mm (millimeters)
	1,000,000 μ = 1×10^{6} μ (micrometers)
	1×10^{9} nm (nanometers)

1 mps (meters/second) =	196.9 fpm (feet/minute)
	3.6 kmph (kilometers/hour)
	2.237 mph (miles/hour)
1 mph (mile/hour) =	88 fpm (feet/minute)
	1.61 kmph (kilometers/hour)
	0.447 mps (meters/second)
1 kt (nautical mile) =	6,076.1 ft (feet)
	1.852 km (kilometers)
	1.15 mi (statute miles)
	2,025.4 yds (yards)
1 mi (statute mile) =	5,280 ft (feet)
	1.609 km (kilometers)
	1,609.3 m (meters)
	0.869 nt (nautical miles)
	1,760 yds (yards)
1 mCi (millicurie) =	0.001 Ci (curie)
	3.7×10^{10} radioactive disintegrations/second
	3.7×10^{10} Bq (becquerel)
1 mm Hg (millimeter of mercury) =	1.316×10^{-3} atm (atmospheres)
	0.535 in H_2O (inches of water)
	1.33 mb (millibars)
	133.32 pascals
	1 torr
	0.0193 psia (pounds/square inch - absolute)
1 min (minute) =	6.94×10^{-4} days
	0.0167 hrs (hours)
	60 sec (seconds)
	9.92×10^{-5} weeks
	1.90×10^{-6} yrs (years)
1 nt (newton) =	1×10^{5} dynes
1 nt-m (newton-meter) =	1.00 j (joules)
	2.78×10^{-4} watt-hours
1 ppm (parts/million-volume) =	1.00 ml/m^3 (milliliters/cubic meter)
1 ppm[wt] (parts/million-weight) =	1.00 mg/kg (milligrams/kilogram)
1 pascal =	9.87×10^{-6} atm (atmospheres)
	4.015×10^{-3} in H_2O (inches of water)
	0.01 mb (millibars)
	7.5×10^{-3} mm Hg (millimeters of mercury)

DEFINITIONS, CONVERSIONS, AND CALCULATIONS

1 lbs (pound) =	453.59 gms (grams) 16 oz (ounces)
1 lbs/ft^3 (pounds/cubic foot) =	16.02 gms/l (grams/liter)
1 lbs/in^3 (pounds/cubic inch) =	27.68 gms/cc (grams/cubic centimeter) 1,728 lbs/ft^3 (pounds/cubic foot)
1 psi (pounds/square inch) =	0.068 atm (atmospheres) 27.67 in H_2O (inches of water) 68.85 mb (millibars) 51.71 mm Hg (millimeters of mercury) 6,894.76 pascals
1 qt (quart) =	946.4 cc (cubic centimeters) 57.75 in^3 (cubic inches) 0.946 ℓ (liters)
1 rad =	100 ergs/gm (ergs/gram) 0.01 Gy (gray) 1 rem [unless modified through division by an appropriate factor, such as Q and/or N]
1 rem =	1 rad [unless modified through division by an appropriate factor, such as Q and/or N]
1 Sv (sievert) =	1 Gy (gray) [unless modified through division by an appropriate factor, such as Q and/or N]
1 cm^2 (square centimeter) =	1.076×10^{-3} ft^2 (square feet) 0.155 in^2 (square inches) 1×10^{-4} m^2 (square meters)
1 ft^2 (square foot) =	2.296×10^{-5} acres 929.03 cm^2 (square centimeters) 144 in^2 (square inches) 0.0929 m^2 (square meters)
1 m^2 (square meter) =	10.76 ft^2 (square feet) 1,550 in^2 (square inches)
1 mi^2 (square mile) =	640 acres 2.79×10^7 ft^2 (square feet) 2.59×10^6 m^2 (square meters)
1 torr =	1.33 mb (millibars)
1 watt =	3.41 BTU/hr (British Thermal Units/hour) 1.341×10^{-3} hp (horsepower) 1.00 j/sec (joules/second)

1 watt-hour =	3.412 BTUs (British Thermal Units)
	859.8 cal (calories)
	3,600 j (joules)
	35.53 liter-atmospheres
1 week =	7 days
	168 hrs (hours)
	10,080 min (minutes)
	6.048×10^5 sec (seconds)
	0.0192 yrs (years)
1 yr (year) =	365.25 days
	8,766 hrs (hours)
	5.26×10^5 min (minutes)
	3.16×10^7 sec (seconds)
	52.18 weeks

Appendix C

Conversion Factors Listing By Unit Category

Units of Length

1 cm (centimeter) =	0.0328 ft (feet)
	0.394 in (inches)
	10,000 microns (micrometers)
	100,000,000 Å = 10^8 Å (Ångstroms)
1 ft (foot) =	30.48 cm (centimeters)
	12 in (inches)
	0.3048 m (meters)
	1.65×10^{-4} nt (nautical miles)
	1.89×10^{-4} mi (statute miles)
1 in (inch) =	2.54 cm (centimeters)
	1,000 mils
1 km (kilometer) =	3,280.8 ft (feet)
	0.54 nt (nautical miles)
	0.6214 mi (statute miles)
1 m (meter) =	1×10^{10} Å (Ångstroms)
	100 cm (centimeters)
	3.28 ft (feet)
	39.37 in (inches)
	1×10^{-3} km (kilometers)
	1,000 mm (millimeters)
	1,000,000 μ = 1×10^6 μ (micrometers)
	1×10^9 nm (nanometers)
1 kt (nautical mile) =	6,076.1 ft (feet)
	1.852 km (kilometers)
	1.15 mi (statute miles)
	2,025.4 yds (yards)
1 mi (statute mile) =	5,280 ft (feet)
	1.609 km (kilometers)
	1,609.3 m (meters)
	0.869 nt (nautical miles)
	1,760 yds (yards)

Units of Area

1 cm^2 (square centimeter) =	1.076×10^{-3} ft^2 (square feet)
	0.155 in^2 (square inches)
	1×10^{-4} m^2 (square meters)
1 ft^2 (square foot) =	2.296×10^{-5} acres
	929.03 cm^2 (square centimeters)
	144 in^2 (square inches)
	0.0929 m^2 (square meters)
1 m^2 (square meter) =	10.76 ft^2 (square feet)
	1,550 in^2 (square inches)
1 mi^2 (square mile) =	640 acres
	2.79×10^{7} ft^2 (square feet)
	2.59×10^{6} m^2 (square meters)

Units of Volume

1 cc (cubic centimeter) =	3.53×10^{-5} ft^3 (cubic feet)
	0.061 in^3 (cubic inches)
	2.64×10^{-4} gal (gallons)
	0.001 ℓ (liters)
	1.00 ml (milliliters)
1 ft^3 (cubic foot) =	28,317 cc (cubic centimeters)
	1,728 in^3 (cubic inches)
	0.0283 m^3 (cubic meters)
	7.48 gal (gallons)
	28.32 ℓ (liters)
	29.92 qts (quarts)
1 in^3 (cubic inch) =	16.39 cc (cubic centimeters)
	16.39 ml (milliliters)
	5.79×10^{-4} ft^3 (cubic feet)
	1.64×10^{-5} m^3 (cubic meters)
	4.33×10^{-3} gal (gallons)
	0.0164 ℓ (liters)
	0.55 fl oz (fluid ounces)
1 m^3 (cubic meter) =	1,000,000 cc = 10^6 cc (cubic centimeters)
	35.31 ft^3 (cubic feet)
	61,023 in^3 (cubic inches)
	264.17 gal (gallons)
	1,000 ℓ (liters)

1 yd^3 (cubic yard) =	201.97 gal (gallons)
	764.55 ℓ (liters)
1 gal (gallon) =	3,785 cc (cubic centimeters)
	0.134 ft^3 (cubic feet)
	231 in^3 (cubic inches)
	3.785 ℓ (liters)
1 ℓ (liter) =	1,000 cc (cubic centimeters)
	1 dm^3 (cubic decimeters)
	0.0353 ft^3 (cubic feet)
	61.02 in^3 (cubic inches)
	0.264 gal (gallons)
	1,000 ml (milliliters)
	1.057 qts (quarts)
1 qt (quart) =	946.4 cc (cubic centimeters)
	57.75 in^3 (cubic inches)
	0.946 ℓ (liters)

Units of Mass

1 gm (gram) =	0.001 kg (kilograms)
	1,000 mg (milligrams)
	1,000,000 ng = 10^6 ng (nanograms)
	2.205×10^{-3} lbs (pounds)
1 kg (kilogram) =	1,000 gms (grams)
	2.205 lbs (pounds)
1 lbs (pound) =	453.59 gms (grams)
	16 oz (ounces)

Units of Time

1 day =	24 hrs (hours)
	1,440 min (minutes)
	86,400 sec (seconds)
	0.143 weeks
	2.738×10^{-3} yrs (years)
1 hr (hour) =	0.0417 days
	60 min (minutes)
	3,600 sec (seconds)
	5.95×10^{-3} weeks
	1.14×10^{-4} yrs (years)

1 min (minute) =	6.94×10^{-4} days
	0.0167 hrs (hours)
	60 sec (seconds)
	9.92×10^{-5} weeks
	1.90×10^{-6} yrs (years)
1 week =	7 days
	168 hrs (hours)
	10,080 min (minutes)
	6.048×10^{5} sec (seconds)
	0.0192 yrs (years)
1 yr (year) =	365.25 days
	8,766 hrs (hours)
	5.26×10^{5} min (minutes)
	3.16×10^{7} sec (seconds)
	52.18 weeks

Units of the Measure of Temperature

°C (degree Celsius) =	$\left[\left(5/9\right)(°F - 32°)\right]$
1°C (expressed as an interval) =	1.8°F = $\left[9/5\right]$°F (degrees Fahrenheit)
	1.8°R (degrees Rankine)
	1.0 K (degrees Kelvin)
°F (degree Fahrenheit) =	$\left[\left(9/5\right)(°C) + 32°\right]$
1°F (expressed as an interval) =	0.556°C = $\left[5/9\right]$°C (degrees Celsius)
	1.0°R (degrees Rankine)
	0.556 K (degrees Kelvin)

Units of Force

1 dyne =	1×10^{-5} nt (newtons)
1 nt (newton) =	1×10^{5} dynes

Units of Work or Energy

1 BTU (British Thermal Unit) =	252 cal (calories)
	1,055.06 j (joules)
	10.41 liter-atmospheres
	0.293 watt-hours
1 cal (calorie) =	3.97×10^{-3} BTUs (British Thermal Units)
	4.18 j (joules)
	0.0413 liter-atmospheres
	1.163×10^{-3} watt-hours
1 ev (electron volt) =	1.602×10^{-12} ergs
	1.602×10^{-19} j (joules)
1 erg =	1 dyne-centimeter
	1×10^{-7} j (joules)
	2.78×10^{-11} watt-hours
1 j (joule) =	9.48×10^{-4} BTUs (British Thermal Units)
	0.239 cal (calories)
	10,000,000 ergs = 1×10^{7} ergs
	9.87×10^{-3} liter-atmospheres
	1.00 nt-m (newton-meters)
1 kcal (kilocalorie) =	3.97 BTUs (British Thermal Units)
	1,000 cal (calories)
	4,186.8 j (joules)
1 kw-hr (kilowatt-hour) =	3,412.14 BTU (British Thermal Units)
	3.6×10^{6} j (joules)
	859.8 kcal (kilocalories)
1 nt-m (newton-meter) =	1.00 j (joules)
	2.78×10^{-4} watt-hours
1 watt-hour =	3.412 BTUs (British Thermal Units)
	859.8 cal (calories)
	3,600 j (joules)
	35.53 liter-atmospheres

Units of Power

1 hp (horsepower) =	745.7 j/sec (joules/sec)
1 kw (kilowatt) =	56.87 BTU/min (British Thermal Units/minute)
	1.341 hp (horsepower)
	1,000 j/sec (joules/sec)
1 watt =	3.41 BTU/hr (British Thermal Units/hour)
	1.341×10^{-3} hp (horsepower)
	1.00 j/sec (joules/second)

Units of Pressure

1 atm (atmosphere) =	1.013 bars
	10.133 newtons/cm^2 (newtons/square centimeter)
	33.90 ft. of H_2O (feet of water)
	101.325 kp (kilopascals)
	1,013.25 mb (millibars)
	14.70 psia (pounds/square inch - absolute)
	760 torr
	760 mm Hg (millimeters of mercury)
1 bar =	0.987 atm (atmospheres)
	1×10^{6} dynes/cm^2 (dynes/square centimeter)
	33.45 ft. of H_2O (feet of water)
	1×10^{5} pascals [nt/m^2] (newtons/square meter)
	750.06 torr
	750.06 mm Hg (millimeters of mercury)
1 inch of water =	1.86 mm Hg (millimeters of mercury)
	249.09 pascals
	0.0361 psi (lbs/in^2)
1 mm Hg (millimeter of mercury) =	1.316×10^{-3} atm (atmospheres)
	0.535 in H_2O (inches of water)
	1.33 mb (millibars)
	133.32 pascals
	1 torr
	0.0193 psia (pounds/square inch - absolute)
1 pascal =	9.87×10^{-6} atm (atmospheres)
	4.015×10^{-3} in H_2O (inches of water)
	0.01 mb (millibars)
	7.5×10^{-3} mm Hg (millimeters of mercury)

1 psi (pounds/square inch) =	0.068 atm (atmospheres)
	27.67 in H_2O (inches of water)
	68.85 mb (millibars)
	51.71 mm Hg (millimeters of mercury)
	6,894.76 pascals
1 torr =	1.33 mb (millibars)

Units of Velocity or Speed

1 fps (feet/second) =	1.097 kmph (kilometers/hour)
	0.305 mps (meters/second)
	0.01136 mph (miles/hour)
1 mps (meters/second) =	196.9 fpm (feet/minute)
	3.6 kmph (kilometers/hour)
	2.237 mph (miles/hour)
1 mph (mile/hour) =	88 fpm (feet/minute)
	1.61 kmph (kilometers/hour)
	0.447 mps (meters/second)

Units of Density

1 gm/cc (grams/cubic centimeter) =	62.43 lbs/ft^3 (pounds/cubic foot)
	0.0361 lbs/in^3 (pounds/cubic inch)
	8.345 lbs/gal (pounds/gallon)
1 lbs/ft^3 (pounds/cubic foot) =	16.02 gms/ℓ (grams/liter)
1 lbs/in^3 (pounds/cubic inch) =	27.68 gms/cc (grams/cubic centimeter)
	1,728 lbs/ft^3 (pounds/cubic foot)

Units of Concentration

1 ppm (parts/million-volume) =	1.00 ml/m^3 (milliliters/cubic meter)
1 ppm(wt) (parts/million-weight) =	1.00 mg/kg (milligrams/kilogram)

Radiation & Dose Related Units

1 Bq (becquerel) =	1 radioactive disintegration/second
	2.7×10^{-11} Ci (curie)
	2.7×10^{-8} mCi (millicurie)
1 Ci (curie) =	3.7×10^{10} radioactive disintegrations/second
	3.7×10^{10} Bq (becquerel)
	1,000 mCi (millicurie)

DEFINITIONS, CONVERSIONS, AND CALCULATIONS

1 Gy (gray) =	1 j/kg (joule/kilogram)
	100 rad
	1 Sv (sievert) — [unless modified through division by an appropriate factor, such as Q and/or N]
1 mCi (millicurie) =	0.001 Ci (curie)
	3.7×10^{10} radioactive disintegrations/second
	3.7×10^{10} Bq (becquerel)
1 rad =	100 ergs/gm (ergs/gm)
	0.01 Gy (gray)
	1 rem — [unless modified through division by an appropriate factor, such as Q and/or N]
1 rem =	1 rad — [unless modified through division by an appropriate factor, such as Q and/or N]
1 Sv (sievert) =	1 Gy (gray) — [unless modified through division by an appropriate factor, such as Q and/or N]

INDEX

DEFINITIONS, CONVERSIONS, AND CALCULATIONS